THEORETICAL
STATISTICAL
OPTICS

THEORETICAL STATISTICAL OPTICS

Olga Korotkova

University of Miami, USA

World Scientific

NEW JERSEY · LONDON · SINGAPORE · BEIJING · SHANGHAI · HONG KONG · TAIPEI · CHENNAI · TOKYO

Published by

World Scientific Publishing Co. Pte. Ltd.
5 Toh Tuck Link, Singapore 596224
USA office: 27 Warren Street, Suite 401-402, Hackensack, NJ 07601
UK office: 57 Shelton Street, Covent Garden, London WC2H 9HE

British Library Cataloguing-in-Publication Data
A catalogue record for this book is available from the British Library.

THEORETICAL STATISTICAL OPTICS

Copyright © 2022 by World Scientific Publishing Co. Pte. Ltd.

All rights reserved. This book, or parts thereof, may not be reproduced in any form or by any means, electronic or mechanical, including photocopying, recording or any information storage and retrieval system now known or to be invented, without written permission from the publisher.

For photocopying of material in this volume, please pay a copying fee through the Copyright Clearance Center, Inc., 222 Rosewood Drive, Danvers, MA 01923, USA. In this case permission to photocopy is not required from the publisher.

ISBN 978-981-123-497-2 (hardcover)
ISBN 978-981-123-498-9 (ebook for institutions)
ISBN 978-981-123-499-6 (ebook for individuals)

For any available supplementary material, please visit
https://www.worldscientific.com/worldscibooks/10.1142/12230#t=suppl

Desk Editor: Nur Syarfeena Binte Mohd Fauzi

Typeset by Stallion Press
Email: enquiries@stallionpress.com

To my family...

Preface

Statistical optics is an interdisciplinary field within the domain of natural sciences encompassing partially differential equations, theory of random processes, as well as physical, environmental and biological optics. Since its formation into a stand-alone area, statistical optics has experienced perpetual expansion, mainly due to the fact that understanding of the nature and the properties of light and those of the media it interacts with has also substantially grown over time. At present, this field is abundant and developing with acceleration, which makes it rather impossible for a single scientist or even a research group to deeply comprehend the variety of underlying phenomena. In addition, optical engineering technologies are currently introduced at unprecedented rates making scientists revisit and enhance the methods for light manipulation on a yearly, if not a monthly, basis.

Nevertheless, there are theoretical foundations in statistical optics that seem to withstand the passage of time and the technological evolution. These are primarily the laws and the principles based on the equations of mathematical physics, and electromagnetics, in particular, as well as the rigorous statistical analysis of the observable optical phenomena. The aim of this monograph is to overview the most popular theories, principles, techniques and experiments relating to random light and/or random media, their description and interaction, and to highlight some of the related technologies.

The selection of topics for this text was made with a purpose of blending the overview of the well-established theories, also belonging to famous texts such as *Statistical Optics* by Joseph W. Goodman, *Introduction to the Theories of Coherence and Polarization of Light* by Emil Wolf, and others, with the developments that have appeared very recently, such as structured light coherence, polarization and ghost imaging techniques, light interaction with non-classic turbulence, including the double-diffusing oceanic turbulence, and non-stationary pulses, among others. The attempt is also made in providing the reader with the complete characterization of the electromagnetic random light, including, for example, its description in terms of energy, linear and angular momentum as well as the discussion of the three-dimensional polarization states. While being a theoretically oriented monograph, it could not avoid inclusion of a concise review of the four most important experiments of statistical optics: the classic Young, the Michelson and the Hanbury Brown and Twiss interferometric procedures, and a stand-alone random phase conjugation phenomenon. The other objective for the selection of the material for the book was to share with the reader the knowledge acquired by the author through active research involvement in all these areas. Indeed, a large number of sections are based on the results obtained by the author, her students and collaborators, which would hopefully make the presentation more inspiring.

The monograph is intended for graduate students, scientists and engineers interested in the theory of statistical optics and some of its applied facets such as imaging, communications and sensing by means of optical fields. It requires some introductory knowledge of the Fourier analysis, partial differential equations, random processes, electromagnetics and physical optics but is meant to be mostly self-explanatory. Otherwise, it includes all necessary citations to more elementary and classic texts. Several sections contain material currently being at the very edge of science which can be therefore used as a starting point for new research endeavors.

A large number of people have implicitly participated in the development of this text. First of all, I would like to acknowledge the most enthusiastic support from my academic advisors: Larry Andrews,

Ronald Phillips, Aristide Dogariu and Emil Wolf, who taught me how to select good problems and to develop enough intuition to deal with them, how to write, collaborate, present, and just love science. I would also like to extend my gratitude to my former graduate students: Serkan Sahin, Zhisong Tong, Nathan Farwell, Jia Li and Xi Chen, who shared with me their fresh thinking, broad range of skills and positive attitude towards any new ideas. We have spent countless hours trying to see "the big picture" and to scrutinize all these tiny details ... With the special feeling I address the contribution of the undergraduate and high school students with whom I happened to enjoy working in the last several years: Gordon Martinez-Piedra, Jasmine Rodriguez, Daniel Montaño, Arianna Soresi, Sean McDuffie and Mohammad Najjar. It was always an invaluable and thrilling experience to explain to them the complex phenomena "from scratch".

I am also indebted to my colleagues who have become good friends over the years. Svetlana Avramov-Zamurovic, Charles Nelson, and Reza Malek-Madani, thank you for welcoming me infinitely many times for work and great conversations at the US Naval Academy. I am also grateful for my close connection to the galaxy of scientists raised or supported by Emil Wolf: Greg Gbur, Scott Carney, Miguel Alonso, Taco Visser, Daniel James and Ari Friberg — it never feels lonely in optics because of this common set of problems we are trying to solve, and these old jokes we share. Franco Gori, Massimo Santarsiero and Riccardo Borghi — I thank you for setting the example of excellence in every published piece of work — it is always a pleasure to learn from you and work with you. Tero Setälä — I am thankful to you for the continued support of our collaboration and for your unique ability to crystallize any rough idea. I also acknowledge the contributions from several scientists who visited the University of Miami for an extended time: Italo Toselli, for sharing his deep insight into turbulence modeling (and teaching me to sail), Elena Shchepakina, for introducing me to fascinating nonlinear optical phenomena, Zhangrong Mei, for dedication in joint development of important classes of beams with structured coherence, Fei Wang, for converting our theoretical research into experimental in a course of a single year.

I am also grateful to the Deep Turbulence MURI team and other collaborations in Dayton, OH — especially Mikhail Vorontsov, David Voelz, Xifeng Xiao, Milo Hyde and Santasri Bose-Pillai for our joint work on untangling the complex turbulence–light interactions and random light simulations. Yangian Cai — thank you for the enormous number of projects carried out together on so many different topics; Chaoliang Ding, Liuzhan Pan and Daomu Zhao — for our joint work on random pulses and other areas, JinRen Yao — for our very recent, fruitful collaboration on natural water turbulence.

Very special thanks are also sent to my family who made sure the book is completed in a reasonable time, to my close friends who were understanding and supportive, and to two handsome golden retrievers, Roma and Paris, who observed the thrills and the pains of this writing process from the first page to the last.

I do hope this "light reading" will be enriching and stimulating for the curious reader.

Olga Korotkova
18 May 2021

Contents

Preface   vii

1. Introduction   1
   - 1.1. Random Nature of Light 1
   - 1.2. History of Statistical Optics 8
     - 1.2.1. Environmental scattering 8
     - 1.2.2. Light as a random electromagnetic wave ... 10
     - 1.2.3. Foundation of random process theory 14
   - 1.3. Mathematical Preliminaries 16
     - 1.3.1. Fourier transform 16
     - 1.3.2. Analytic signal 25
   - 1.4. Statistical Preliminaries 28
     - 1.4.1. Linear processes 28
     - 1.4.2. Random variables and processes 29
     - 1.4.3. Types of random processes 33
     - 1.4.4. Spectral content of a wide-sense stationary process 38
     - 1.4.5. Gaussian random process 41

2. Statistical Characterization of Optical Fields   45
   - 2.1. Deterministic Optical Fields 45
     - 2.1.1. The Maxwell equations 45
     - 2.1.2. Energy 47

		2.1.3.	Linear and angular momentum	51
		2.1.4.	Polarization	55
	2.2.	Scalar Stationary Optical Fields		58
		2.2.1.	Cross-spectral density	58
		2.2.2.	One- and two-point properties	62
	2.3.	Electromagnetic Beam-like Fields		63
		2.3.1.	Cross-spectral density matrix	63
		2.3.2.	One-point properties	64
		2.3.3.	Two-point properties	69
	2.4.	Electromagnetic General Fields		74
	2.5.	Model Sources		80
		2.5.1.	Schell and quasi-homogeneous models	80
		2.5.2.	Homogeneous model for spherical shell sources	84

3. Famous Experiments and Phenomena Relating to Random Light 87

	3.1.	Young's Interference Experiment	87
	3.2.	Michelson Interference Experiment	90
	3.3.	Hanbury Brown and Twiss Interference Experiment	92
	3.4.	BackScatter Amplification Effect	93

4. Free-space Propagation of Stationary Light 97

	4.1.	Deterministic Light		97
		4.1.1.	Scalar theory	97
		4.1.2.	Electromagnetic theory	99
	4.2.	Stationary Light		101
		4.2.1.	3×3 cross-spectral density tensor propagation	101
		4.2.2.	The van Cittert–Zernike theorem	104
	4.3.	Source Correlation-induced Changes		106
		4.3.1.	The Huygens–Fresnel integral	106
		4.3.2.	Examples	107

5. Structured Light Coherence — 113

- 5.1. One-dimensional Sources 113
  - 5.1.1. Bochner's theorem method 113
  - 5.1.2. Examples 117
  - 5.1.3. Sliding function method 118
  - 5.1.4. Examples 121
- 5.2. Two-dimensional, Scalar Sources and Beams 128
  - 5.2.1. Bochner's theorem method 128
  - 5.2.2. Examples: Uniform correlations, radial symmetry 130
  - 5.2.3. Examples: Uniform correlations, Cartesian symmetry 133
  - 5.2.4. Examples: Uniform correlations, no symmetry 136
  - 5.2.5. Examples: Twisted correlations 141
  - 5.2.6. Examples: Separable phases 144
  - 5.2.7. Sliding function method 145
  - 5.2.8. Example 147
- 5.3. Other Methods, Models, Statistics 149

6. Light Interaction with Devices of Polarization Optics — 153

- 6.1. Jones Calculus 153
- 6.2. Stokes–Mueller Calculus 157
  - 6.2.1. Mueller matrix 157
  - 6.2.2. Stokes parameters' determination 158
  - 6.2.3. Mueller matrix determination 159
- 6.3. Two-point Stokes–Mueller Calculus 164
  - 6.3.1. Two-point Mueller matrix 164
  - 6.3.2. Analytic example: Spatial light modulator 165

7. Image Formation with Random Light — 169

- 7.1. Classic Imaging Systems 169
  - 7.1.1. Linear system approach 169

	7.1.2.	Isoplanatic systems, homogeneous illumination 172
	7.1.3.	Coherent and incoherent imaging systems . . 179
	7.1.4.	ABCD matrices 182
	7.1.5.	Generalized Huygens–Fresnel integral 185
7.2.	Polarization Imaging 187
	7.2.1.	Linear system approach 187
	7.2.2.	Isoplanatic polarimetric systems 193
7.3.	Two-point Resolution 196
	7.3.1.	Imaging by structured illumination 196
	7.3.2.	Rayleigh resolution criterion 198
7.4.	Ghost Imaging 204

8. Light Scattering from Three-Dimensional Media — 209

8.1.	The Scattering Phenomenon 209
8.2.	Potential Scattering for Scalar Fields 211
	8.2.1.	The first Born approximation 211
	8.2.2.	Far-zone approximation 214
	8.2.3.	Scattering matrix 215
	8.2.4.	Random incident field and/or scatterer . . . 218
	8.2.5.	Deterministic mode representation of scatterers 221
	8.2.6.	Pair-scattering matrix 222
8.3.	Potential Scattering for Electromagnetic Fields . . . 223
	8.3.1.	Deterministic incident field and scatterer . . 223
	8.3.2.	Far-zone approximation 225
	8.3.3.	Random incident field and scatterer 227
8.4.	Examples of Scattering from Deterministic Media . . 230
	8.4.1.	Spherically symmetric media 230
	8.4.2.	Hard-edged ellipsoids, cylinders, parallelepipeds 233
	8.4.3.	Deterministic collections of scatterers 237
	8.4.4.	Effect of random incident field 238
8.5.	Examples of Scattering from Random Media 239
	8.5.1.	Gaussian-correlated particle 239

| | 8.5.2. Scatterers with structured correlations . . . . 242 |
| | 8.5.3. Random collections of scatterers . . . . 246 |

9. Light Interaction with Turbulence　　253

- 9.1. Phenomenon of Optical Turbulence 254
  - 9.1.1. Classic and non-classic turbulence 254
  - 9.1.2. Major turbulence parameters 255
  - 9.1.3. Obukhov–Corrsin power spectra 257
- 9.2. Atmospheric Turbulence 259
  - 9.2.1. Classic model 259
  - 9.2.2. Non-classic extensions 260
- 9.3. Oceanic Turbulence 264
  - 9.3.1. Classic model 264
  - 9.3.2. Natural Earth's water turbulence 266
- 9.4. Bio-tissue Turbulence 272
- 9.5. Methods for Light–turbulence Interaction 274
  - 9.5.1. Extended Huygens–Fresnel method 274
  - 9.5.2. Convolution method 277
  - 9.5.3. Other methods 281
- 9.6. Behavior of Light Beams in Turbulence 282
  - 9.6.1. General phenomena 282
  - 9.6.2. Probability density functions of intensity 284

10. Non-stationary Pulse Ensembles　　287

- 10.1. Theory of Quasi-stationary Pulses 287
- 10.2. Mathematical Models 289
  - 10.2.1. Laser-based examples 289
  - 10.2.2. Gaussian Schell-model pulses 290
- 10.3. Propagation of Pulse Ensembles in Dispersive Media 293
- 10.4. Structured Pulse Coherence 295
  - 10.4.1. Bochner's theorem method 295
  - 10.4.2. Sliding function method 298

Appendix A Natural Water Parameters Varying with
  $\langle T \rangle$ and $\langle S \rangle$ 299

Bibliography   305

Chapter 1

Introduction

1.1 Random Nature of Light

Optical radiation belongs to a relatively narrow but very important portion of the electromagnetic spectrum spanning wavelengths from 380 μm to 750 μm. The boundaries of this interval are fuzzy since the sensitivity of optical detectors, including the most important of them, the human eye, does not decrease to zero sharply. Within the domain of classical theory optical radiation exhibits wave-like properties and in the quantum treatment it is viewed as a collection of light quanta – photons. The wave nature of light has been established via spatial (Young, 1804) and temporal (Michelson and Morley, 1887) interference experiments manifesting itself on free-space propagation, while the particulate nature of light was confirmed with the help of the photoelectric effect (Stoletov, 1888; Einstein, 1905; Millikan, 1916) based on the light–matter interaction process.

We first briefly overview the quantum properties of light. A photon is characterized by four physical quantities: *wavelength*, *energy*, *linear momentum* and *spin angular momentum*. The wavelength defines the photon's color. Photons in the optical range are detectable by the human eye since they have enough energy to change the bond structure of molecules in the retina's rhodopsin via light absorption mechanism. Hence, the optical portion of the electromagnetic spectrum is also termed *visible*. Remarkably, the important process of photosynthesis also occurs in the optical range since the photons of light are capable of exciting the molecules of chlorophyll. Under the assumptions of relativistic theory, a photon of light with wavelength λ_p propagating in vacuum with speed $c = 299,792,458$ (m/s) carries energy

$$E_p = \frac{hc}{\lambda_p}, \qquad (1.1)$$

where $h = 6.62607150(81) \times 10^{-34}$ (J·s) is the Plank constant (Plank, 1901). Equation (1.1) is known as the *Plank–Einstein relation*. The photon's energy can also be expressed via its frequency, $f_p = c/\lambda_p$, as

$$E_p = h f_p. \qquad (1.2)$$

The linear momentum vector of light is defined as

$$\mathbf{p}_p = \frac{h \mathbf{k}_p}{2\pi}, \qquad (1.3)$$

where $\mathbf{k}_p$ is the three-dimensional wave vector specifying the direction of photon's propagation, while its length relates to photon's wavelength as $k_p = |\mathbf{k}_p| = 2\pi/\lambda_p$. Combination of Eqs. (1.1) and (1.3) results in relations

$$p_p = h/\lambda_p, \quad E_p = p_p c, \qquad (1.4)$$

where $p_p = |\mathbf{p}_p|$.

The spin angular momentum of a photon being independent of its frequency can have quantized magnitude $h/\sqrt{2\pi}$ and helicity (component measured along its direction of motion) $\pm h$. The positive and negative helicities are termed right-handed and left-handed, respectively, corresponding to two circular polarization states, with clockwise and countreclockwise directions.

The aforementioned properties of light are pertinent to individual photons and are studied within the domain of *quantum optics*. On the other hand, the subject of *classical optics* treats the photon distribution – the wavefront – as the whole entity, regarding it as a continuous electromagnetic field.

Electromagnetic waves may be classified as *deterministic* and *random*. The amplitude and the phase of a deterministic wavefront can be uniquely specified at any given position and time instant.

Although very useful in concept and for performing theoretical calculations, the deterministic optical waves do not exist: even the stabilized laser radiation involves a certain degree of spatio-temporal randomness. Similarly, a seemingly smooth edge of a material object diffracting a laser beam is typically rough if compared with the beam's wavelength, leading to spatial wave randomization. Moreover, a strictly deterministic optical detection process cannot be achieved.

Random nature of light fields is the consequence of either spontaneous emission of photons by their sources or interaction of previously emitted radiation with random media. Such interaction is broadly termed *scattering*, encompassing reflection from a rough surface, interaction with a localized particle or a collection of particles, diffraction from a rough aperure, etc. If the scattered field is characterized within the scattering volume, this interaction process is also referred to as *propagation* through a random medium. Even though emission and scattering of the electromagnetic radiation are two fundamentally different physical processes, their mathematical treatments share a number of similarities. *Statistical optics*, being an extension of physical optics, brings together the toolsets of partial differential equations, statistics and optical phemonena, for characterization of random processes involved in light radiation, evolution and scattering. The statistical description of random light fields is, most of the time, the only possibility because the enormous number of participating photons, each having certain, often unknown, initial conditions, precludes application of any of the deterministic treatments. In the rest of this chapter, after overviewing the history of scientific developments within the field of statistical optics, we introduce the important mathematical and statistical apparati, such as the Fourier transform, the analytic signal representation and the theory of random processes, essential for clear understanding of the monograph.

Statistical description of a random optical wave is based on the information about its correlation (similarity on average) with its own versions, and other (reference) waves taken at various sets of positions and/or times. In statistical optics, the preference is usually given to light properties that are directly observable whether by the eye or by any manmade optical detector (Wolf, 1954). For example, a stationary optical field can be completely or partially correlated at a pair of positions in space. The directly measurable quantity associated with such a correlation, known as the *spatial degree of coherence*, is of fundamental importance in statistical optics for predictions of various light properties evolving on propagation. Similarly, on considering non-stationary pulse trains, one may define the directly measurable *temporal degree of coherence*, which characterizes typical discrepancies between any two pulse realizations. As yet another example, the components of an optical vector-field at a certain position in space might fluctuate randomly with time and be partially correlated. The *degree of polarization* taking into account the auto-correlations and the cross-correlations among the pairs of the optical field components in a single expression provides a very useful characteristic of the wavefield's vectorial content. This quantity cannot be directly observed but can be evaluated from a set of several values of measured average intensities. The statistical characterization of the stationary, random optical fields in terms of observable quantities is introduced in Chapter 2.

In dealing with various statistical descriptors of random optical fields, it appears important to distinguish among one-, two- and three-dimensional wavefronts in terms of their geometry as well as the number of non-trivial electric-field components. Consider for instance, an optical beam with one-dimensional geometry, i.e., a highly directional optical wavefront distributed along one transverse direction but invariant along the other transverse direction. In this case, one or two components of the electric field can be responsible for such a disturbance. If a single non-trivial electric field component is present, the wavefront is polarized and the scalar approximation is applicable. Otherwise, for two non-trivial electric field components the treatment must be based on two-dimensional vectors (for deterministic fields) and 2×2 correlation matrices (for random fields). The opposite example is a circularly symmetric, polarized beam, being one-dimensional polarimetrically and two-dimensional geometrically. The theoretical foundation and experimental techniques for dealing with one- and two-dimensional fields, either geometrically or polarimetrically, are currently fairly well-established.

Optical fields of random nature that require three-dimensional treatment geometrically and/or electromagnetically are less studied and some of the underlying concepts still appear controversial. In the most general case, all three components of the electric field could be present in a volume and, hence, its description must be made with the help of the three-dimensional vectors (for deterministic fields) or 3×3 correlation matrices (for random fields). The examples of such optical fields are those occuring in the focal regions of lenses with high numerical apertures and in the near-field regions, i.e., the layers of subwavelength thickness in the vicinity of sources and scatterers. Another important example of a genuinely three-dimensional field is the black-body radiation, i.e., a field that came to equilibrium after sufficiently many interactions with the inner walls of a perfectly absorbing material (Blomstedt et al., 2017). In a number of situations, the electromagnetic nature of two- and three-dimensional fields is ignored and the treatment is made under scalar approximation.

We, however, stress that the outlined classification is only pertinent to a fixed line, plane or volume: as light propagates from these domains, its geometric and/or polarimetric dimensionality can be adjusted. For instance, an electric field having two transverse polarization components in a source plane and propagating into a half-space will acquire the third (longitudinal) component in the intermediate zone from the source but will restore (locally) its two-component structure in the far zone. The rigorous statistical description of optical fields originating from more complex domains than planes and spherical volumes, such as two- and three-dimensional curves and three-dimensional surfaces, has not yet been satisfactorily developed. Similarly, light interacting with media may or may not preserve its initial geometrical and/or polarimetric dimensionality. For instance, a geometrically two-dimensional light beam having one non-trivial polarization component may acquire the second polarization component on interaction with a sufficiently thick layer of a biological tissue.

As we have already mentioned, one of the cornerstones of statistical optics is the idea of light characterization in terms of observable quantities. This makes one use only those of the theoretically available light statistics that can be directly measured and study only the phenomena that can be practically accessible. Chapter 3 presents four fundamental experiments that have built our current understanding of random light and media: the Young–Zernike double-slit spatial coherence interferometer, the Michelson–Morley time-coherence interferometer, the Hanbury Brown–Twiss high-order correlation interferometer and phase conjugation experiment with a random medium.

Perhaps the most unexpected revelation regarding random light is the ability of its source correlations to control its other properties on evolution in free space, in media and in optical systems. The phenomena of source correlation-induced changes in diffraction rate of the radiated beam, in its spectral and polarization states and in its orbital angular momentum flux density are summarized in Chapter 4.

One of the fundamental problems of statistical optics is modeling and synthesizing light sources with prescribed spatio-temporal statistical properties. It is well known by now how from the knowledge of an optical source's statistics it is possible to determine and even tune the variety of physical properties of the radiated field in free space and rather occurately predict the results of its interaction with media. In particular, one of the currently rapidly growing areas of statistical optics, *structured light coherence*, uses specific profiles of the source correlations (whether temporal or spatial) for delivering produced radiation to specific locations in space and/or at specific moments in time. A detailed discussion of this subject relating to spatial structuring is presented in Chapter 5.

A large subarea of statistical optics deals specifically with vectorial (polarimetric) properties of electric fields, their evolution in free space and their changes on interaction with optical systems and media. In Chapter 6, we present Jones–Mueller calculus that allows for efficient treatment of interaction of random light with deterministic and random devices of polarization optics. This analysis is based on the well-established paraxial treatment of fluctuating electric fields and considers one- and two-point transformations.

Another important subject of statistical optics is characterization of interaction of random light with various optical systems, for example, broadly used classic imaging systems. Such systems employ stationary light as the carrier of information about an object interacting with it in one plane (object plane) and forming its image in another plane (image plane). Depending on the nature of illumination, of the object and of the system itself (deterministic or random), image formation is in general a very complex process, which can be optimized in terms of similarity of the object and the image, largely using the methods of statistical optics for adjusting the illumination's statistics. In Chapter 7, we offer a detailed analysis of scalar and a brief introduction to polarimetric imaging systems. We also explain the theoretical aspects of ghost image formation with thermal light, a technique that utilizes the Hanbury Brown and Twiss interferometry of high-order field correlations.

Since the statistical state of light can be altered on interaction with random media, the other fundamental theme of statistical optics is the characterization of random processes governing the media's refractive index distributions. In this respect, two types of light–medium interaction problems are distinguished: light *scattering from* a random collection of discrete particles and *light propagation in* an extended, continuously fluctuating medium. The practically important examples of the scattering process include light passage in a relatively thick, turbulence-free atmospheric layer with high concentration of water droplets or aerosoles; light trespassing a still ocean water column densely populated with plankton; and light interacting with relatively hard biological tissues. On the contrary, the examples of light propagation in a continuous medium may include those in particle-free atmospheric/oceanic turbulence and in soft bio-tissues. Obviously, discrete scatterers may also be embedded in a continuous random medium, in this case constituting a very complex, analytically untreatable random process. Further, depending on their chemical composition, random scatterers or media can introduce either scalar or polarimetric changes in light waves. For instance, clear-air or clear-water (scatterer-free) turbulence are known to act on all the electric field components in the same manner, but bio-tissues are known to be randomly birefrigent and, hence, to modify the vectorial light properties. For scalar interactions, one uses various correlation functions in the refractive index and in vectorial interactions 2×2 or 3×3 matrices with elements that are scalar correlation functions of the refractive index are employed. The classic approach for light scattering from particles and their collections is presented in Chapter 8 and light propagation in extended random (turbulent-like) media is discussed in Chapter 9.

Chapters 2-9 of this text discuss statistically stationary optical fields, i.e., those in which different monochromatic frequency components are uncorrelated. Relatively recently, non-stationary optical fields, and their particular kind, cyclostationary fields (also known as random optical pulses) became of acute interest. In such fields, there exists a partial correlation among different frequency components. A brief discussion of such pulsed optical radiation including its characterization, shaping and propagation, occupies Chapter 10 and concludes the monograph.

1.2 History of Statistical Optics

1.2.1 *Environmental scattering*

Historically, the very first field of study that paved the path to development of the modern statistical optics was sunlight scattering from natural random media, and, in particular, from water droplets and aerosoles suspended in the air. Indeed, optical phenomena manifesting themselves primarily in formation of certain intensity or color patterns in the sky or in water columns have fascinated scientists and philosophers since ancient times. It is believed, on the basis of a Latin translation from Arabic, that Claudius Ptolemy (140 AD) was interested in refraction and scattering phenomena, in general, and atmospheric scattering, in particular. A millennium later, an 11th century Arabic scientist Al Hasen of Basra has claimed that the brightness of the sky could be attributed to the reflection of sunlight from the airborne particles. A British philosopher Robert Grosseteste (1220 AD) has stated that the rainbow was the result of sunlight interaction with layers in a "watery cloud", not, however, recognizing the importance of the individual water droplets. Somewhat later, another British scholar Roger Bacon bridged this gap: he explained the rainbow phenomenon as the result of sunlight's interaction with individual water particles. Later, a German scholar Theodoric of Frieberg (Dietrich, 1304–1310) then found experimentally that the rainbow was formed by internal reflection within each droplet, and managed to rigorously explain the formation of the primary and the secondary bows. During the same epoch, a Persian scientist Kamal al-Din Hasan ibn Ali ibn Hasan al-Farisi (1267–1319) came to a similar conclusion independently. Both scientists have only managed to reproduce the rainbow in the laboratory conditions but have not given its theoretical explanation. The first ray diagrams of the rainbow formation belong to a Croatian scholar Marco Antonio de Dominis (1560–1624).

During the Renaissance period, numerous experiments on scattering by particulate media were conducted by Leonardo Da Vinci (1452–1519) who would illuminate smoke and water spray and observe the produced patterns by projecting them onto an opaque screen. He is also believed to be the first to describe and systematically use in his paintings (e.g., *Madonna of the Rocks*) the effect of atmospheric scattering on the appearance of distant objects. This technique is currently known as *aerial perspective*, in which the objects' colors are mixed with the dominant colors of the atmosphere, blue during the day and red or orange at sunset and sunrise. Since the human brain evaluates ranges to distant objects embedded in the atmosphere from their color patterns, the aerial perspective substantially enriches the information carried by the image.

The first recorded systematic experiments on scattering by atmospheric aerosoles belong to John Tyndall (1820–1893) [Tyndall (1869)] (see Fig. 1.1). Using Tyndall's results, Lord Rayleigh (1842–1919) (also published under the name John Strutt), has developed the quantitative theoretical treatment of light scattering by non-absorbing (dielectric) spherical particles, explained some natural color phenomena by light scattering from small particles and derived the famous λ^{-4} scattering law. In particular, he has then concluded that the sky is blue due to predominantly scattered light, while sunrises and sunsets are reddish due to the difference of the incident white light (sunlight) and the scattered blue light. He has

John Tyndall (1820–1893)   Lord Rayleigh (1842–1919)   Gustav Mie (1868–1957)

Fig. 1.1 Scientists who revolutionized our understanding of particulate scattering.

also outlined an optical method for mass determination of very small particles (Strutt, 1871a,b, 1881, 1899).

A theory describing light scattering from particles of different sizes, having spherical symmetry and hard edges, was introduced by a German scholar Gustav Mie (1868–1967) (Mie, 1908). Although directly applicable only to a single sphere, Mie's theory has later rigorously explained the color schemes formed in natural rainbows (Nussenzveig, 1977) and other atmopsheric phenomena, such as haloes (Yaacoub et al., 2019) (see Fig. 1.2). Currently, it serves as the basis for various numerical models of light scattering valid in particular regions of the Earth's atmosphere and oceans.

1.2.2 Light as a random electromagnetic wave

An Italian scientist Francesco Maria Grimaldi (1618–1663) is believed to be the first to record his observations of optical diffraction of white light through small openings and to suggest that light is a fluid that exhibits wave-like motion. Sometime later, an English physicist Robert Hooke (1635–1703) conducted experiments on the colors produced in flakes of mica, soap bubbles and films of oil on water, concluding that color is related to the thickness of the scatterers. Isaac Newton (1643–1727) was the first to observe rings, now carrying

Introduction 11

Fig. 1.2 A rare atmospheric sun halo due to scattering from Cirrus clouds taken on April 11, 2020, in Miami, Florida. The Sun was in its zenith and had to be blocked by palm leaves in order to obtain a clear image. Courtesy of Ryder Arafet.

his name, that are the result of interference pattern created by the reflection of light between two adjucent surfaces, spherical and flat. He also was among the first to put forward the corpuscular (photon) theory of light. However, the true cornerstone for the development of statistical optics was the double-slit experiment of English scientist Thomas Young (1773–1829) that provided support for the wave theory of light by demonstrating its spatial interference (Young, 1804, 1807) (see Fig. 1.3).

Although the phenomenon of light's polarization had been discovered much earlier, the convenient manner in which polarization properties of a random wave can be described and measured as a four-dimensional vector was introduced by Gabriel Stokes (1819–1903) and now carries his name (Stokes, 1852) (see Fig. 1.4). A decade later, James Clerk Maxwell (1820–1879) established the theoretical background for the electromagnetic radiation by means of the famous set of differential equations. Albert Michelson (1852–1931)

12 *Theoretical Statistical Optics*

Thomas Young (1773–1829)   Albert Michelson (1852–1931)   Frits Zernike (1888–1966)

Fig. 1.3 Scientists who explained the light interference phenomenon.

Gabriel Stokes (1819–1903)   Hans Mueller (1900–1965)   Subrahmanyan Chandrasekhar (1903–1987)

Fig. 1.4 Scientists who developed understanding of polarization in random light.

and Edward Morley (1838–1923) have proposed an interferometer, currently known as Michelson interferometer, that illustrated the temporal interference phenomenon of light waves (Michelson and Morley, 1887). However, neither Young nor Michelson considered the results of light interference as the outcome of their statistical similarity. On the other hand, French physicist Emile Verdet (1824–1866) as early as in 1865 has established that the vibrations in sunlight are statistically similar only at very closely situated points but the similarity vanishes as separation increases to about an order of magnitude as compared with a typical light wavelength (Verdet, 1865). This finding was not of much use for a number of decades until Pieter Hendrik van Cittert (1889–1959) (van Cittert, 1934) and Frits Zernike (1888–1966) in (Zernike, 1938), both working in Netherlands, revealed that there exists a close relation between light randomness and its ability to form sharp fringe patterns on self-interference. Moreover, their theory has led to an important discovery that on passing in vacuum light becomes less random, explaining a variety of natural phenomena, such as twinkling of stars. Indeed, radiated from completely random sources and traveling through very long distances, starlight comes to the Earth's atmopshere spatially coherent. Atmospheric turbulent layers then readily destroy its phase, and, as a consequence, its intensity. The van Cittert–Zernike theory has also led to a discovery by a Czech-American scientist Emil Wolf (1921–2018) that light's spectral composition can change on its passage in vacuum (Wolf, 1986).

The interaction of electromagnetic light with devices of polarization optics has been theoretically descibed by an American scientist Robert Clark Jones (1916–2004) (Jones, 1941) and a Swiss-American scholar Hans Mueller (1900–1965) (Mueller, 1948). The radiative transfer theory involving electromagentic radiation (described by Stokes vectors) was developed by an Indian physicist Subrahmanyan Chandrasekhar (1903–1987) (Chandrasekhar, 1960). Scattering theory of random scalar and electromagnetic fields that incorporated the information about source coherence properties was put forward in Jannson et al. (1988). Based on this development, important phenomena pertinent to random illumination scattered by (elastic) deterministic and random media, such as frequency shifts, were later predicted (Wolf et al., 1989).

The unification of coherence and polarization theories of light was completed in Italy (Gori et al., 1998) and the United States (Wolf, 2003) shortly after publication of a seminal paper by a British-Canadian physicist Daniel F. V. (James, 1994). This theory made it possible to fully appreciate the interplay between these two local optical properties, and its consequences for light evolution in free space and interaction with media.

1.2.3 Foundation of random process theory

The origins of the probability theory can be traced back to the 17th century. In 1654, in the correspondence between French mathematicians Pierre Fermat (1607–1665) and Blaise Pascal (1623–1662), the concept of probability has been introduced in connection with the outcomes in a gambling problem. An independent work by an Italian mathematician Gerolamo Cardano (1501–1576) offers another discussion on probabilty (Cardano, 1565). In 1713, the results of an Italian scholar Jacob Bernoulli (1654–1705) on combinatorics and the law of large numbers were first published (Bernoulli, 1713) (see Fig. 1.5). During the 18th century important contributions on the normal distribution and the central limit theorem were made by Abraham de Moivre (1667–1754) (de Moivre, 1718) in France and Carl Friedrich Gauss (1777–1855) in Germany (Gauss, 1809).

During the 19th century the bayesian interpretation of probability was developed by a French mathematician Pierre-Simon Laplace (1749–1827), substantial contributions to the probability theory, such as the weak law of large numbers were made by Pafnuty Chebyshev (1821–1894) in Russia and the important distribution of random variables was introduced by Siméon Denis Poisson (1781–1840) in France, now carrying his name. During that time, development of the theories of probability and random processes were also greatly stimulated by the rapid development of

Jacob Bernoulli
(1654–1705)

Andrej Markov
(1856–1922)

Andrey Kolmogorov
(1903–1987)

Fig. 1.5 Scientists who developed the theories or random variables and processes.

statistical mechanics and the kinetic theory of gases. In 1859, a Scottish physicist James Clerk Maxwell (1831–1879) directly used the propability laws for description of random velocity field of a gas. The first statistical account of a Brownian motion (Wiener-type random process) appeared in 1880 due to a Danish astronomer Thorvald Thiele (1838–1910) and the second important account was made by a French mathematician Louis Jean-Batiste Alphose Bachelier (1870–1946) in his 1900 thesis *The Theory of Speculation*.

The first half of the twentieth century can be viewed as the golden age for the conceptual development of random processes. In Sweden in 1903, Filip Lundberg (1876–1965) published the thesis in which he introduced the Poisson random process. In 1905, in Great Britain, a bio-statistician Karl Pearson (1857–1936) was the first to discuss random walk restricting the analysis to the plane. Random walk in higher dimensions was analyzed by a Hungerian-Swiss-American mathematician George Pólya (1887–1985) in 1919–1921. In 1906, a Russian mathematician Andrey Andreevich Markov (1856–1922) introduced a discrete random process named after him, *Markov chain*. The assumptions leading to this process are widely used in modern statistical optics for a simplified treatment of light–turbulence interaction problems.

In 1925, Paul Levy (1886–1971) in France and, in 1929, Andrey Nikolaevich Kolmogorov (1903–1987) in the Soviet Union independently published monographs in which they attempted to combine the propability theory and the measure theory. In early 1930s, the modern rigorous foundations of the theory of random processes were developed in the Soviet Union. In particular, Alexander Yakovlevich Khintchin (1894–1959) rigorously introduced the notion of the random process, and A. N. Kolmogorov introduced axiomatic approach to probability. In the 1940s, a Japanese mathematician Kiyosi Ito (1915–2008) developed the theories of stochastic calculus and of the stochastic differential equations. Throughout the 1940s the theory of homogeneous turbulence, perhaps the most complex random process occuring in nature, was developed by Kolmogorov (1941a) and his student Alexander Mikhailovich Obukhov (1918–1989).

1.3 Mathematical Preliminaries

1.3.1 *Fourier transform*

The application of integral transforms, and in particular, the Fourier transform, is onmipresent in optics and in statistical optics, in particular. The Fourier theory stipulates that it is possible to represent a well-behaving function defined over a domain of complex numbers as a linear superposition of harmonic contributions oscillating at different frequences (Debnath and Bhatta, 2015). Let us first consider a time-varying signal, $\mathcal{F}(t)$, and form its one-dimensional Fourier transform, $F(\omega)$, as

$$F(\omega) = FT[\mathcal{F}(t)] = \int_{-\infty}^{\infty} \mathcal{F}(t) e^{i\omega t} dt. \tag{1.5}$$

Then the inverse Fourier transform

$$\mathcal{F}(t) = FT^{-1}[F(\omega)] = \frac{1}{2\pi} \int_{-\infty}^{\infty} F(\omega) e^{-i\omega t} dt \tag{1.6}$$

reconstructs $\mathcal{F}(t)$ from $F(\omega)$. We will refer to t as the direct-space variable and to ω as the Fourier-space variable. For a signal distributed in time, the Fourier transform variable is termed *angular frequency*. The only requirement for a function to have a Fourier transform is its absolute integrability

$$\int_{-\infty}^{\infty} |\mathcal{F}(t)| dt < \infty. \tag{1.7}$$

On using, instead of time t and angular frequency ω, a Cartesian coordinate x and a *spatial frequency*, say κ, respectively, one can also form the spatial Fourier transform pair. We will discuss the spatial counterpart later on in the section on introducing Fourier transforms in higher dimensions.

The choice of the coefficients in front and of the positive/negative signs in the complex exponential kernel of the transform pair is arbitrary. However, the product of the coefficients appearing in front of the (one-dimensional) direct and inverse transforms should be $1/2\pi$ and the signs in the kernel should be different for the direct and the inverse transforms. In addition, if the arguments of the exponential

Introduction

functions have coefficients $\pm 2\pi$, then no coefficients in front of either of the transforms are needed.

The basic properties of the Fourier transform directly follow from its definition. Let $F(\omega)$ and $G(\omega)$ be Fourier transforms of functions $\mathcal{F}(t)$ and $\mathcal{G}(t)$, respectively, and let a and b be complex-valued constants. Then the following properties are readily verified:

(1) *Linearity*:
$$FT[a\mathcal{F}(t) + b\mathcal{G}(t)] = aF(\omega) + bG(\omega). \tag{1.8}$$

(2) *Hermiticity*: For real $f(t)$, the positive and the negative frequency components are related as
$$F(\omega) = F^*(-\omega), \tag{1.9}$$

where star denotes complex conjugate. The functions obeying such a relation are referred to as Hermitian.

(3) *Differentiation*:
$$FT[\mathcal{F}^{(n)}(t)] = (i\omega)^n F(\omega). \tag{1.10}$$

(4) *Scaling*:
$$FT[\mathcal{F}(at)] = \frac{1}{a} F\left(\frac{\omega}{a}\right). \tag{1.11}$$

(5) *Direct variable shift*:
$$FT[\mathcal{F}(t+a)] = e^{ia\omega} F(\omega). \tag{1.12}$$

(6) *Fourier variable shift*:
$$FT[e^{at}\mathcal{F}(t)] = F(\omega + ia). \tag{1.13}$$

Fourier transform of a function rapidly fluctuating in the direct space has a wide Fourier transform, i.e., it involves sufficiently high frequencies. On the other hand, if a function is slowly varying in the direct space, its Fourier transform is a very narrow function centered around zero.

In statistical optics, two operations, *convolution* and *correlation*, are frequently used for a variety of purposes. Convolution $\mathcal{Q}(\tau)$ of two functions, say, $\mathcal{F}(t)$ and $\mathcal{G}(t)$, is defined via integral

$$\mathcal{Q}(\tau) = [\mathcal{F} \circledast \mathcal{G}](\tau) = \int_{-\infty}^{\infty} \mathcal{F}(t)\mathcal{G}(\tau - t)dt, \qquad (1.14)$$

and their cross-correlation $\mathcal{C}(\tau)$ is defined as

$$\mathcal{C}(\tau) = [\mathcal{F} \otimes \mathcal{G}](\tau) = \int_{-\infty}^{\infty} \mathcal{F}^*(t)\mathcal{G}(\tau + t)dt. \qquad (1.15)$$

On passage from the direct/Fourier space to the Fourier/direct space, the convolution of two functions becomes the product of their direct/inverse Fourier transforms. The opposite is also true: the product of two functions in one domain becomes their convolution in the other. These relations are generally referred to as the *convolution theorem*. More precisely,

$$Q(\omega) = F(\omega)G(\omega), \qquad (1.16)$$

and

$$FT[\mathcal{F}(t)\mathcal{G}(t)] = \frac{1}{2\pi} F(\omega) \circledast G(\omega), \qquad (1.17)$$

where Q is the Fourier transform of $\mathcal{Q}$.

Convolution can also be used in measurements of a signal, say, $\mathcal{F}(t)$, by a measurement device with resolution function $\mathcal{G}(t)$. The obtained data is then convolution $\mathcal{Q}(\tau)$ of these two distributions. If resolution function $\mathcal{G}(t)$ is known and $\mathcal{Q}(\tau)$ is recorded then from their knowledge one can reconstruct the original function $\mathcal{F}(t)$. Indeed, by the convolution theorem,

$$F(\omega) = \frac{Q(\omega)}{G(\omega)}, \qquad (1.18)$$

and $\mathcal{F}(t)$ is determined by taking the inverse Fourier transform of both parts of the expression above:

$$FT(t) = \mathcal{F}^{-1}\left[\frac{Q(\omega)}{G(\omega)}\right]. \qquad (1.19)$$

The process of extraction of a true distribution from a distribution modified by a measured device is known as *deconvolution*. This procedure is widely used in pulse processing and imaging systems for extracting the information about an object from the recorded signal/image and the knowledge of the system's properties.

One can also prove a pair of statements regarding the properties of correlation of two functions, being similar to the convolution theorem, which are of particular importance for statistical optics. Namely, on passage from the direct to the Fourier space, the correlation of two functions leads to the product of their Fourier transforms, with the first term being conjugated. Also, the Fourier transform of the product of two functions, the first of which is conjugated, equals to the scaled cross-correlation of their Fourier transforms. More precisely,

$$C(\omega) = F^*(\omega)G(\omega), \tag{1.20}$$

$$FT[\mathcal{F}^*(t)\mathcal{G}(t)] = \frac{1}{2\pi}F(\omega) \otimes G(\omega), \tag{1.21}$$

where $C(\omega)$ is the Fourier transform of correlation function $\mathcal{C}(\tau)$.

Let us separately consider the special case of an *auto-correlation*, i.e., when $\mathcal{F}(t) = \mathcal{G}(t)$, defining it as

$$\mathcal{A}(\tau) = [\mathcal{F} \otimes \mathcal{F}](\tau) = \int_{-\infty}^{\infty} \mathcal{F}^*(t)\mathcal{F}(\tau + t)dt. \tag{1.22}$$

The auto-correlation characterizes the function's similarity with its own shifted version. Equations (1.21) and (1.22) imply that

$$\begin{aligned}\mathcal{A}(\tau) &= \frac{1}{2\pi}\int_{-\infty}^{\infty} A(\omega)e^{-i\omega\tau}d\omega \\ &= \frac{1}{2\pi}\int_{-\infty}^{\infty}[F^*(\omega)F(\omega)]e^{-i\omega\tau}d\omega \\ &= FT^{-1}[|F(\omega)|^2], \end{aligned} \tag{1.23}$$

where $A(\omega)$ is the Fourier transform of $\mathcal{A}(t)$. Relation (1.23) expresses the *Wienner theorem*. Quantity $|F(\omega)|^2$ is known as the *energy spectrum*, i.e., it characterizes the distribution of energy among the frequencies. Since the auto-correlation function can be directly measured, it provides access to the energy spectrum of the signal. Further, if the auto-correlation function is found at $\tau = 0$, then, using the Wienner theorem, we find that

$$\mathcal{A}(0) = \int_{-\infty}^{\infty} |\mathcal{F}(t)|^2 dt$$
$$= \int_{-\infty}^{\infty} |F(\omega)|^2 d\omega. \quad (1.24)$$

The relation above is the *generalized Parseval theorem*, i.e., its version applicable to Fourier transforms. It implies that if $\mathcal{F}(t)$ is the amplitude of a wave, then its intensity $|\mathcal{F}(t)|^2$, integrated over all values of t, which is just its total power, is equal to energy spectrum $|F(\omega)|^2$ integrated over all the frequencies ω. Geometrically, the areas under curves $|\mathcal{F}(t)|^2$ and $|F(\omega)|^2$ are equal and this theorem can be regarded as the energy conservation law in dual domains. Later in its chapter, we will come back to this discussion in connection with the spectral characterization of random processes.

The Fourier transforms in higher dimensions are frequently required in statistical optics for description of various spatio-temporal field correlations. In general, for n dimensions, the Fourier transform pair for function $F(\mathbf{r})$, where $\mathbf{r} = (r_1, r_2, \ldots, r_n)$, has the forms

$$\mathfrak{F}(\boldsymbol{\kappa}) = \int F(\mathbf{r}) e^{i\mathbf{r}\cdot\boldsymbol{\kappa}} d\mathbf{r} \quad (1.25)$$

and

$$F(\mathbf{r}) = \frac{1}{(2\pi)^n} \int \mathfrak{F}(\boldsymbol{\kappa}) e^{-i\mathbf{r}\cdot\boldsymbol{\kappa}} d\boldsymbol{\kappa}, \quad (1.26)$$

with $\boldsymbol{\kappa} = (\kappa_1, \kappa_2, \ldots, \kappa_n)$ being an n-dimensional vector in the Fourier space. Integration in these expressions is performed over the n-dimensional real or complex space. On introducing the temporal and the spatial Fourier transforms, we have specifically set and will use throughout the text the following selection of fonts: $F(\mathbf{r}, \omega)$ (space and angular frequency), $\mathcal{F}(\mathbf{r}, t)$ (space–time) and $\mathfrak{F}(\boldsymbol{\kappa}, \omega)$ (spatial frequency – angular frequency).

An important particular case is the three-dimensional spatial Fourier transform. Let us consider function $F(\mathbf{r}) = F(x, y, z)$, where x, y and z are the Cartesian coordinates of a three-dimensional position vector **r**. One then can define the three-dimensional Fourier transform, $\mathfrak{F}(\boldsymbol{\kappa})$, as

$$\mathfrak{F}(\boldsymbol{\kappa}) = \int_{-\infty}^{\infty} F(\mathbf{r})e^{i\mathbf{r}\cdot\boldsymbol{\kappa}}d\mathbf{r}, \qquad (1.27)$$

where $\boldsymbol{\kappa} = (\kappa_x, \kappa_y, \kappa_z)$ is a spatial frequency vector, and integration extends over the whole three-dimensional space. Conversely,

$$F(\mathbf{r}) = \frac{1}{(2\pi)^3} \int \mathfrak{F}(\boldsymbol{\kappa}) e^{-i\mathbf{r}\cdot\boldsymbol{\kappa}} d\boldsymbol{\kappa}. \qquad (1.28)$$

Here, $d\mathbf{r} = dxdydz$, $d\boldsymbol{\kappa} = d\kappa_x d\kappa_y d\kappa_z$, $\mathbf{r}\cdot\boldsymbol{\kappa}$ is the scalar product.

If function $F(\mathbf{r})$ has spherical symmetry, i.e., if $F(\mathbf{r}) = F(r) = F(\sqrt{x^2+y^2+z^2})$, then it is possible to relate its three-dimensional Fourier transform to the one-dimensional Fourier transform with respect to radial variable r, which is known as the *spherical Fourier transfrom*. In fact, in the spherical coordinate system with polar angle θ and azimuthal angle ϕ, we have $d\mathbf{r} = r^2\sin\phi\, dr d\phi d\theta$ and $\mathbf{r}\cdot\boldsymbol{\kappa} = \kappa r\cos\phi$ (without loss of generality, if vector $\boldsymbol{\kappa}$ coincides with the z-axis). Hence,

$$\mathfrak{F}(\boldsymbol{\kappa}) = \int F(\mathbf{r})e^{i\mathbf{r}\cdot\boldsymbol{\kappa}}d\mathbf{r}$$

$$= \int_0^{\infty} dr \int_0^{\pi} d\phi \int_0^{2\pi} d\theta F(r)r^2 \sin\phi e^{i\kappa r\cos\phi}$$

$$= 2\pi \int_0^{\infty} dr F(r)r^2 \int_0^{\pi} d\phi \sin\phi e^{i\kappa r\cos\phi}. \qquad (1.29)$$

Using the fact that

$$\frac{\partial}{\partial\phi}(e^{i\kappa\cos\phi}) = i\kappa r\sin\phi e^{i\kappa r\cos\phi}, \qquad (1.30)$$

we then get the result

$$\mathfrak{F}(\boldsymbol{\kappa}) = 2\pi \int_0^{\infty} dr F(r)r^2 \left[\frac{e^{i\kappa r\cos\phi}}{i\kappa r}\right]_{\phi=0}^{\pi}$$

$$= 4\pi \int_0^{\infty} F(r)r^2 \frac{\sin\kappa r}{\kappa r} dr$$

$$= 4\pi \int_0^{\infty} F(r)r^2 \mathrm{sinc}(\kappa r) dr, \qquad (1.31)$$

where sinc$(x) = \sin x/x$. The spherical Fourier transform is widely used in problems relating to light radiation/scattering from the spherically symmetric sources/particles.

The two-dimensional Fourier transform is obtained in a similar way if $F(\mathbf{r}) = F(x, y)$, and $\boldsymbol{\kappa} = (\kappa_x, \kappa_y)$:

$$\mathfrak{F}(\boldsymbol{\kappa}) = \int F(\mathbf{r}) e^{i\mathbf{r}\cdot\boldsymbol{\kappa}} d\mathbf{r}, \tag{1.32}$$

where integration is performed over a plane, and, conversely,

$$F(\mathbf{r}) = \frac{1}{(2\pi)^2} \int \mathfrak{F}(\boldsymbol{\kappa}) e^{-i\mathbf{r}\cdot\boldsymbol{\kappa}} d\boldsymbol{\kappa}, \tag{1.33}$$

with $d\mathbf{r} = dxdy$, $d\boldsymbol{\kappa} = d\kappa_x d\kappa_y$.

Further, if function $F(\mathbf{r})$ defined in two dimensions has polar symmetry, i.e., if $F(\mathbf{r}) = F(r) = F(\sqrt{x^2 + y^2})$, then it is possible to relate its two-dimensional Fourier transform with the one-dimensional version taken with respect to the radial variable r. In polar coordinate system with radius r and angle θ, we have $d\mathbf{r} = rdrd\theta$, $\mathbf{r}\cdot\boldsymbol{\kappa} = \kappa r \cos\theta$ and, hence,

$$\mathfrak{F}(\boldsymbol{\kappa}) = \int_{r=0}^{\infty} \int_{\theta=0}^{2\pi} F(r) e^{-i r\kappa \cos\theta} r d\theta dr. \tag{1.34}$$

After separating the integrals over θ and r and on setting

$$J_0(x) = \int_0^{2\pi} e^{-ix\cos\theta} d\theta, \tag{1.35}$$

J_0 being the zero-order Bessel function, we arrive at the expression

$$\mathfrak{F}(\boldsymbol{\kappa}) = \int_0^{\infty} r F(r) J_0(r\kappa) dr. \tag{1.36}$$

This version of the Fourier transfrom, is known as the zero-order *Hankel transform*, being of great importance in problems with axial symmetry.

The situation becomes more complex when function $F(\mathbf{r})$ expressed in polar coordinates has angular dependence. To illustrate this point, we will outline the important special case of a two-dimensional function defined in polar coordinates that is separable in radial and polar variables:

$$F(r, \theta) = F_r(r) F_\theta(\theta), \qquad (1.37)$$

where function $F_\theta(\theta)$ is periodic with period 2π. Then the Fourier transform reduces to trigonometric-Bessel series:

$$F(\kappa) = a_0 \pi \int_0^\infty F_r(r) J_0(\kappa r) r \, dr$$

$$+ \sum_{m=1}^\infty 2 a_m \pi \cos(m\theta) (-i)^m \int_0^\infty F_r(r) J_m(\kappa r) r \, dr$$

$$+ \sum_{m=1}^\infty 2 b_m \pi \sin(m\theta) (-i)^m \int_0^\infty F_r(r) J_m(\kappa r) r \, dr, \quad (1.38)$$

where J_m is the mth order Bessel function and a_m and b_m are defined as

$$a_m = \frac{1}{\pi} \int_{-\pi}^\pi F_\theta(\theta) \cos(m\theta) d\theta, \quad (m = 0, 1, 2, \ldots) \qquad (1.39)$$

and

$$b_m = \frac{1}{\pi} \int_{-\pi}^\pi F_\theta(\theta) \sin(m\theta) d\theta, \quad (m = 1, 2, \ldots). \qquad (1.40)$$

Let us now briefly review the definition and the properties of the delta function and establish its relation with the Fourier transforms. Momentarily, we switch back to the one-dimensional $t-\omega$ pair. One-dimensional *Dirac-delta function* or, simply, *δ-function* $\delta(t)$ is a generalized function (distribution) which satisfies the following two conditions:

$$\delta(t) = 0, \quad t \neq 0, \qquad (1.41)$$

and

$$\int_{-\infty}^\infty \delta(t) \mathcal{F}_t(t) dt = f(0), \qquad (1.42)$$

for any smooth *test function* $\mathcal{F}_t(t)$. Alternatively, the delta-function can be defined by means of a *δ-sequence* of ordinary functions, i.e., functions $\delta_n(t)$ such that $\int_{-\infty}^{\infty} \delta_n(t)dt = 1$ and $\delta_n(t)$ should tend to $\delta(t)$ as $n \to \infty$. For example, we can choose a Gaussian sequence:

$$\delta_n(t) = \frac{n}{\pi} \exp[-n^2 t^2]. \tag{1.43}$$

The delta-function obeys the following properties:

(1) *Sifting*:

$$\int_{-\infty}^{\infty} \delta(t-a)\mathcal{F}(t)dt = \mathcal{F}(a). \tag{1.44}$$

(2) *Scaling*:

$$\delta(bt) = \frac{\delta(t)}{|b|}. \tag{1.45}$$

(3) *Composition*:

$$\delta[\mathcal{F}(t)] = \sum_a \frac{\delta(t-a)}{|\mathcal{F}'(a)|}, \tag{1.46}$$

where $t = a$ is the zero of $\mathcal{F}(t)$, and prime denotes the derivative.

(4) *Differentiation*:

$$\int_{-\infty}^{\infty} \frac{d^n \delta(t)}{dt^n} \mathcal{F}(t) dt = (-1)^n \frac{d^n \mathcal{F}(t)}{dt^n}\bigg|_{t=0}. \tag{1.47}$$

(5) Delta-function is *even*:

$$\delta(-t) = \delta(t). \tag{1.48}$$

Let us now establish the relationship between the delta-function and the Fourier transform. Any well-behaved function $\mathcal{F}(t)$ can be first represented by the pair of nested Fourier transforms, direct and

Introduction

inverse:

$$\mathcal{F}(t) = \frac{1}{2\pi} \int_{-\infty}^{\infty} d\omega e^{-i\omega t} \int_{-\infty}^{\infty} du \mathcal{F}(u) e^{i\omega u}$$
$$= \int_{-\infty}^{\infty} du \mathcal{F}(u) \left[\frac{1}{2\pi} \int_{-\infty}^{\infty} e^{-i\omega(t-u)} d\omega \right], \quad (1.49)$$

where u is the integration variable. Then the part of the formula in the square brackets may be recognized as delta-function $\delta(t-u)$:

$$\delta(t) = \frac{1}{2\pi} \int_{-\infty}^{\infty} 1 \cdot e^{-i\omega t} d\omega, \quad (1.50)$$

implying that delta-function $\delta(t)$ and identity function $\mathcal{F}(t) = 1$ are the Fourier transform pair. Say, if a process in real space or time is of the form of an impulse, then its Fourier transform is equally distributed among all the frequencies, constituting *white noise*.

Similarly, for the analysis of optical fields varying in space, the delta-function can be extended to higher (n) dimensions by the formula:

$$\delta^{(n)}(\mathbf{r}) = \frac{1}{(2\pi)^n} \int 1 \cdot e^{-i\boldsymbol{\kappa} \cdot \mathbf{r}} d\boldsymbol{\kappa}, \quad (1.51)$$

where integration is performed over the n-dimensional real space.

As we will see in the subsequent chapters, the temporal Fourier transform is typically applied for bringing the analysis of stationary optical signals from the space–time to the space–frequency domain for considerable simplification of theoretical calculations. It is also used as an analytical tool for characterizing cyclo-stationary pulse trains in the frequency domain. On the other hand, the spatial Fourier transform is frequently used for quantitaive analysis of imaging systems and random media. The spatial frequencies provide the insight into the presence and strength of certain spatial scales involved in free-space light evolution and light–matter interactions.

1.3.2 Analytic signal

It follows from the basic properties of the Fourier transform that a real-valued function of time, say $\mathcal{F}^{(R)}(t)$, has both positive and negative frequency components. While mathematically acceptable, negative frequencies must be avoided in optics. This can be done by introducing the complex-valued *analytic signal*, say $\mathcal{F}^{(A)}(t)$, associated with $\mathcal{F}^{(R)}(t)$ as

$$\mathcal{F}^{(A)}(t) = \mathcal{F}^{(R)}(t) + i\left[\mathcal{F}^{(R)}(t) \circledast \frac{1}{\pi t}\right], \quad (1.52)$$

where, as before, ⊛ stands for convolution. Indeed, let us first choose the Fourier transform of $\mathcal{F}^{(A)}(t)$ to be defined by piecewise continuous function

$$FT\left[\mathcal{F}^{(A)}(t)\right] = F^{(A)}(\omega)$$

$$= \begin{cases} 2F^{(R)}(\omega), & \omega > 0, \\ F^{(R)}(\omega), & \omega = 0, \\ 0, & \omega < 0, \end{cases}$$

$$= F^{(R)}(\omega) \cdot 2H_s(\omega), \quad (1.53)$$

where $F^{(R)}(\omega)$ is the Fourier transform of $\mathcal{F}^{(R)}(t)$ and $H_s(\omega)$ is a Heaviside step function in the Fourier domain. It only contains the non-negative frequency components of $f^{(R)}(t)$. Next, the Hermiticity of $F(\omega)$ implies

$$F^{(R)}(\omega) = \begin{cases} \dfrac{1}{2}F^{(A)}(\omega), & \omega > 0, \\ F^{(A)}(\omega), & \omega = 0, \\ \dfrac{1}{2}F^{(A)*}(-\omega), & \omega < 0. \end{cases} \quad (1.54)$$

Finally, on using the fact that the inverse Fourier transform of $F^{(R)}(\omega)$ is $\mathcal{F}^{(R)}(t)$ and that of $H_s(\omega)$ is $\delta(t) + i/(\pi t)$, $\delta(t)$ being the Dirac delta-function, Eq. (1.52) is confirmed.

In fact, the real and the imaginary parts of the analytic signal, $\mathcal{F}^{(R)}(t)$ and $\mathcal{F}^{(I)}(t)$, are related as the Hilbert transform pair (Gabor,

1946):

$$\mathcal{F}^{(I)}(t) = \frac{1}{\pi} Pr \int_{-\infty}^{\infty} \frac{\mathcal{F}^{(R)}(t')}{t'-t} dt',$$
$$\mathcal{F}^{(R)}(t) = -\frac{1}{\pi} Pr \int_{-\infty}^{\infty} \frac{\mathcal{F}^{(I)}(t')}{t'-t} dt', \quad (1.55)$$

where Pr denotes the *principle value* of a contour integral. These expressions are also known in physics as the Kramers–Kronig relations and in mathematics as the Sokhotski–Plemelj theorem.

To give a simple example, let us consider a monochromatic real signal

$$\mathcal{F}^{(R)}(t) = a\cos(\omega t), \quad (1.56)$$

where a is a non-trivial real constant. Its spectrum is immediately found by involving Euler's formula:

$$\mathcal{F}^{(R)}(t) = \frac{a}{2}\left[e^{i\omega t} + e^{-i\omega t}\right], \quad (1.57)$$

and contains both positive and negative frequency components. The corresponding analytic signal takes the form

$$\mathcal{F}^{(A)}(t) = a[\cos(\omega t) + i\sin(\omega t)] = ae^{i\omega t}, \quad (1.58)$$

in which the negative frequency component vanishes and the positive one doubles. The analytic signal expressed in polar coordinates becomes

$$\mathcal{F}^{(A)}(t) = a(t)e^{i\varphi(t)}, \quad (1.59)$$

where $a(t)$ is the *instantaneous amplitude* or the *envelop* and $\varphi(t)$ is the *instantaneous phase* or *phase angle*. Further, the *instantaneous angular frequency* is then defined as $\omega(t) = d\varphi(t)/dt$.

If the signal is quasi-monochromatic, i.e., has a small spread $\Delta\omega$ around its central frequency $\bar{\omega}$: $\Delta\omega \ll \bar{\omega}$, then it may be written as

$$\mathcal{F}^{(R)}(t) = a(t)\cos[\bar{\omega}t - \varphi(t)]. \quad (1.60)$$

In this case, the choice of $a(t)$ and $\varphi(t)$ is not unique since the left and the right sides of Eq. (1.60) have a different number of functions. Such ambiguity is removed if one defines the analytic signal as

$$\mathcal{F}^{(A)}(t) = a(t)\exp[i\varphi(t)]\exp[i\bar{\omega}t]. \tag{1.61}$$

Its envelope $a(t)\exp[i\varphi(t)]$ varies with time much slower than the monochromatic signal $\exp[i\bar{\omega}t]$ at the mean angular frequency.

1.4 Statistical Preliminaries

1.4.1 *Linear processes*

In physics and optics, in particular, the description of all the phenomena of interest is typically made with the help of a *dynamical process*, characterizing spatio-temporal evolution of a system isolated from the rest of the universe. The process starts from an *initial state* which occurs at some moment in time (or a set of moments) and converts to a *final state* at a later moment (or set of moments) by means of a *dynamical system*. Depending on the problem of interest, typically two out of three entities (system, initial state, final state) are known and one is unknown. If the initial state and the dynamical system are known, then the problem of finding the final state is termed *direct*. Finding the initial state from the dynamical system and the final state, or the dynamical system from the initial and final states are known as *inverse* problems. For example, finding the properties of scattered light from the knowledge of those of the illumination and of the scattering medium is a direct problem while finding the properties of the medium from those of the illumination and of the scattered light is an inverse problem.

Dynamical systems can be broadly classified into *linear* and *nonlinear*. If a linear combination of two initial states enters a linear system, then the return is the linear combination of the states passed through the system independently. This can be mathematically expressed as

$$L(af + bg) = aL(f) + bL(g), \tag{1.62}$$

where L denotes a generic linear system, f and g are any two inputs and a and b are arbitrary, generally complex constants. On the other hand, a nonlinear system will not generally return such a combination. This book will be solely confined to linear systems and, hence, linear processes. The states of the process can generally be of any complexity, being real or complex scalars, vectors, tensors; discrete or continuous distributions of such quantities in space, etc. In statistical optics, one is typically interested in spatial and/or temporal distributions of an optical field, being complex-valued vector functions.

All natural and man-made processes can also be classified as *deterministic* or *stochastic* (*random*) (Papoulis and Pillai, 2002). For deterministic processes, the initial state and the system must be characterized with certainty, leading to a completely predictable final state. For random processes, the initial state and/or the system might be random, i.e., either not predictable at all or predictable only to some extent. Then the final state of the random process is always random. Caution must be taken in dealing with *chaotic* processes, that are intrinsically deterministic and nonlinear, but with a behavior that is so difficult to predict that they seem to act as random. Examples of optical chaotic systems are a semiconductor or a fiber laser amplifier.

The most interesting problems of statistical optics arise when averages of a random process governing light fluctuations are measurable with values being invariant if the experiment is repeated. An example of this is the formation of fringes with a constant contrast in the two-slit experiment with direct sunlight. Another set of interesting and practically useful problems appear when the randomness of the source radiating a light field is used for sensing or mitigation of complex and random media. The most successful technologies based on such interactions are the Optical Coherence Tomography (OCT) and the Free-Space Optical (FSO) laser communications.

1.4.2 *Random variables and processes*

We begin by introducing the concept of a *discrete random variable*, say, X, i.e., one which may take any value from a given, finite set of values, called a *state space* $\bar{X}_n$: $\{X_1, X_2, \ldots, X_n, \ldots, X_N\}$ with a fixed probability, say $P(X_n)$. We will assume for now that $\bar{X}_n$ is a subset of real numbers, without any loss of generality. Axiomatically, the *probability* of attaining a certain value from the state space can be defined by imposing three conditions (Papoulis and Pillai, 2002):

(1) certainty: $P(X_n) = 1$ if X_n is the only possible state;
(2) non-negativity: $P(X_n) \geq 0$ for any n;
(3) mutual exclusiveness: $P(X_{n_1} \text{ or } X_{n_2}) = P(X_{n_1}) + P(X_{n_2})$.

A map $p(X_n)$ from the state space to the set of probabilities is known as the discrete *Probability Density Function* (PDF). Two random variables are called *identically distributed* if they have the same PDF. The simplest example of a PDF is *uniform*, for which the probabilities of N outcomes are the same, $P(X_n) = 1/N$, $n = 1, \ldots, N$. Another classic example is the *Bernoulli* PDF, for which $N = 2$ and the two probabilities are $P(X_1) = \rho$ and $P(X_2) = 1 - \rho$, $0 \leq \rho \leq 1$.

Empirically, the probabilities of random variables are determined by forming a sequence of experiments (trials) and analyzing the results of the outcomes. For a variable with state space $\bar{X}_n$, $n = 1, \ldots, N$, the experiment is performed as a sequence of trials $\bar{T}_m$: $\{T_1, T_2, \ldots, T_m, \ldots, T_M\}$ in which only one element of the state space is randomly drawn at a time. The experimentally obtained random samples of state variables constitute possible *realizations* $\bar{x}_m = \{x_1, x_2, \ldots, x_m, \ldots x_M\}$. While the beginning of the trial sequence appears random, for large number of trials M, the number of occurrences of each state variable tends to a certain number, say $P(X_n)$, that can be empirically obtained as a limit:

$$P(X_n) = \lim_{M \to \infty} \frac{P_n}{M}, \qquad (1.63)$$

where P_n is the number of times state variable X_n appears in trial sequence $\bar{T}_m$. It is also typically assumed that the result of any trial is independent from the results of all previous trials and that the state space and the probabilities are the same for any trial.

Function $p(X_n)$ accounting for the distribution of probabilities occuring in all possible states is the experimental PDF of the random variable.

If a random variable X takes any value in a given interval of real numbers, $X \in [a, b]$, $a, b \in (-\infty, +\infty)$, with fixed probabilities, then it is called a *continuous random variable* and $[a, b]$ becomes its state space. Then its PDF $p_X(x)$ also becomes a continuous function defined on this interval. The most famous example of a continuous random variable is Gaussian:

$$p_X(x) = \frac{1}{\sigma\sqrt{2\pi}} \exp\left[-\frac{1}{2}\left(\frac{x - x_m}{\sigma}\right)^2\right], \quad x \in (-\infty, +\infty), \quad (1.64)$$

where parameters x_m and σ characterize the mean and the variance of the distribution. Widely used probability density function models of the instantaneous intensity of laser light after interacting with turbulent media are discussed in Chapter 9.

A random variable may change with time, taking on the values from its state space. In such cases, one introduces a *random process*, say $\mathcal{X}(t)$. In addition, the value of the random variable may also fluctuate in space forming a *random field*. Both time advances and spatial increments may be either discrete or continuous. In theoretical, classical statistical optics, we are primarily concerned with light fields governed by continuous random processes that are also undergoing continuous fluctuations in space. For instance, an analytic signal describing the fluctuations in one of the components of the electric field represents an important example of the continuous spatio-temporal random process discussed in this book.

Let us fix for a moment a spatial position in the random field and only consider temporal fluctuations of the process. At this position, the complete characterization of random process $\mathcal{X}(t)$ can only be achieved if the nth order *joint* probability density functions $p_n = p_n(\mathcal{X}_1, \mathcal{X}_2, \ldots, \mathcal{X}_n; t_1, t_2, \ldots, t_n)$, describing probability that $\mathcal{X}(t)$ takes value $\mathcal{X}_1$ at time instant t_1 and so on, up to taking value $\mathcal{X}_n$ at moment t_n, simultaneously for all $n = 1, \ldots, \infty$. Obviously, in a typical practical situation it is never a possibility. Another theoretically legitimate method, listing all possible individual realizations of the process, i.e., its complete ensemble of realizations, $\bar{x}(t) = \{x_1(t), x_2(t), \ldots, x_m(t), \ldots, x_M(t)\}$, for all allowed time instants t, is also impractical. That is why, in statistics, and in statistical optics in particular, most of the methods for theoretical work and experimental measurements of a random process rely on the knowledge of its *statistical moments* of the first several orders which provide information about the random process averaged with respect to some subset of its characteristics.

There are two fundamentally different ways of defining the statistical moments, via *ensemble averaging* and *time averaging*. The nth order joint moment of the former type is based on integral:

$$\langle w_E(\mathcal{X}(t_1), \mathcal{X}(t_2), \ldots, \mathcal{X}(t_n)) \rangle_E$$
$$= \int \int \cdots \int w_E(\xi_1, \xi_2, \ldots, \xi_n)$$
$$\times p_n(\xi_1, \xi_2, \ldots, \xi_n, t_1, t_2, \ldots, t_n) d\xi_1 d\xi_2 \ldots d\xi_n, \quad (1.65)$$

where subscript E stands for the average taken over the ensemble of realizations. Here, the weighting function w_E can be arbitrary but is typically chosen to be the product of all its arguments, i.e.,

$$w_E(\xi_1, \xi_2, \ldots, \xi_n) = \xi_1 \cdot \xi_2 \cdot \ldots \cdot \xi_n. \quad (1.66)$$

With such a choice of w_E, the first two moments, the average value and the two-point correlation function, are given by the expressions

$$\langle \mathcal{X}(t_1) \rangle_E = \int \xi p_1(\xi, t_1) d\xi, \quad (1.67)$$

and

$$\langle \mathcal{X}(t_1) \mathcal{X}(t_2) \rangle_E = \int \int \xi_1 \xi_2 p_2(\xi_1, \xi_2, t_1, t_2) d\xi_1 d\xi_2. \quad (1.68)$$

The time averaging procedure is defined by selecting particular realizations, say $x_{m_1}(t), \ldots, x_{m_M}(t)$, choosing a sufficiently long time interval T, and performing integration and limit of the form

$$\langle w_T(x(t_1), x(t_2), \ldots, x(t_M))\rangle_T$$
$$= \lim_{T\to\infty} \frac{1}{T} \int_{-T/2}^{T/2} w_T(x_{m_1}(t+t_1), x_{m_2}(t+t_2), \ldots, x_{m_M}(t+t_m))dt, \tag{1.69}$$

where w_T can be an arbitrary function, but is typically chosen as product

$$w_T(x_{m_1}(t), x_{m_2}(t), \ldots, x_{m_M}(t)) = x_{m_1}(t)x_{m_2}(t)\cdots x_{m_M}(t). \tag{1.70}$$

Then the long time average and the two-point correlation function become

$$\langle x(t_1)\rangle_T = \lim_{T\to\infty} \frac{1}{T} \int_{-T/2}^{T/2} x_{m_1}(t)dt, \tag{1.71}$$

and

$$\langle x(t_1)x(t_2)\rangle_T = \lim_{T\to\infty} \frac{1}{T} \int_{-T/2}^{T/2} x_{m_1}(t+t_1)x_{m_2}(t+t_2)dt. \tag{1.72}$$

1.4.3 *Types of random processes*

After listing formal definitions of statistical moments, we now briefly outline the types of random processes used in this book. One such classification can be introduced on the basis of the process' *stationarity*, i.e., with respect to constraints imposed on the behavior of its statistical moments of different orders. Historically, most of the early studies made within the theories of speckle and optical coherence relied on stationary, or, more practically, wide-sense stationary processes. A typical laser beam passing through a ground-glass diffuser rotating with a constant speed is the best example of a stationary process. Consequently, various extensions into non-stationary random processes have been made: laser light interaction with optical turbulence has required the development of the framework for dealing with non-stationary processes with stationary increments (Tatarskii,

1971), and the analysis of random pulse trains stipulated the adoption of the cyclostationary process theory (Gardner et al., 2006) for the optical waves. While the stationary optical fields will dominate our discussions throughout the text, the other aforementioned types of non-stationary processes will be considered in the last two chapters, respectively.

A *statistically stationary* random process is defined as being independent from the origin of time, viz., its joint statistical moments of any order n must have time translation symmetry:

$$p_n(\mathcal{X}_1, \mathcal{X}_2, \ldots, \mathcal{X}_n, t_1, t_2, \ldots, t_n)$$
$$= p_n(\mathcal{X}_1, \mathcal{X}_2, \ldots, \mathcal{X}_n, t_1 + \tau, t_2 + \tau, \ldots, t_n + \tau), \quad (1.73)$$

for all values of time lag τ. Such processes are the idealization and cannot exist in practice since they must carry an infinite amount of energy.

A *wide-sense statistically stationary* random process was introduced in order to circumvent the unrealizability of a strictly stationary process. For such a process, it is only required that its average value is a constant and the two-instant correlation function depends on the time difference:

$$\langle \mathcal{X}(t_1) \rangle_E = \text{constant}; \quad \langle \mathcal{X}(t_1)\mathcal{X}(t_2) \rangle_E = \mathcal{G}(t_1 - t_2), \quad (1.74)$$

for some function $\mathcal{G}$, while no requirement is set for higher-order moments.

A *non-stationary process with stationary increments* requires that difference $\mathcal{X}(t_1) - \mathcal{X}(t_2)$ is stationary, for all time instants t_1 and t_2. In particular, the processes whose average values change linearly with time belong to this category.

A *cyclostationary process* must have the joint statistical moments of any order n obeying relation (1.73) *for some values* of time lag τ (and not all values as for the stationary process) (Davis, 2007). The smallest value of τ is associated with the period of the process. Realizations of cyclostationary processes can be visualized as periodic pulse trains in which individual pulse events may differ from each other in duration, magnitude, shape, or the combination of these properties.

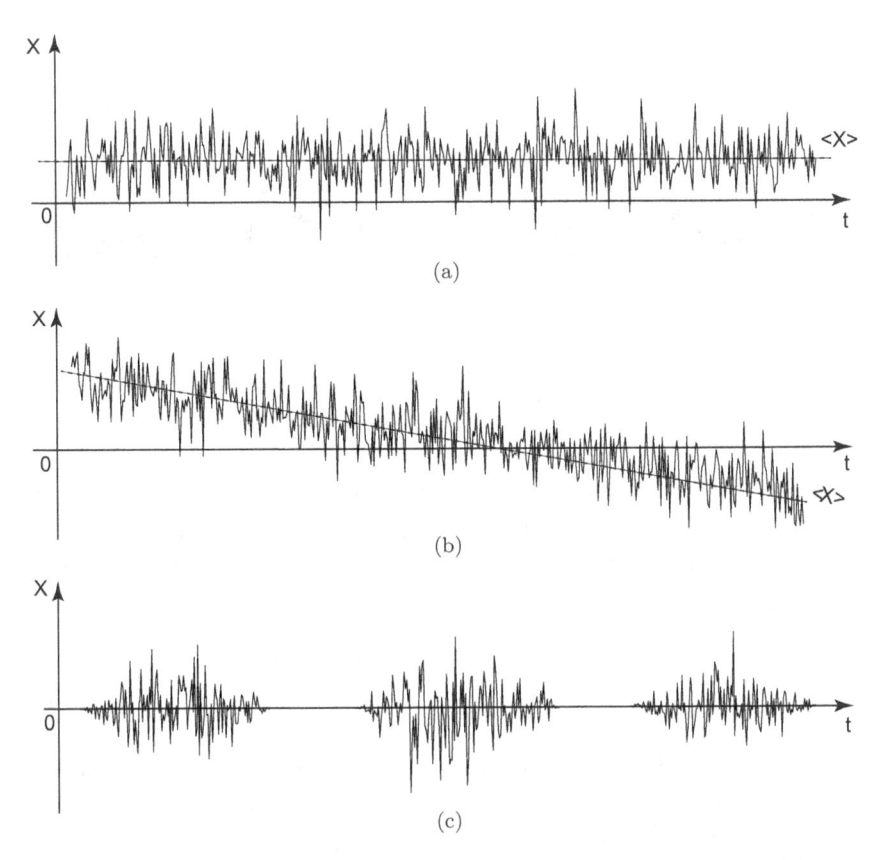

Fig. 1.6 Typical realizations of random processes: (a) stationary; (b) non-stationary with stationary increments; (c) cyclostationary.

Figure 1.6 shows realizations of a stationary process, a non-stationary process with stationary increments and a cyclostationary process. While in the first two subfigures the averages are shown explicitly, in the last subfigure it is trivial.

Another classification of random processes can be obtained in terms of equivalence of their ensemble and time-averaged statistical moments. Such classification carries enormous practical importance since it addresses the question whether the analytic properties of the process carried in the statistical ensemble of its realizations can be inferred solely from its single realization, since it is frequently the only one available experimentally.

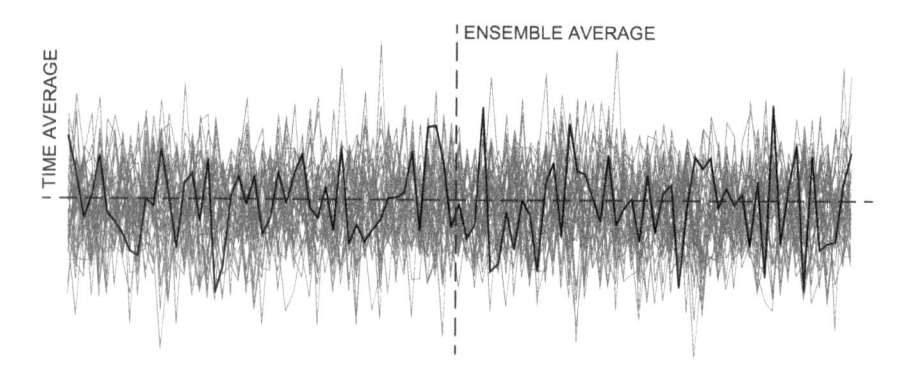

Fig. 1.7 Illustrating the concept of ergodicity.

Ergodic random processes require that for the moments of any order the time averages obtained from a single realization and the ensemble averages calculated for all realizations at a fixed set of times are equal:

$$\langle w_T[x(t_1), x(t_2), \ldots, x(t_n)] \rangle_T = \langle w_E[\mathcal{X}(t_1), \mathcal{X}(t_2), \ldots, \mathcal{X}(t_n)] \rangle_E, \quad (1.75)$$

for arbitrary weighting functions w_T (or w_E), and any realization m. If the process is ergodic, it is automatically stationary and wide-sense stationary. Because the opposite is not always true, the ergodicity requirement must be imposed for characterizing the process by measuring its long-time averages. As long as the process is ergodic, the complete estimation of its probability density functions of various orders can be made from its single realization.

Figure 1.7 illustrates the concept of ergodicity as applied to the average value of the process. Twenty realizations of a generic ergodic process were generated in a computer simulation: 19 realizations are given in gray color and the 20th is highlighted by black color. The horizontal line denotes the time average obtained from a single realization. The vertical line is drawn by using all 20 realizations at a single, while generic, time instant, and is used for finding the ensemble average. In our example, the two estimates lead to the same average, making processes ergodic.

The concept of the *cycloergodic random process* is needed for ensuring the stable measurements of cyclostationary processes. In contrast with stationary processes, no insight into the statistics of a non-stationary process can be generally made from analyzing its single realization or even a set of realizations. However, if the process is cyclostationary, imposing the cycloergodicity condition solves the problem: one can still consistently estimate the probability density functions of all orders from a single realization of the process. The rigorous mathematical formulation of the cycloergodicity conditions is beyond the scope of this book but may be found elsewhere (Gardner et al., 2006). Intuitively, within the validity of these conditions the averaging taken across a number of periods of the process is sufficient for obtaining consistent estimates of the statistical moments at individual positions within a single period.

Since the optical fields are complex-valued, we briefly state here that all the concepts discussed in the preceding text regarding real-valued processes may be readily generalized to the complex domain. Let

$$\mathcal{Z}(t) = \mathcal{X}(t) + i\mathcal{Y}(t) \qquad (1.76)$$

be a complex-valued random process. It is practically useful to assume that its real part $\mathcal{X}(t)$ and imaginary part $\mathcal{Y}(t)$ constitute an analytic signal. The realizations, the statistical moments, the probability density functions, etc. of such a process must be complex-valued as well. For instance, the definition of a statistical moment must now rely on complex conjugates. For example, the nth order joint statistical moment is defined by the formula

$$\langle \mathcal{Z}^*(t_1)\mathcal{Z}^*(t_2)\cdots\mathcal{Z}(t_{n-1})\mathcal{Z}(t_n)\rangle_E$$
$$= \int\int\cdots\int \chi_1^*\chi_2^*\cdots\chi_{n-1}\chi_n$$
$$\times p_n(\chi_1,\chi_2,\ldots,\chi_n;t_1,t_2\ldots t_n)d^2\chi_1 d^2\chi_2\ldots d^2\chi_n, \qquad (1.77)$$

where χ_i $(i = 1,\ldots n)$ are the complex variables of integration and $d^2\chi_i$, $(i = 1,\ldots n)$ are the differentials in the complex plane.

1.4.4 Spectral content of a wide-sense stationary process

As we already stated, any absolutely integrable function $\mathcal{F}(t)$ possesses the Fourier transform, viz., integral in Eq. (1.5) converges. Then function

$$A(\omega) = |F(\omega)|^2$$
$$= \left| \int_{-\infty}^{\infty} \mathcal{F}(t) e^{i\omega t} dt \right|^2 \tag{1.78}$$

represents the *energy spectrum* of $\mathcal{F}(t)$, i.e., gives the distribution of the average energy per angular frequency ω. On the other hand, if inequality

$$\lim_{T \to \infty} \frac{1}{T} \int_{-T/2}^{T/2} \mathcal{F}^2(t) dt < \infty \tag{1.79}$$

holds, where T is a sufficiently large time interval, the power of signal $\mathcal{F}(t)$ is finite, but it might not have the Fourier transform. Then consider instead a version of $\mathcal{F}(t)$ having finite support:

$$\mathcal{F}_c(t) = \begin{cases} \mathcal{F}(t), & -\dfrac{T}{2} \leq t \leq \dfrac{T}{2}, \\ 0, & t < -\dfrac{T}{2}, t > \dfrac{T}{2}. \end{cases} \tag{1.80}$$

The latter function does have the Fourier transform. On taking the limit as $T \to \infty$, we define *power spectrum* $S(\omega)$ of $\mathcal{F}(t)$ as a limit:

$$S(\omega) = \lim_{T \to \infty} \frac{1}{T} \left| \int_{-\infty}^{\infty} \mathcal{F}_c(t) e^{i\omega t} dt \right|^2. \tag{1.81}$$

We now turn to random (real) processes for which, in general, the limit in Eq. (1.81) might not exist. This is the case, for instance, for the realizations of a strictly stationary process. Therefore, for a random process $\mathcal{X}(t)$, which we assume to be ergodic, the energy spectrum and the power spectrum can be defined by the expressions

$$A_X(\omega) = \left\langle \left| \int_{-\infty}^{\infty} x(t)e^{i\omega t} dt \right|^2 \right\rangle_T,$$
$$S_X(\omega) = \lim_{T \to \infty} \frac{1}{T} \left\langle \left| \int_{-\infty}^{\infty} x_c(t)e^{i\omega t} dt \right|^2 \right\rangle_T, \qquad (1.82)$$

where $x(t)$ is its original realization and $x_c(t)$ is its truncated version.

On the other hand, the *auto-correlation function* of random process $\mathcal{X}(t)$

$$\mathcal{W}_X(t_1, t_2) = \langle x(t_1)x(t_2) \rangle_T = \langle \mathcal{X}(t_1)\mathcal{X}(t_2) \rangle_E, \qquad (1.83)$$

defined by the double integral (see Eqs. (1.68) and (1.72)), determines the time-averaged degree of its self-similarity at time instances t_1 and t_2. For a complex-valued process $\mathcal{Z}(t)$, formula (1.83) generalizes to form

$$\mathcal{W}_Z(t_1, t_2) = \langle z^*(t_1)z(t_2) \rangle_T = \langle \mathcal{Z}^*(t_1)\mathcal{Z}(t_2) \rangle_E. \qquad (1.84)$$

In cases when a random process is at least wide-sense statistically stationary, the autocorrelation function and the power spectrum are simply related as the Fourier transform pair, viz.,

$$S_Z(\omega) = FT[\mathcal{W}_Z(\tau)], \quad \mathcal{W}_Z(\tau) = FT^{-1}[S_Z(\omega)], \qquad (1.85)$$

where $\tau = t_2 - t_1$. The Fourier duality of these two functions is known as the *Wiener–Khintchin theorem* (for random processes) whose practical significance is manifested in a relatively simple measurement procedure of $\mathcal{W}_Z(\tau)$ and the straightforward evaluation of $S_Z(\omega)$ from it.

Another pair of functions important in the theory of optical random processes is the *auto-covariance function* $\mathcal{B}_Z(t_1, t_2)$ and the structure function $\mathcal{D}_Z(t_1, t_2)$ defined by the formulas

$$\begin{aligned}\mathcal{B}_Z(t_1, t_2) &= \langle [z(t_1) - \langle z(t_1)\rangle_T][z(t_2) - \langle z(t_2)\rangle_T]\rangle_T \\ &= \mathcal{W}_Z(t_1, t_2) - \langle z(t_1)\rangle_T \langle z(t_2)\rangle_T,\end{aligned} \quad (1.86)$$

and

$$\begin{aligned}\mathcal{D}_Z(t_1, t_2) &= \langle [z(t_1) - z(t_2)]^2\rangle_T \\ &= \langle z^2(t_1)\rangle_T + \langle z^2(t_2)\rangle_T - 2\mathcal{W}_Z(t_1, t_2),\end{aligned} \quad (1.87)$$

respectively. If structure function $\mathcal{D}_Z(t_1, t_2)$ depends only on time delay $\tau = t_2 - t_1$, it might be of great importance, even for certain processes that are not wide-sense stationary. This is the case for processes with stationary increments. For instance, atmospheric fluctuations in temperature, and, hence, in the refractive index, can be well-modeled for short time intervals as a process with stationary increments because the mean temperature slowly changes during the day–night cycle. Hence, for atmospheric applications, the knowledge of the refractive-index structure functions is a must (Tatarskii, 1971).

It can be readily verified that the structure function relates to the power spectrum by the formula

$$\mathcal{D}_Z(\tau) = 2\int_{-\infty}^{\infty} S_Z(\omega)[1 - \cos(\omega\tau)]d\omega. \quad (1.88)$$

Another group of functions frequently used in statistical optics are the *cross-correlation functions*, defined for two processes, $\mathcal{Z}_1$ and $\mathcal{Z}_2$, as

$$\mathcal{W}_{Z_1 Z_2}(t_1, t_2) = \langle z_1^*(t_1) z_2(t_2)\rangle_T, \quad (1.89)$$

and, in particular,

$$\mathcal{W}_{Z_1 Z_2}(\tau) = \langle z_1^*(t) z_2(t + \tau)\rangle_T. \quad (1.90)$$

Random processes are called *jointly wide-sense stationary* when

$$\mathcal{W}_{Z_1 Z_2}(t_1, t_2) = \mathcal{W}_{Z_1 Z_2}(\tau), \quad (1.91)$$

i.e., it depends only on time difference $\tau = t_1 - t_2$. The counterpart of the cross-correlation function in frequency domain is defined as

$$W_{Z_1 Z_2}(\omega) = \lim_{T \to \infty} \frac{1}{T} \langle FT[z(t_1)]_c^*(\omega) FT[z(t_2)]_c(\omega) \rangle_E, \quad (1.92)$$

and is termed the *cross-spectral density function*, where $FT[z(t)]_c$ is the Fourier transform of a truncated version of the process. The cross-spectral density function is a measure of similarity between the two processes at a fixed frequency. It obeys the quasi-Hermiticity property:

$$W_{Z_1 Z_2}(\omega) = W_{Z_2 Z_1}^*(\omega), \quad (1.93)$$

and, in particular, for any real-valued processes $\mathcal{X}$ and $\mathcal{Y}$:

$$W_{XY}(-\omega) = W_{XY}(\omega). \quad (1.94)$$

The *generalized Wiener–Khintchin theorem* relates $W_{Z_1 Z_2}(\omega)$ and $\mathcal{W}_{Z_1 Z_2}(\tau)$ as a Fourier transform pair:

$$W_{Z_1 Z_2}(\omega) = FT[\mathcal{W}_{Z_1 Z_2}(\tau)], \quad \mathcal{W}_{Z_1 Z_2}(\tau) = FT^{-1}[W_{Z_1 Z_2}(\omega)]. \quad (1.95)$$

We will see in the following chapters that these relations play a crucial part in characterization of random optical fields.

1.4.5 *Gaussian random process*

Since the complete characterization of a random process generally involving specification of the joint probability density functions of any order is an impossible task, it appears convenient to work with processes, for which the knowledge of the statistical moments of low orders is sufficient. As an example of such a random process, we consider (a real-valued) Gaussian process, for which at any finite set of time instants, say $t_1, \ldots, t_n$, the n-dimensional random

vector
$$\vec{\mathcal{X}} = \begin{bmatrix} \mathcal{X}(t_1) \\ \mathcal{X}(t_2) \\ ... \\ \mathcal{X}(t_n) \end{bmatrix} \quad (1.96)$$

has the multi-variate Gaussian joint probability density function:

$$p_n(\vec{\mathcal{X}}) = \frac{1}{(\sqrt{2\pi})^n \sqrt{\overleftrightarrow{C}}} \exp\left[-\frac{1}{2}[\vec{\mathcal{X}} - \langle\vec{\mathcal{X}}\rangle]^T \overleftrightarrow{C}^{-1} [\vec{\mathcal{X}} - \langle\vec{\mathcal{X}}\rangle]\right], \quad (1.97)$$

where superscript T stands for matrix transposition,

$$\langle\vec{\mathcal{X}}\rangle = \begin{bmatrix} \langle\mathcal{X}(t)\rangle \\ \langle\mathcal{X}(t)\rangle \\ ... \\ \langle\mathcal{X}(t)\rangle \end{bmatrix}, \quad (1.98)$$

is the n-dimensional vector of the average values while $\overleftrightarrow{C}$ is the $n \times n$ covariance matrix with elements

$$C_{lp}^2 = \langle[\mathcal{X}(t_{(l)}) - \langle\mathcal{X}(t_{(l)})\rangle][\mathcal{X}(t_{(p)}) - \langle\mathcal{X}(t_{(p)})\rangle]\rangle,$$
$$(l, p = 1, ..., n). \quad (1.99)$$

In these expressions, time averages are assumed but subscript T is omitted.

A remarkable feature of Gaussian processes is that their statistical moments of any order can be expressed via their second-order moments. For instance, the fourth-order moment at time instants t_1, t_2, t_3 and t_4 becomes

$$\langle\mathcal{X}(t_1)\mathcal{X}(t_2)\mathcal{X}(t_3)\mathcal{X}(t_4)\rangle = \mathcal{W}_X(t_2, t_1)\mathcal{W}_X(t_4, t_3)$$
$$+ \mathcal{W}_X(t_3, t_1)\mathcal{W}_X(t_4, t_2)$$
$$+ \mathcal{W}_X(t_3, t_2)\mathcal{W}_X(t_4, t_1), \quad (1.100)$$

where $\mathcal{W}$ is the auto-correlation function. This implies that if a Gaussian process is wide-sense stationary, it is stationary automatically.

Introduction

A random process is called a *complex Gaussian process* if its real and imaginary parts are joint Gaussian processes. Suppose $\mathcal{Z}^{(A)}(t)$ is an analytic signal representation of the real-valued Gaussian process $\mathcal{X}(t)$, i.e., $\mathcal{Z}^{(A)}(t) = \mathcal{X}(t) + i\mathcal{Y}(t)$. Then $\mathcal{Y}(t)$ is also the Gaussian random process, because of relations (1.55) and, hence, $\mathcal{Z}^{(A)}(t)$ itself is the complex Gaussian process. Nevertheless, the converse is not true, i.e., not every complex Gaussian process is the analytic signal for some real Gaussian process.

The complex Gaussian process is called *circular* if for any quad of time instances the following relations hold:

$$\langle \mathcal{X}(t_m) \rangle = \langle \mathcal{Y}(t_m) \rangle = 0, \qquad (m = 1, 2, 3, 4);$$
$$\langle \mathcal{X}(t_m)\mathcal{X}(t_n) \rangle = \langle \mathcal{X}(t_m)\mathcal{X}(t_n) \rangle, \qquad (m, n = 1, 2, 3, 4); \qquad (1.101)$$
$$\langle \mathcal{X}(t_m)\mathcal{Y}(t_n) \rangle = -\langle \mathcal{X}(t_m)\mathcal{Y}(t_n) \rangle, \qquad (m, n = 1, 2, 3, 4).$$

Analytic signal $\mathcal{Z}^{(A)}(t)$ of a real Gaussian process $\mathcal{X}(t)$ with zero mean can be shown to be a circular Gaussian process (Goodman, 2000). The fourth-order moment of $\mathcal{Z}^{(A)}(t)$ then reduces to the sum

$$\mathcal{W}^{(4)}(t_1, t_2, t_3, t_4) = \langle \mathcal{Z}^{(A)*}(t_1)\mathcal{Z}^{(A)*}(t_2)\mathcal{Z}^{(A)}(t_3)\mathcal{Z}^{(A)}(t_4) \rangle_T$$
$$= \mathcal{W}_Z(t_3, t_1)\mathcal{W}_Z(t_4, t_2) + \mathcal{W}_Z(t_3, t_2)\mathcal{W}_Z(t_4, t_1). \qquad (1.102)$$

This formula is of great significance in statistical optics since, if evaluated at $t_1 = t_2 = t_3 = t_4$, it relates the average intensity of a fluctuating field with the contrast (variance) of fluctuation. Also, for $t_1 = t_3$ and $t_2 = t_4$, it relates intensity–intensity fluctuations with the auto-correlation function. A similar relation can be written with respect to spatial arguments.

Among other examples of random processes that are of particular importance for some areas of statistical optics are Poisson processes, Markov chains and the Wienner processes. However, this monograph will not explicitly employ them.

Chapter 2

Statistical Characterization of Optical Fields

In this chapter, after overviewing the description and properties of deterministic electromagnetic fields, based on the Maxwell theory, we will give the complete statistical characterization of wide-sense stationary electromagnetic fields, discussing in detail their energy, linear and angular momentum, as well as their polarization and coherence states. This material is followed by introduction of major mathematical models to be used later for illustration of various phenomena relating to random light.

2.1 Deterministic Optical Fields

2.1.1 *The Maxwell equations*

We first consider the most general set of equations governing deterministic electromagnetic fields on their free-space evolution and interaction with matter. Let the time-varying electric and magnetic field vectors at point $\mathbf{r}$ of the three-dimensional space and at time instant t be $\vec{\mathcal{E}}(\mathbf{r},t)$ and $\vec{\mathcal{B}}(\mathbf{r},t)$, respectively. Each of these field vectors has three components, in general. In the System International (SI) of units (kilogramm, meter, second) the set of four macroscopic Maxwell's equations, expressed in space–time domain, has the following form (Jackson, 1998):

$$\nabla \times \vec{\mathcal{E}}(\mathbf{r},t) = -\dot{\vec{\mathcal{B}}}(\mathbf{r},t), \qquad (2.1)$$

$$\nabla \times \vec{\mathcal{H}}(\mathbf{r},t) = \vec{\mathcal{J}}_c(\mathbf{r},t) + \dot{\vec{\mathcal{D}}}(\mathbf{r},t), \qquad (2.2)$$

$$\nabla \cdot \vec{\mathcal{B}}(\mathbf{r},t) = 0, \qquad (2.3)$$

$$\nabla \cdot \vec{\mathcal{D}}(\mathbf{r},t) = \rho_c(\mathbf{r},t). \qquad (2.4)$$

Here, $\nabla \times$ and $\nabla \cdot$ are curl and divergence operators, dot stands for time derivative, while electric displacement field $\vec{\mathcal{D}}$, magnetizing field $\vec{\mathcal{H}}$, electric current density $\vec{\mathcal{J}}_c$ and electric charge density ρ_c describe interaction of the field with the material. The individual equations (2.1)–(2.4) describe the Faraday induction law, the Ampère law, the electric and the magnetic Gauss laws, respectively.

We will assume for now that the field is quasi-monochromatic, i.e., has a sufficiently narrow spectrum, $\Delta\lambda \ll \lambda_0$, where $\Delta\lambda$ is the spectral width and λ_0 is the central wavelength. For quasi-monochromatic fields and homogeneous and isotropic materials in the absence of either polarization or magnitization, the following set of *material* or *constitutive* relations hold:

$$\vec{\mathcal{J}}_c(\mathbf{r},t) = \sigma_m \vec{\mathcal{E}}(\mathbf{r},t), \quad \vec{\mathcal{D}}(\mathbf{r},t) = \varepsilon_m \vec{\mathcal{E}}(\mathbf{r},t), \quad \vec{\mathcal{B}}(\mathbf{r},t) = \mu_m \vec{\mathcal{H}}(\mathbf{r},t). \qquad (2.5)$$

The first of these equations is also known as Ohm's law. Here, σ_m is specific conductivity, ε_m is dialectric constant (or electric permittivity) and μ_m is magnetic permeability of the material. These material quantities may be functions of frequency, describing dispersion, but we omit this dependence for brevity. In particular, in vacuum, denoted by subscript 0,

$$\sigma_0 = 0, \quad \varepsilon_0 = 8.8541878128(13) \times 10^{-12} \frac{F}{m}, \quad \mu_0 \approx 4\pi \times 10^{-7} \frac{H}{m}. \qquad (2.6)$$

On applying material relations (2.5) in Maxwell's equations (2.1)–(2.4) and using several vector calculus identities, it is possible to decouple the equations governing the evolution of the electric and the magnetic fields (Born and Wolf, 1999). In the former case,

$$\nabla^2 \vec{\mathcal{E}}(\mathbf{r},t) - \varepsilon_m \mu_m \ddot{\vec{\mathcal{E}}}(\mathbf{r},t) + (\nabla \ln \mu_m) \times \nabla \times \vec{\mathcal{E}}(\mathbf{r},t)$$
$$+ \nabla(\vec{\mathcal{E}}(\mathbf{r},t) \cdot \nabla \ln \varepsilon_m) = 0, \qquad (2.7)$$

where ∇ denotes gradient and ∇^2 stands for Laplacian. A similar equation holds for magnetic field $\vec{\mathcal{B}}(\mathbf{r},t)$.

For interaction with materials in which ε_m and μ_m vary sufficiently with spatial position, the gradient terms vanish and the six equations (three for the electric field and three for the magnetic field) decouple to six wave equations, for individual components of the electric field,

$$v_m^2 \nabla^2 \vec{\mathcal{E}}(\mathbf{r},t) = \ddot{\vec{\mathcal{E}}}(\mathbf{r},t), \qquad (2.8)$$

and the similar equations for the magnetic field components. Here,

$$v_m = (\varepsilon_m \mu_m)^{-1/2} \qquad (2.9)$$

is the speed of the wave in the medium. In vacuum, v_m reduces to c.

For monocromatic electric fields oscillating at angular frequency ω,

$$\vec{\mathcal{E}}(\mathbf{r},t) = \mathrm{Re}[\vec{E}(\mathbf{r},\omega)e^{-i\omega t}], \qquad (2.10)$$

and Eqs. (2.8) reduce to three Helmholz equations

$$\nabla^2 \vec{E}(\mathbf{r},\omega) = k^2 n^2 \vec{E}(\mathbf{r},\omega), \qquad (2.11)$$

and similar equations are obeyed by $\vec{B}(\mathbf{r},\omega)$. Here, $k = 2\pi/\lambda = \omega/c$ is the wave number of light and $n = c/v_m$ is the *refractive index* of the medium. Here, we switch from cursive to straight font since $\vec{E}(\mathbf{r},\omega)$ can be regarded as a field in space–frequency domain with a single Fourier component.

2.1.2 *Energy*

Consider a closed region of a three-dimensional space with charge distribution $\rho_c(\mathbf{r},t)$ and current density distribution $\vec{\mathcal{J}}_c(\mathbf{r},t)$. On a microscopic level, the Lorentz force acting by the electromagnetic field on charge q moving with velocity vector $\vec{v}(t)$ has the form

$$\vec{\mathcal{F}}_L(\mathbf{r},t) = q[\vec{\mathcal{E}}(\mathbf{r},t) + \vec{v}(t) \times \vec{\mathcal{B}}(\mathbf{r},t)]. \qquad (2.12)$$

Since magnetic field $\vec{\mathcal{B}}$ is orthogonal to the charge's direction of motion, the mechanical power applied to the charge only depends on contribution from the electric field vector:

$$\vec{\mathcal{G}}_L(\mathbf{r},t) = q\vec{v}(t) \cdot \vec{\mathcal{E}}(\mathbf{r},t). \tag{2.13}$$

On integrating power $\vec{\mathcal{G}}_L$ over the volume, and recognizing that

$$q\vec{v}(t) = \vec{\mathcal{J}}_c(\mathbf{r},t)d\mathbf{r}, \tag{2.14}$$

we find that the mechanical power has two contributions:

$$\begin{aligned}\vec{\mathcal{G}}_L(t) &= \int \vec{\mathcal{J}}_c(\mathbf{r},t) \cdot \vec{\mathcal{E}}(\mathbf{r},t)d\mathbf{r} \\ &= \int [\nabla \times \vec{\mathcal{H}}(\mathbf{r},t) - \vec{\mathcal{D}}(\mathbf{r},t)] \cdot \vec{\mathcal{E}}(\mathbf{r},t)d\mathbf{r} \\ &= -\int \nabla \cdot [\vec{\mathcal{E}}(\mathbf{r},t) \times \vec{\mathcal{H}}(\mathbf{r},t)]d\mathbf{r} \\ &\quad - \int [\vec{\mathcal{H}}(\mathbf{r},t) \cdot \vec{\mathcal{B}}(\mathbf{r},t) + \vec{\mathcal{E}}(\mathbf{r},t) \cdot \vec{\mathcal{D}}(\mathbf{r},t)]d\mathbf{r}, \end{aligned} \tag{2.15}$$

where we have used Eq. (2.2) on passage to the second line as well as vector identity $\nabla \cdot (\vec{\mathcal{E}} \times \vec{\mathcal{H}}) = \vec{\mathcal{H}} \cdot (\nabla \times \vec{\mathcal{E}}) - \vec{\mathcal{E}} \cdot (\nabla \times \vec{\mathcal{H}})$ and Eq. (2.1) on passage to the third line. Now we write

$$\vec{\mathcal{G}}_L(t) = -\vec{\mathcal{G}}_P(t) - \vec{\mathcal{G}}_D(t), \tag{2.16}$$

which can be viewed as the conservation of the electromagneric power law. On using the Gauss theorem, we find that the first term becomes

$$\begin{aligned}\vec{\mathcal{G}}_P(t) &= \int \nabla \cdot [\vec{\mathcal{E}}(\mathbf{r},t) \times \vec{\mathcal{H}}(\mathbf{r},t)]d\mathbf{r} \\ &= \oint [\vec{\mathcal{E}}(\mathbf{r},t) \times \vec{\mathcal{H}}(\mathbf{r},t)] \cdot d\mathbf{s}, \end{aligned} \tag{2.17}$$

where the surface integral is taken over the boudary of the region, s being the differential area. Further, using the last two material relations (2.5) we find that the second contribution in Eq. (2.16) takes the form

$$\vec{\mathcal{G}}_D(t) = \int [\vec{\mathcal{H}}(\mathbf{r},t) \cdot \vec{\mathcal{B}}(\mathbf{r},t) + \vec{\mathcal{E}}(\mathbf{r},t) \cdot \vec{\mathcal{D}}(\mathbf{r},t)] d\mathbf{r}$$

$$= \frac{1}{2} \frac{\partial}{\partial t} \int [\varepsilon_m |\vec{\mathcal{E}}(\mathbf{r},t)|^2 + \mu_m |\vec{\mathcal{H}}(\mathbf{r},t)|^2] d\mathbf{r}. \quad (2.18)$$

The integrand in the second line of Eq. (2.17) is the *Poynting vector*:

$$\vec{\mathcal{S}}_P(\mathbf{r},t) = \vec{\mathcal{E}}(\mathbf{r},t) \times \vec{\mathcal{H}}(\mathbf{r},t), \quad (2.19)$$

and the quantity

$$\vec{\mathcal{S}}_D(\mathbf{r},t) = \frac{1}{2}[\vec{\mathcal{H}}(\mathbf{r},t) \cdot \vec{\mathcal{B}}(\mathbf{r},t) + \vec{\mathcal{E}}(\mathbf{r},t) \cdot \vec{\mathcal{D}}(\mathbf{r},t)] \quad (2.20)$$

obtained from the integrand of Eq. (2.18) is regarded as the *electromagnetic field energy density*. We finally mention that the energy conservation law (2.16) can also be expressed as

$$\vec{\mathcal{G}}_P(t) = -\oint \vec{\mathcal{S}}_P(\mathbf{r},t) \cdot d\mathbf{s}, \quad (2.21)$$

where $\vec{\mathcal{G}}_P$ is the net power being the combination of mechanical power and that due to the electromagnetic fields.

The derivation above was carried out in space–time domain. Optical signals oscillate at frequencies on the order of 10^{15} Hz and, hence, their phases cannot be directly assessed. Let us consider a scalar monochromatic signal oscillating at a fixed frequency ω (see also Eq. (2.10)),

$$\mathcal{U}(\mathbf{r},t) = \text{Re}[U(\mathbf{r},\omega)e^{-i\omega t}]. \quad (2.22)$$

Then the scalar product of the field with itself appearing in the energy-related expressions takes the form

$$\mathcal{U}(\mathbf{r},t) \cdot \mathcal{U}(\mathbf{r},t) = \frac{1}{2} U^*(\mathbf{r},\omega) \cdot U(\mathbf{r},\omega) + \frac{1}{4}[U(\mathbf{r},\omega) \cdot U(\mathbf{r},\omega)e^{-i2\omega t}$$
$$+ U^*(\mathbf{r},\omega) \cdot U^*(\mathbf{r},\omega)e^{+i2\omega t}]. \quad (2.23)$$

After averaging the field over one oscillation cycle with period $T = 2\pi/\omega$, only the first term in Eq. (2.23) survives. Indeed, we then define the *average intensity* of a scalar, monochromatic field by the expression

$$\begin{aligned}\mathcal{I}(\mathbf{r},\omega) &= 2\langle \mathcal{U}(\mathbf{r},t)\cdot\mathcal{U}(\mathbf{r},t)\rangle \\ &= \frac{2}{T}\int_{-T/2}^{T/2}\mathcal{U}(\mathbf{r},t)\cdot\mathcal{U}(\mathbf{r},t)dt \\ &= U^*(\mathbf{r},\omega)U(\mathbf{r},\omega) \\ &= |U(\mathbf{r},\omega)|^2. \end{aligned} \qquad (2.24)$$

Similarly, for two different fields, say $\mathcal{U}_1(\mathbf{r},t)$ and $\mathcal{U}_2(\mathbf{r},t)$, the averaged scalar product can be shown to reduce to

$$\langle \mathcal{U}_1(\mathbf{r},t)\cdot\mathcal{U}_2(\mathbf{r},t)\rangle = \frac{1}{2}\mathrm{Re}[U_1^*(\mathbf{r},\omega)\cdot U_2(\mathbf{r},\omega)]. \qquad (2.25)$$

For monochromatic electromagnetic fields oscillating at frequency ω with electric field given in Eq. (2.10), and similar magnetic fields, we find that the average mechanical power becomes

$$\langle \vec{G}_L(\omega)\rangle = \frac{1}{2}\int \langle \vec{J}_c^*(\mathbf{r},\omega)\cdot\vec{E}(\mathbf{r},\omega)\rangle d\mathbf{r}. \qquad (2.26)$$

In order to determine the frequency-dependent G_P and G_D, we first express the first two Maxwell equations (2.1) and (2.2) as

$$\nabla\times\vec{E}(\mathbf{r},\omega) = i\omega\vec{B}(\mathbf{r},\omega),\quad \nabla\times\vec{H}(\mathbf{r},\omega) = \vec{J}_c(\mathbf{r},\omega) - i\omega\vec{D}(\mathbf{r},\omega), \qquad (2.27)$$

and then find that

$$\langle \vec{G}_L(\omega)\rangle = -\langle \vec{G}_P(\omega)\rangle - \langle \vec{G}_D(\omega)\rangle, \qquad (2.28)$$

where

$$\langle \vec{G}_P(\omega)\rangle = \oint \vec{S}_P(\mathbf{r},\omega)\cdot d\mathbf{s} \qquad (2.29)$$

with average Poynting vector

$$\vec{S}_P(\mathbf{r},\omega) = \frac{1}{2}\vec{E}(\mathbf{r},\omega)\times\vec{H}^*(\mathbf{r},\omega), \qquad (2.30)$$

and

$$\langle \vec{G}_D(\omega)\rangle = 2i\omega \int [\vec{T}_E(\mathbf{r},\omega) - \vec{T}_M(\mathbf{r},\omega)]d\mathbf{r} \quad (2.31)$$

with the electric and the magnetic energy densities

$$\vec{T}_E(\mathbf{r},\omega) = \frac{1}{4}\vec{D}^*(\mathbf{r},\omega)\cdot\vec{E}(\mathbf{r},\omega),$$

$$\vec{T}_M(\mathbf{r},\omega) = \frac{1}{4}\vec{H}^*(\mathbf{r},\omega)\cdot\vec{B}(\mathbf{r},\omega). \quad (2.32)$$

For materials with linear response, $\vec{D}(\mathbf{r},\omega) = \varepsilon_m(\mathbf{r},\omega)\vec{E}(\mathbf{r},\omega)$ and $\vec{B}(\mathbf{r},\omega) = \mu_m(\mathbf{r},\omega)\vec{H}(\mathbf{r},\omega)$, the two energy densities further simplify to

$$\vec{T}_E(\mathbf{r},\omega) = \frac{-i}{4}\varepsilon_m^{(i)}(\mathbf{r},\omega)|\vec{E}(\mathbf{r},\omega)|^2,$$

$$\vec{T}_M(\mathbf{r},\omega) = \frac{-i}{4}\mu_m^{(i)}(\mathbf{r},\omega)|\vec{H}(\mathbf{r},\omega)|^2, \quad (2.33)$$

where $\varepsilon_m^{(i)}$ and $\mu_m^{(i)}$ stand for the imaginary parts of electric permitivity and magnetic permeability addresing the medium's absorption or gain. To arrive at Eq. (2.33), we have taken the real parts of the complex numbers at the last step. The z-component of the Poynting vector, $S_z(\mathbf{r},\omega) = S(\mathbf{r},\omega)$, can be shown to have the form

$$S_z(\mathbf{r},\omega) = \frac{k}{2\mu_0\omega}[E_x^*(\mathbf{r},\omega)E_x(\mathbf{r},\omega) + E_y^*(\mathbf{r},\omega)E_y(\mathbf{r},\omega)]. \quad (2.34)$$

The constant factor on the right-hand side is often omitted in theoretical calculations.

2.1.3 Linear and angular momentum

In order to characterize the linear and the angular momenta, we use the material relations (2.5) for vacuum in the macroscopic form of Maxwell's equations (2.1)–(2.4). This leads us to the following equations expressed on the microscopic level:

$$\nabla \times \vec{\mathcal{E}}(\mathbf{r},t) = -\dot{\vec{\mathcal{B}}}(\mathbf{r},t), \tag{2.35}$$

$$\nabla \times \vec{\mathcal{B}}(\mathbf{r},t) = \mu_0 \vec{\mathcal{J}}_c(\mathbf{r},t) + \mu_0\varepsilon_0 \dot{\vec{\mathcal{E}}}(\mathbf{r},t), \tag{2.36}$$

$$\nabla \cdot \vec{\mathcal{B}}(\mathbf{r},t) = 0, \tag{2.37}$$

$$\nabla \cdot \vec{\mathcal{E}}(\mathbf{r},t) = \rho_c(\mathbf{r},t)/\varepsilon_0. \tag{2.38}$$

Then for a closed three-dimensional region with charge density $\rho_c(\mathbf{r},t)$ and current density $\vec{\mathcal{J}}_c(\mathbf{r},t)$ distributions, the net Lorentz force takes the form

$$\vec{\mathcal{F}}_L(t) = \int [\rho_c(\mathbf{r},t)\vec{\mathcal{E}}(\mathbf{r},t) + \vec{\mathcal{J}}_c(\mathbf{r},t) \times \vec{\mathcal{B}}(\mathbf{r},t)]d\mathbf{r}. \tag{2.39}$$

Using Eqs. (2.35)–(2.38) in Eq. (2.39), we find that

$$\vec{\mathcal{F}}_L(t) = \int [\vec{\mathcal{F}}_P(\mathbf{r},t) - \vec{\mathcal{F}}_D(\mathbf{r},t)]d\mathbf{r}, \tag{2.40}$$

where

$$\vec{\mathcal{F}}_P(\mathbf{r},t) = \varepsilon_0 \vec{\mathcal{E}}(\mathbf{r},t)[\nabla \cdot \vec{\mathcal{E}}(\mathbf{r},t)] - \varepsilon_0 \vec{\mathcal{E}}(\mathbf{r},t) \times [\nabla \times \vec{\mathcal{E}}(\mathbf{r},t)]$$
$$+ \frac{1}{\mu_0}\vec{\mathcal{B}}(\mathbf{r},t)[\nabla \cdot \vec{\mathcal{B}}(\mathbf{r},t)] - \frac{1}{\mu_0}\vec{\mathcal{B}}(\mathbf{r},t) \times [\nabla \times \vec{\mathcal{B}}(\mathbf{r},t)], \tag{2.41}$$

and

$$\vec{\mathcal{F}}_D(\mathbf{r},t) = \varepsilon_0 \frac{\partial}{\partial t}[\vec{\mathcal{E}}(\mathbf{r},t) \times \vec{\mathcal{B}}(\mathbf{r},t)]. \tag{2.42}$$

Function $\vec{\mathcal{F}}_P(\mathbf{r},t)$ can be expressed in the component form as

$$\vec{\mathcal{F}}_P(\mathbf{r},t) = \partial_j \mathcal{T}_{ij}$$
$$= \partial_j \left\{ -\left[\varepsilon_0 \mathcal{E}_i \mathcal{E}_j + \frac{1}{\mu_0}\mathcal{B}_i\mathcal{B}_j - \frac{1}{2}\delta_{ij}\left(\varepsilon_0 \mathcal{E}_k\mathcal{E}_k + \frac{1}{\mu_0}\mathcal{B}_k\mathcal{B}_k\right)\right]\right\}, \tag{2.43}$$

where $\overleftrightarrow{\mathcal{T}}(\mathbf{r},t) = \{\mathcal{T}_{ij}(\mathbf{r},t)\}$, $(i,j = x,y,z)$ is known as the *Maxwell stress tensor* while the Einstein summation notation is implied and the Kronecker symbol δ_{ij} is used. In vector notation

$$\langle \overleftrightarrow{T}(\mathbf{r},t)\rangle = -\varepsilon_0 \vec{\mathcal{E}}^\dagger \vec{\mathcal{E}} + \frac{\varepsilon_0}{2}\text{Tr}[\vec{\mathcal{E}}^\dagger \vec{\mathcal{E}}]\overleftrightarrow{I}^{(3)} - \frac{1}{\mu_0}\vec{\mathcal{B}}^\dagger \vec{\mathcal{B}}$$
$$+ \frac{1}{2\mu_0}\text{Tr}[\vec{\mathcal{B}}^\dagger \vec{\mathcal{B}}]\overleftrightarrow{I}^{(3)}, \qquad (2.44)$$

where $I^{(3)}$ is a unit 3×3 matrix and dagger stands for Hermitian adjoint.

In particular, for monochromatic fields the components of the cycle-averaged Maxwell stress tensor can be shown to take the form (Kim and Gbur, 2009)

$$\langle T_{ij}(\mathbf{r},\omega)\rangle = -\frac{1}{2}\left[\varepsilon_0 E_i^* E_j + \frac{1}{\mu_0}B_i^* B_j\right.$$
$$\left. -\frac{1}{2}\delta_{ij}\left(\varepsilon_0 E_k^* E_k + \frac{1}{\mu_0}B_k^* B_k\right)\right]. \qquad (2.45)$$

Alternatively, the whole tensor $\overleftrightarrow{T}$ can also be written as

$$\langle \overleftrightarrow{T}(\mathbf{r},\omega)\rangle = -\frac{\varepsilon_0}{2}\vec{E}^\dagger \vec{E} + \frac{\varepsilon_0}{4}\text{Tr}[\vec{E}^\dagger \vec{E}]\overleftrightarrow{I}^{(3)} - \frac{1}{2\mu_0}\vec{B}^\dagger \vec{B}$$
$$+ \frac{1}{4\mu_0}\text{Tr}[\vec{B}^\dagger \vec{B}]\overleftrightarrow{I}^{(3)}. \qquad (2.46)$$

Since a force vector is the time derivative of the linear momentum vector, expressing Eq. (2.40) in terms of the linear momenta yields

$$\dot{\vec{\mathcal{P}}}_P(t) = \dot{\vec{\mathcal{P}}}_L(t) + \dot{\vec{\mathcal{P}}}_D(t). \qquad (2.47)$$

The relation between the components of $\vec{\mathcal{P}}_P$ and $\mathcal{T}_{ij}$ becomes

$$[\dot{\vec{\mathcal{P}}}_P(t)]_i = -\int \partial_j \mathcal{T}_{ij}(\mathbf{r},t)d\mathbf{r}$$
$$= -\oint \mathcal{T}_{ij}(\mathbf{r},t)ds_j, \qquad (2.48)$$

where in the last step the divergence theorem is applied, and the integral is taken over the boundary of the region, while ds_j represents component j of the differential area element $d\mathbf{s}$. In view of Eq. (2.48), the Maxwell stress tensor can be regarded as the *linear momentum flux density*.

Further, on using Eqs. (2.40), (2.42) and (2.47) we make association of the vector

$$\vec{\mathcal{P}}_D(t) = \int \varepsilon_0 \vec{\mathcal{E}}(\mathbf{r},t) \times \vec{\mathcal{B}}(\mathbf{r},t) d\mathbf{r} \tag{2.49}$$

with the *total linear momentum* of the field in the region. This suggests that the *linear momentum density* can be defined as

$$\vec{\mathcal{P}}_D(\mathbf{r},t) = \varepsilon_0 \vec{\mathcal{E}}(\mathbf{r},t) \times \vec{\mathcal{B}}(\mathbf{r},t). \tag{2.50}$$

On the other hand, the torque induced by the Lorentz force (2.39) becomes

$$\vec{\mathcal{Q}}_L(t) = \int \mathbf{r} \times [\rho_c(\mathbf{r},t)\vec{\mathcal{E}}(\mathbf{r},t) + \vec{\mathcal{J}}_c(\mathbf{r},t) \times \vec{\mathcal{B}}(\mathbf{r},t)]d\mathbf{r}$$

$$= \int \mathbf{r} \times [\vec{\mathcal{F}}_P(\mathbf{r},t) - \vec{\mathcal{F}}_D(\mathbf{r},t)]d\mathbf{r}. \tag{2.51}$$

Since the torque vector is the time derivative of the angular momentum vector, we express (2.51) as

$$\dot{\vec{\mathcal{L}}}_P(t) = \dot{\vec{\mathcal{L}}}_L(t) + \dot{\vec{\mathcal{L}}}_D(t), \tag{2.52}$$

where the first term on the right can be written as

$$\dot{\vec{\mathcal{L}}}_P(t) = \int \mathbf{r} \times \vec{\mathcal{F}}_P(\mathbf{r},t) d\mathbf{r},$$

$$= -\int \nabla \cdot [\mathbf{r} \times \overleftrightarrow{\mathcal{T}}(\mathbf{r},t)] d\mathbf{r}$$

$$= -\oint [\mathbf{r} \times \overleftrightarrow{\mathcal{T}}(\mathbf{r},t)] \cdot d\mathbf{s}, \tag{2.53}$$

and the second, as

$$\dot{\vec{\mathcal{L}}}_D(t) = \int \mathbf{r} \times \vec{\mathcal{F}}_D(\mathbf{r},t) d\mathbf{r}$$

$$= \int \mathbf{r} \times \left[\varepsilon_0 \vec{\mathcal{E}}(\mathbf{r},t) \times \vec{\mathcal{B}}(\mathbf{r},t)\right] d\mathbf{r}. \tag{2.54}$$

In view of these relations, the integrand in the last line of Eq. (2.53) can be associated with the *angular momentum flux density* tensor of the field:

$$\overleftrightarrow{\mathcal{L}}_P(\mathbf{r},t) = \mathbf{r} \times \overleftrightarrow{\mathcal{T}}(\mathbf{r},t), \qquad (2.55)$$

and the integrand in the second line of Eq. (2.54) can be regarded as the *angular momentum density* vector of the field:

$$\vec{\mathcal{L}}_D(\mathbf{r},t) = \mathbf{r} \times [\varepsilon_0 \vec{\mathcal{E}}(\mathbf{r},t) \times \vec{\mathcal{B}}(\mathbf{r},t)]. \qquad (2.56)$$

In the following chapters, we will be using the zz-component of the cycle-averaged angular momentum flux density of an optical beam (Kim and Gbur, 2012)

$$\langle L_P(\boldsymbol{\rho},\omega)\rangle = \langle L^{(\mathrm{sp})}(\boldsymbol{\rho},\omega)\rangle + \langle L^{(\mathrm{orb})}(\boldsymbol{\rho},\omega)\rangle, \qquad (2.57)$$

where spin $\langle L^{(\mathrm{sp})}\rangle$ and orbital $\langle L^{(\mathrm{orb})}\rangle$ contributions take the forms

$$\langle L^{(\mathrm{sp})}(\boldsymbol{\rho},\omega)\rangle = \frac{\varepsilon_0}{2k}\mathrm{Im}[E_x^* E_y - E_y^* E_x], \qquad (2.58)$$

$$\langle L^{(\mathrm{orb})}(\boldsymbol{\rho},\omega)\rangle = \frac{\varepsilon_0}{2k}\mathrm{Im}\{y E_y^* \partial_x E_y - x E_x^* \partial_y E_x$$
$$+ x[\partial_y E_y^*]E_y - y[\partial_x E_x^*]E_x\}$$
$$= \frac{\varepsilon_0}{2k}\mathrm{Im}\{E_x^* \partial_\theta E_x + E_y^* \partial_\theta E_y\}. \qquad (2.59)$$

Here, ∂_θ is the partial derivative with respect to polar angle and $\boldsymbol{\rho} = x\hat{x} + y\hat{y}$ is the two-dimensional vector in the plane orthogonal to z direction.

2.1.4 *Polarization*

Vectorial nature of a beam-like field can also be characterized by the set of its polarization properties. Let the transverse part of the electric field $\vec{\mathcal{E}}$ at a point with position vector $\mathbf{r}$ at time instant t propagating along the positive z axis in free space have the

form
$$\vec{\mathcal{E}}(\mathbf{r},t) = \vec{E}(\mathbf{r},\omega)\exp[i(kz - \omega t)]. \tag{2.60}$$

Here, $\vec{E}(\mathbf{r},\omega)$ is a complex-valued vector that can be expressed in the Cartesian coordinate system as

$$\begin{aligned}\vec{E}(\mathbf{r},\omega) &= E_x(\mathbf{r},\omega)\hat{x} + E_y(\mathbf{r},\omega)\hat{y} \\ &= a_x(\mathbf{r},\omega)\exp[i\varphi_x(\mathbf{r},\omega)]\hat{x} + a_y(\mathbf{r},\omega)\exp[i\varphi_y(\mathbf{r},\omega)]\hat{y},\end{aligned} \tag{2.61}$$

with a_i and φ_i, $(i = x, y)$ being the amplitudes and the phases of the two time-independent electric-field components, E_x and E_y, respectively. Let us now take the real part of vector $\vec{\mathcal{E}}(\mathbf{r},t)$ in Eq. (2.60):

$$\vec{\mathcal{E}}^{(R)}(\mathbf{r},t) = \text{Re}[\vec{\mathcal{E}}(\mathbf{r},t)] = \vec{E}(\mathbf{r},\omega)\cos(kz - \omega t). \tag{2.62}$$

Combining the two components of vector $\vec{\mathcal{E}}^{(R)}$ in Eq. (2.62) by elimination of argument $(kz - \omega t)$ and setting $\varphi(\mathbf{r},\omega) = \varphi_y(\mathbf{r},\omega) - \varphi_x(\mathbf{r},\omega)$ yields the elliptical quadratic form

$$\frac{E_x^2(\mathbf{r},\omega)}{a_x^2(\mathbf{r},\omega)} + \frac{E_y^2(\mathbf{r},\omega)}{a_y^2(\mathbf{r},\omega)} - 2\frac{E_x(\mathbf{r},\omega)}{a_x(\mathbf{r},\omega)}\frac{E_y(\mathbf{r},\omega)}{a_y(\mathbf{r},\omega)}\cos\varphi(\mathbf{r},\omega) = \sin^2\varphi(\mathbf{r},\omega), \tag{2.63}$$

that generally describes the polarization state of deterministic light.

In particular, the elliptical polarization state degenerates to *linear* state if $\varphi(\mathbf{r},\omega) = n\pi$, $n = 0, \pm 1, \pm 2, \ldots$, i.e., the electric-field components oscillate exactly in phase or out of phase. The other special case is *circular* polarization, when $a_x(\mathbf{r},\omega) = a_y(\mathbf{r},\omega)$ and $\varphi(\mathbf{r},\omega) = \pm(n + 1/2)\pi$, i.e., the electric field components oscillate with phase delay being effectively $\pm\pi/2$. In addition, the *elliptical* (and in the special case, circular) polarization state is characterized by its helicity (spin): left and right polarization corresponding to $\varphi(\mathbf{r},\omega) > 0$ and $\varphi(\mathbf{r},\omega) < 0$, respectively.

When the ellipse is not degenerate, it can be uniquely characterized by three parameters: the orientation angle $\psi(\mathbf{r},\omega)$, measured as

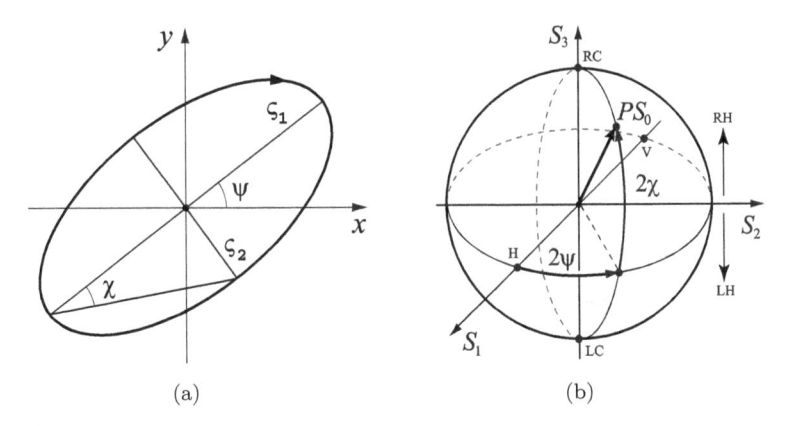

Fig. 2.1 Polarization properties of optical beams. (a) polarization ellipse. (b) Poincaré sphere; letter labels stand for the polarization states: H — horizontal, V — vertical, RC — right circular, LC — left circular, HR — righthandedness, HL — lefthandedness.

the smallest angle between the positive x-axis and the direction of the major semi-axis:

$$\psi(\mathbf{r},\omega) = \frac{1}{2} \arctan\left(\frac{2a_x(\mathbf{r},\omega)a_y(\mathbf{r},\omega)\cos\varphi(\mathbf{r},\omega)}{a_x^2(\mathbf{r},\omega) - a_y^2(\mathbf{r},\omega)}\right), \quad 0 \leq \psi \leq \pi, \tag{2.64}$$

and the magnitudes of the major and minor semi-axes, $\zeta_+(\mathbf{r},\omega)$ and $\zeta_-(\mathbf{r},\omega)$:

$$\zeta_\pm(\mathbf{r},\omega) = [a_x^2(\mathbf{r},\omega)\cos^2\psi(\mathbf{r},\omega) + a_y^2(\mathbf{r},\omega)\sin^2\psi(\mathbf{r},\omega)$$
$$\pm a_x(\mathbf{r},\omega)a_y(\mathbf{r},\omega)\sin[2\psi(\mathbf{r},\omega)]\cos\varphi(\mathbf{r},\omega)]^{1/2}. \tag{2.65}$$

The shape of the polarization ellipse is typically characterized by ellipticity $\epsilon(\mathbf{r},\omega)$ or azimuthal angle $\chi(\mathbf{r},\omega)$ (see Fig. 2.1(a)):

$$\epsilon(\mathbf{r},\omega) = \frac{\zeta_+(\mathbf{r},\omega)}{\zeta_-(\mathbf{r},\omega)}, \quad 1 \leq \psi \leq \infty;$$
$$\chi(\mathbf{r},\omega) = \arctan\epsilon(\mathbf{r},\omega), \quad -\pi/4 \leq \chi(\mathbf{r},\omega) \leq \pi/4, \tag{2.66}$$

where for circular polarization $\epsilon(\mathbf{r},\omega) = 1$ and $\chi(\mathbf{r},\omega) = \pm\pi/4$ and for linear polarization $\epsilon(\mathbf{r},\omega) = 0$ and $\chi(\mathbf{r},\omega) = 0$.

The polarization properties of optical beams can be alternatively described by the set of four Stokes parameters frequently used in experimental measurements (Stokes, 1852). For deterministic beams, they are defined as

$$\begin{aligned} S_0(\mathbf{r},\omega) &= a_x^2(\mathbf{r},\omega) + a_y^2(\mathbf{r},\omega), \\ S_1(\mathbf{r},\omega) &= a_x^2(\mathbf{r},\omega) - a_y^2(\mathbf{r},\omega), \\ S_2(\mathbf{r},\omega) &= 2a_x(\mathbf{r},\omega)a_y(\mathbf{r},\omega)\cos\varphi(\mathbf{r},\omega), \\ S_3(\mathbf{r},\omega) &= 2a_x(\mathbf{r},\omega)a_y(\mathbf{r},\omega)\sin\varphi(\mathbf{r},\omega). \end{aligned} \qquad (2.67)$$

One can readily verify that

$$S_0^2(\mathbf{r},\omega) = S_1^2(\mathbf{r},\omega) + S_2^2(\mathbf{r},\omega) + S_3^2(\mathbf{r},\omega). \qquad (2.68)$$

The Stokes parameters are related to those of the polarization ellipse as

$$\begin{aligned} S_1(\mathbf{r},\omega) &= S_0(\mathbf{r},\omega)\cos 2\chi(\mathbf{r},\omega)\cos 2\psi(\mathbf{r},\omega), \\ S_2(\mathbf{r},\omega) &= S_0(\mathbf{r},\omega)\cos 2\chi(\mathbf{r},\omega)\sin 2\psi(\mathbf{r},\omega), \\ S_3(\mathbf{r},\omega) &= 2S_0(\mathbf{r},\omega)\sin 2\psi(\mathbf{r},\omega). \end{aligned} \qquad (2.69)$$

If Stokes parameters S_1, S_2 and S_3 of a deterministic beam are used as the mutually orthogonal axes of the three-dimensional Cartesian coordinate system, then in view of relation (2.68) all possible polarization states occupy the surface of a spherical shell with radius S_0, known as the Poincaré sphere (Brosseau, 1998). Figure 2.1(b) shows the Poincaré sphere and the parameters of the polarization ellipse ψ and χ as well as the particular polarization states: horizontal (H) (x-linear), vectical V (y-linear), right-hand circular (RC), left-hand circular (LC) and handedness (RH or LH).

2.2 Scalar Stationary Optical Fields

2.2.1 *Cross-spectral density*

We will now consider a realization of a wide-sense stationary, scalar, quasi-monochromatic optical field, $\mathcal{U}(\mathbf{r},t)$, that may represent the analytic signal (symbol not explicitly shown) associated with one of its electric (or magnetic) field components. Then according to

Eq. (2.8), on evolution in vacuum $\mathcal{U}(\mathbf{r}, t)$ must be a solution of wave equation

$$\nabla^2 \mathcal{U}(\mathbf{r}, t) = \frac{1}{c^2} \frac{\partial^2 \mathcal{U}(\mathbf{r}, t)}{\partial t^2}. \qquad (2.70)$$

As we have already discussed, it appears impossible to measure the phase of the optical disturbance directly. Instead, one can theoretically and experimentally characterize the optical field by means of observable correlation functions, taken at one or several points in time and space (Wolf, 1954). The most practically accessible such correlation is the *mutual coherence function* defined as (Mandel and Wolf, 1995)

$$\mathcal{W}(\mathbf{r}_1, \mathbf{r}_2, t_1, t_2) = \langle \mathcal{U}^*(\mathbf{r}_1, t_1) \mathcal{U}(\mathbf{r}_2, t_2) \rangle_T, \qquad (2.71)$$

where time average is implied.

We will assume for the immediately following discussion that the optical field obeys *wide-sense stationarity* condition: the average of the field is independent of the origin of time and the mutual coherence function only depends on time difference. The latter function can then be expressed as

$$\mathcal{W}(\mathbf{r}_1, \mathbf{r}_2, \tau) = \langle \mathcal{U}^*(\mathbf{r}_1, t) \mathcal{U}(\mathbf{r}_2, t + \tau) \rangle_T, \qquad (2.72)$$

where $\tau = t_1 - t_2$ is the time delay. It immediately follows from Eqs. (2.70) and (2.72) that the mutual coherence function obeys a pair of wave equations, widely known as *Wolf equations* (Wolf, 1955):

$$\nabla_1^2 \mathcal{W}(\mathbf{r}_1, \mathbf{r}_2, \tau) = \frac{1}{c^2} \frac{\partial^2 \mathcal{W}(\mathbf{r}_1, \mathbf{r}_2, \tau)}{\partial \tau^2},$$
$$\nabla_2^2 \mathcal{W}(\mathbf{r}_1, \mathbf{r}_2, \tau) = \frac{1}{c^2} \frac{\partial^2 \mathcal{W}(\mathbf{r}_1, \mathbf{r}_2, \tau)}{\partial \tau^2}, \qquad (2.73)$$

where subindex of Laplacian denotes the spatial argument with respect to which it is taken.

Two important quantities are based on the mutual coherence function. The *average intensity* of the optical field is given by the expression

$$\mathcal{I}(\mathbf{r}) = \mathcal{W}(\mathbf{r}, \mathbf{r}, 0). \qquad (2.74)$$

This quantity is directly detectable by an unaided eye or by a variety of man-made detectors, such as photographic film, or a CCD camera. Also, the *degree of coherence* of the optical field is the mutual coherence function normalized by the average intensities as

$$\gamma(\mathbf{r}_1, \mathbf{r}_2, \tau) = \frac{\mathcal{W}(\mathbf{r}_1, \mathbf{r}_2, \tau)}{\sqrt{\mathcal{I}(\mathbf{r}_1)}\sqrt{\mathcal{I}(\mathbf{r}_2)}}. \tag{2.75}$$

While the average intensity is real and non-negative, the degree of coherence is complex-valued and may take values anywhere in the unit circle of the complex plane, including the boundary. As we will show in Chapter 3, the degree of coherence can be measured by field self-interference, while its modulus and phase relate to the visibility and location of the produced fringes. For $|\gamma| = 1$, the optical field is called *coherent* and for $\gamma = 0$, it is *incoherent*. At $\tau = 0$ and at $\mathbf{r}_1 = \mathbf{r}_2 = \mathbf{r}$, the degree of coherence (2.75) reduces to its spatial and temporal counterparts, respectively.

The exact solution of the Wolf equations (2.73) being hyperbolic-type partial differential equations is somewhat difficult to obtain, especially for propagation in media. An alternative quantity characterizing the two-point spatial correlations in optical fields can be introduced by taking the one-dimensional Fourier transform of the mutual coherence function with respect to time delay τ:

$$W(\mathbf{r}_1, \mathbf{r}_2, \omega) = \int_{-\infty}^{\infty} \mathcal{W}(\mathbf{r}_1, \mathbf{r}_2, \tau) e^{-i\omega\tau} d\tau. \tag{2.76}$$

This quantity is known as the *cross-spectral density* of the optical field. We stress again that in this text, all functions considered in space–frequency/space–time domain will be denoted in straight/curly font. It can be shown that the cross-spectral density is itself a second-order correlation function (Wolf, 1982)

$$W(\mathbf{r}_1, \mathbf{r}_2, \omega) = \langle U^*(\mathbf{r}_1, \omega) U(\mathbf{r}_2, \omega) \rangle_\omega, \tag{2.77}$$

where the statistical ensemble is taken over the realizations of an equivalent monochromatic optical field oscillating at frequency ω. Taking the one-dimensional Fourier transform of Wolf's equations

(2.73) and using definition (2.77) imply that the cross-spectral density satisfies a pair of Helmholz equations:

$$\nabla_1^2 W(\mathbf{r}_1, \mathbf{r}_2, \omega) + k^2 W(\mathbf{r}_1, \mathbf{r}_2, \omega) = 0,$$
$$\nabla_2^2 W(\mathbf{r}_1, \mathbf{r}_2, \omega) + k^2 W(\mathbf{r}_1, \mathbf{r}_2, \omega) = 0. \tag{2.78}$$

Just like any genuine second-order correlation function, both the mutual coherence function and the cross-spectral density must be (I) quasi-Hermitian, (II) non-negative definite and (III) square-integrable with respect to all arguments (Mandel and Wolf, 1995). For example, for the cross-spectral density, these conditions can be mathematically expressed as

$$W(\mathbf{r}_1, \mathbf{r}_2, \omega) = W^*(\mathbf{r}_2, \mathbf{r}_1, \omega), \tag{2.79}$$

$$\int_D \int_D W(\mathbf{r}_1, \mathbf{r}_2, \omega) f^*(\mathbf{r}_1) f(\mathbf{r}_2) d\mathbf{r}_1 d\mathbf{r}_2 \geq 0, \tag{2.80}$$

$$\int_D \int_D |W(\mathbf{r}_1, \mathbf{r}_2, \omega)| d\mathbf{r}_1 d\mathbf{r}_2 < \infty, \tag{2.81}$$

$$\int_0^\infty |W(\mathbf{r}_1, \mathbf{r}_2, \omega)|^2 d\omega < \infty, \tag{2.82}$$

where D is the domain occupied by the field. In Eq. (2.80), function f is any complex-valued function.

The correlation functions of random processes can be represented via Mercer type expansion (Mercer, 1909), i.e., infinite series of deterministic contributions. For the cross-spectral density, one has:

$$W(\mathbf{r}_1, \mathbf{r}_2, \omega) = \sum_{n=0}^{\infty} \lambda_n(\omega) \phi_n^*(\mathbf{r}_1, \omega) \phi_n(\mathbf{r}_2, \omega), \tag{2.83}$$

where $\lambda_n(\omega)$ are real, non-negative eigenvalues, such that $\lim_{n \to \infty} \lambda_n(\omega) = 0$, and $\phi_n(\mathbf{r}, \omega)$ are orthonormal eigenfunctions satisfying the integral equation

$$\int_D W(\mathbf{r}_1, \mathbf{r}_2, \omega) \phi_n(\mathbf{r}_1, \omega) d\mathbf{r}_1 = \lambda_n(\omega) \phi_n(\mathbf{r}_2, \omega). \tag{2.84}$$

Expansion (2.83) is known as *coherent mode decomposition* since the contributions for each fixed n are fully spatially coherent. Therefore, it provides a valuable insight into the structure of random fields and, moreover, can be used as a practical tool for their modeling, generation and for the analysis of their evolution and interaction with optical systems and media.

2.2.2 One- and two-point properties

The cross-spectral density can be used to evaluate several single-point quantities. For beam-like fields, the *spectral density* is the z-component of the Poynting vector, $S_{P,zz}(\mathbf{r},\omega) = S(\mathbf{r},\omega)$, and can be related to the cross-spectral density as

$$S(\mathbf{r},\omega) = W(\mathbf{r},\mathbf{r},\omega). \tag{2.85}$$

Unlike average intensity $\mathcal{I}(\mathbf{r})$ in Eq. (2.74), it provides information regarding spatial distribution of optical energy at different frequencies.

The total optical intensity is then obtained by the integration of the spectral density over all frequencies:

$$I(\mathbf{r}) = \int_0^\infty S(\mathbf{r},\omega)d\omega. \tag{2.86}$$

In some situations, for instance, for calculating the spectral changes in light beams, it is also convenient to use the normalized version of $S(\mathbf{r},\omega)$,

$$S_N(\mathbf{r},\omega) = \frac{S(\mathbf{r},\omega)}{I(\mathbf{r})}. \tag{2.87}$$

Further, the total energy of the light field can be found by the integration

$$E_{tot} = \int I(\mathbf{r})d\mathbf{r} = \text{const}. \tag{2.88}$$

For scalar fields, the zz component of the *Orbital Angular Momentum* (OAM) *flux density* $L_{zz}(\mathbf{r},\omega) = L^{(\text{orb})}(\mathbf{r},\omega)$ ($L^{(\text{spin})}(\mathbf{r},\omega) = 0$) can be related to the cross-spectral density function as (Kim and Gbur, 2012)

$$L^{(\text{orb})}(\mathbf{r},\omega) = \frac{\varepsilon_0}{k}\text{Im}[(x_1\partial_{y_2} - y_1\partial_{x_2})W(\mathbf{r}_1,\mathbf{r}_2,\omega)]_{\mathbf{r}_1=\mathbf{r}_2}. \tag{2.89}$$

The average OAM flux density per photon $l_d^{(\text{orb})}$ can then be found as

$$l_d^{(\text{orb})}(\mathbf{r},\omega) = \frac{\hbar\omega L^{(\text{orb})}(\mathbf{r},\omega)}{S(\mathbf{r},\omega)}. \qquad (2.90)$$

Further, the total OAM per photon is found by performing integration

$$l_{\text{tot}}^{(\text{orb})}(\omega) = \frac{\hbar\omega \int L^{(\text{orb})}(\mathbf{r},\omega) d^2 r}{\int S(\mathbf{r},\omega) d^2 r}. \qquad (2.91)$$

While $l_{\text{tot}}^{(\text{orb})}$ remains invariant as light propagates, $l_d^{(\text{orb})}$ may exhibit changes in its spatial distribution.

An important two-point quantity derivable from the cross-spectral density is the *spectral degree of coherence* being its normalized version:

$$\mu(\mathbf{r}_1,\mathbf{r}_2,\omega) = \frac{W(\mathbf{r}_1,\mathbf{r}_2,\omega)}{\sqrt{S(\mathbf{r}_1,\omega)}\sqrt{S(\mathbf{r}_2,\omega)}}. \qquad (2.92)$$

Similarly to the time-domain based degree of coherence γ, the magnitude and the phase of the spectral degree of coherence is equal to the visibility and location of the interference fringes, respectively but for fields filtered at a specific value of angular frequency ω.

2.3 Electromagnetic Beam-like Fields

2.3.1 *Cross-spectral density matrix*

Let us now consider a wide-sense statistically stationary beam in which both mutually orthogonal Cartesian components of the electric field $\vec{\mathcal{E}}(\mathbf{r},t) = [\mathcal{E}_x(\mathbf{r},t), \mathcal{E}_y(\mathbf{r},t)]$ fluctuate in space and time. A 2×2 matrix:

$$\begin{aligned}\overleftrightarrow{W}(\mathbf{r}_1,\mathbf{r}_2,\tau) &= \langle \vec{\mathcal{E}}^\dagger(\mathbf{r}_1,t)\cdot\vec{\mathcal{E}}(\mathbf{r}_2,t+\tau)\rangle_T \\ &= \begin{bmatrix}\langle\mathcal{E}_x^*(\mathbf{r}_1,t)\mathcal{E}_x(\mathbf{r}_2,t+\tau)\rangle_T & \langle\mathcal{E}_x^*(\mathbf{r}_1,t)\mathcal{E}_y(\mathbf{r}_2,t+\tau)\rangle_T \\ \langle\mathcal{E}_y^*(\mathbf{r}_1,t)\mathcal{E}_x(\mathbf{r}_2,t+\tau)\rangle_T & \langle\mathcal{E}_y^*(\mathbf{r}_1,t)\mathcal{E}_y(\mathbf{r}_2,t+\tau)\rangle_T\end{bmatrix},\end{aligned}$$
(2.93)

where angular brakets represent the time average, subscript † stands for Hermitian adjoint, · stands for vector or matrix multiplication, τ is time delay, is known as the *beam-coherence-polarization matrix* (Gori et al., 1998). Taking the Fourier transform of matrix (2.93), element by element, yields the 2 × 2 matrix describing the correlations among the field components at two positions, $\mathbf{r}_1$ and $\mathbf{r}_2$, and at angular frequency ω. This matrix is known as the *cross-spectral density matrix* (Wolf, 2003):

$$\overleftrightarrow{W}(\mathbf{r}_1, \mathbf{r}_2, \omega) = \langle \overrightarrow{E}^\dagger(\mathbf{r}_1, \omega) \cdot \overrightarrow{E}(\mathbf{r}_2, \omega) \rangle_\omega$$
$$= \begin{bmatrix} \langle E_x^*(\mathbf{r}_1,\omega)E_x(\mathbf{r}_2,\omega)\rangle_\omega & \langle E_x^*(\mathbf{r}_1,\omega)E_y(\mathbf{r}_2,\omega)\rangle_\omega \\ \langle E_y^*(\mathbf{r}_1,\omega)E_x(\mathbf{r}_2,\omega)\rangle_\omega & \langle E_y^*(\mathbf{r}_1,\omega)E_y(\mathbf{r}_2,\omega)\rangle_\omega \end{bmatrix}.$$
(2.94)

As for scalar fields, $\langle \cdot \rangle_\omega$ stands for the ensemble of monocromatic realizations of the electric field components at this frequency. Each element of correlation matrices (2.93) and (2.94) must obey the square-integrability conditions, just like the scalar correlation function (see Eqs. (2.80) and (2.81)). In addition, they must be quasi-Hermitian:

$$W_{ij}(\mathbf{r}_1, \mathbf{r}_2, \omega) = W_{ji}^*(\mathbf{r}_2, \mathbf{r}_1, \omega),\qquad(2.95)$$

and non-negative definite:

$$\sum_{i,j=x,y} \int_D \int_D f_i^*(\mathbf{r}_1, \omega) W_{ij}(\mathbf{r}_1, \mathbf{r}_2, \omega) f_j(\mathbf{r}_2, \omega) d\mathbf{r}_1 d\mathbf{r}_2 \geq 0, \quad (2.96)$$

where f_i and f_j are arbitrary (possibly complex) functions. Matrices (2.93) and (2.94) are introduced under the assumption of electric *row* vector-field. An alternative definition is conventionally used with the electric *column* vector-field as is the case in polarization optics (Chapter 7) and imaging systems (Chapter 8).

2.3.2 One-point properties

The spectral density of an electromagnetic field is associated with the z component of the Poynting vector, and given by the trace of the cross-spectral density at $\mathbf{r}_1 = \mathbf{r}_2 = \mathbf{r}$, i.e.,

$$S(\mathbf{r}, \omega) = \text{Tr}\overleftrightarrow{W}(\mathbf{r}, \mathbf{r}, \omega), \qquad (2.97)$$

while the proportionality factor is typically ignored in theoretical computations. Its normalized and integrated versions can also be introduced just like for the scalar fields (see Eqs. (2.86) and (2.87)).

As we saw earlier, for electromagnetic beams the zz component of the angular momentum flux density $L_{zz}(\mathbf{r}, \omega)$ has two non-trivial contributions relating to orbital and spin states of photons (Kim and Gbur, 2012). For stationary fields, this relation becomes

$$L_{zz}(\mathbf{r}, \omega) = L^{(\text{sp})}(\mathbf{r}, \omega) + L^{(\text{orb})}(\mathbf{r}, \omega), \qquad (2.98)$$

where the Spin Angular Momentum (SAM) flux density has the form

$$L^{(\text{sp})}(\mathbf{r}, \omega) = \frac{1}{8\pi k}\text{Im}[W_{yx}(\mathbf{r}, \mathbf{r}, \omega) - W_{xy}(\mathbf{r}, \mathbf{r}, \omega)], \qquad (2.99)$$

and the OAM flux density is

$$L^{(\text{orb})}(\mathbf{r}, \omega) = \frac{1}{8\pi k}\text{Im}[y_1\partial_{x_2}W_{yy}(\mathbf{r}_1, \mathbf{r}_2, \omega) - x_1\partial_{y_2}W_{xx}(\mathbf{r}_1, \mathbf{r}_2, \omega)$$
$$+ x_1\partial_{y_1}W_{yy}(\mathbf{r}_1, \mathbf{r}_2, \omega) - y_1\partial_{x_1}W_{xx}(\mathbf{r}_1, \mathbf{r}_2, \omega)]|_{\mathbf{r}_1=\mathbf{r}_2}. \qquad (2.100)$$

This expression takes a particularly concise form in polar coordinates:

$$L^{(\text{orb})}(\mathbf{r}, \omega) = \frac{1}{8\pi k}\text{Im}[\partial_\theta \text{Tr}[\overleftrightarrow{W}(\mathbf{r}_1, \mathbf{r}_2, \omega)]]|_{\mathbf{r}_1=\mathbf{r}_2}. \qquad (2.101)$$

Let us now discuss the polarization properties of a random electromagnetic beam-like field. The cross-spectral density matrix $\overleftrightarrow{W}$ evaluated at coinciding spatial arguments, $\mathbf{r}_1 = \mathbf{r}_2 = \mathbf{r}$, can be uniquely decomposed into a sum of two matrices (Korotkova and Wolf, 2005a):

$$\overleftrightarrow{W}(\mathbf{r}, \mathbf{r}, \omega) = \overleftrightarrow{W}^{(u)}(\mathbf{r}, \mathbf{r}, \omega) + \overleftrightarrow{W}^{(p)}(\mathbf{r}, \mathbf{r}, \omega), \qquad (2.102)$$

where

$$\overleftrightarrow{W}^{(u)}(\mathbf{r}, \mathbf{r}, \omega) = A(\mathbf{r}, \omega)\begin{bmatrix} 1 & 0 \\ 0 & 1 \end{bmatrix},$$

$$\overleftrightarrow{W}^{(p)}(\mathbf{r}, \mathbf{r}, \omega) = \begin{bmatrix} B(\mathbf{r}, \omega) & D(\mathbf{r}, \omega) \\ D^*(\mathbf{r}, \omega) & C(\mathbf{r}, \omega) \end{bmatrix} \qquad (2.103)$$

represent unpolarized and polarized portions of the beam, respectively, at position **r**. Matrix $\overleftrightarrow{W}^{(p)}$ is necessarily a singular matrix, i.e., $\mathrm{Det}[\overleftrightarrow{W}^{(p)}] = 0$, where Det stands for the matrix determinant. Elements of matrices $\overleftrightarrow{W}^{(p)}$ and $\overleftrightarrow{W}^{(u)}$ can be expressed via those of matrix $\overleftrightarrow{W}$ as

$$A(\mathbf{r},\omega) = \frac{1}{2}\left[W_{xx} + W_{yy} - \sqrt{(W_{xx}+W_{yy})^2 + 4|W_{xy}|^2}\right],$$

$$B(\mathbf{r},\omega) = \frac{1}{2}\left[W_{xx} - W_{yy} + \sqrt{(W_{xx}-W_{yy})^2 + 4|W_{xy}|^2}\right],$$

$$C(\mathbf{r},\omega) = \frac{1}{2}\left[W_{yy} - W_{xx} + \sqrt{(W_{xx}-W_{yy})^2 + 4|W_{xy}|^2}\right],$$

$$D(\mathbf{r},\omega) = W_{xy}.$$

(2.104)

The degree of polarization is defined as a ratio of the spectral density of the polarized portion of the beam to that of the total beam:

$$\varrho(\mathbf{r},\omega) = \frac{S^{(p)}(\mathbf{r},\omega)}{S(\mathbf{r},\omega)}$$

$$= \sqrt{1 - \frac{4\mathrm{Det}\overleftrightarrow{W}(\mathbf{r},\mathbf{r},\omega)}{[\mathrm{Tr}\overleftrightarrow{W}(\mathbf{r},\mathbf{r},\omega)]^2}},$$

(2.105)

where the second line in Eq. (2.105) was obtained with the help of Eqs. (2.97), (2.103) and (2.104). The degree of polarization is a real-valued quantity, restricted to the interval $0 \leq \varrho(\mathbf{r}) \leq 1$. In the two limiting cases when $\varrho(\mathbf{r}) = 1$ and $\varrho(\mathbf{r}) = 0$, the beam is called completely polarized and completeley unpolarized. The degree of polarization depends only on the matrix trace and the determinant, and, hence, it remains invariant under rotations, just like the spectral density.

Matrix $\overleftrightarrow{W}^{(p)}$ carries the information about the polarization ellipse of the beam. Being singular, it can be factorized as

$$\overleftrightarrow{W}^{(p)}(\mathbf{r},\mathbf{r},\omega) = \begin{bmatrix} E_x^*(\mathbf{r},\omega)E_x(\mathbf{r},\omega) & E_x^*(\mathbf{r},\omega)E_y(\mathbf{r},\omega) \\ E_y^*(\mathbf{r},\omega)E_x(\mathbf{r},\omega) & E_y^*(\mathbf{r},\omega)E_y(\mathbf{r},\omega) \end{bmatrix},$$

(2.106)

where $E_x(\mathbf{r},\omega)$ and $E_y(\mathbf{r},\omega)$ are the Cartesian components of the equivalent monochromatic electric field at frequency ω. On taking the real parts of the two components, we then find that

$$E_x^{(R)}(\mathbf{r},\omega) = \sqrt{B(\mathbf{r},\omega)}\cos[\omega t + \varphi_x(\mathbf{r},\omega)],$$
$$E_y^{(R)}(\mathbf{r},\omega) = \sqrt{C(\mathbf{r},\omega)}\cos[\omega t + \varphi_y(\mathbf{r},\omega)],$$
(2.107)

where phase lag is defined as

$$\varphi(\mathbf{r},\omega) = \varphi_y(\mathbf{r},\omega) - \varphi_x(\mathbf{r},\omega) = \arg[D(\mathbf{r},\omega)]. \quad (2.108)$$

Eliminating time dependence from Eqs. (2.107) yields the quadratic form

$$C(\mathbf{r},\omega)[E_x^{(R)}(\mathbf{r},\omega)]^2 - 2\mathrm{Re}[D(\mathbf{r};\omega)]E_x^{(R)}(\mathbf{r},\omega)E_y^{(R)}(\mathbf{r},\omega)$$
$$+ B(\mathbf{r},\omega)[E_x^{(R)}(\mathbf{r},\omega)]^2 = (\mathrm{Im}[D(\mathbf{r},\omega)])^2,$$
(2.109)

which characterizes the spectral polarization ellipse at frequency ω. The orientation angle ψ and the magnitudes of the major and minor semi-axes ζ_+ and ζ_- of the ellipse can be related to the elements of the W-matrix as

$$\psi(\mathbf{r},\omega) = \frac{1}{2}\arctan\left(\frac{2\mathrm{Re}[W_{xy}(\mathbf{r},\mathbf{r},\omega)]}{W_{xx}(\mathbf{r},\mathbf{r},\omega) - W_{yy}(\mathbf{r},\mathbf{r},\omega)}\right), \quad (2.110)$$

and

$$\zeta_\pm(\mathbf{r},\mathbf{r},\omega) = \frac{1}{\sqrt{2}}\bigg[\sqrt{[W_{xx}(\mathbf{r},\mathbf{r},\omega) - W_{yy}(\mathbf{r},\mathbf{r},\omega)]^2 + 4|W_{xy}(\mathbf{r},\mathbf{r},\omega)|^2}$$
$$\pm \sqrt{[W_{xx}(\mathbf{r},\mathbf{r},\omega) - W_{yy}(\mathbf{r},\mathbf{r},\omega)]^2 + 4[\mathrm{Re}W_{xy}(\mathbf{r},\mathbf{r},\omega)]^2}\bigg]^{1/2}.$$
(2.111)

The spectral degree of ellipticity and the spectral azimuthal angle of the ellipse are defined by the same expressions as those for deterministic fields in Eq. (2.66) (see Fig. 2.1(a)).

For random beams, the four spectral Stokes parameters at point **r** and angular frequency ω may be defined as

$$S_0(\mathbf{r},\omega) = \langle E_x^*(\mathbf{r},\omega)E_x(\mathbf{r},\omega)\rangle + \langle E_y^*(\mathbf{r},\omega)E_y(\mathbf{r},\omega)\rangle,$$
$$S_1(\mathbf{r},\omega) = \langle E_x^*(\mathbf{r},\omega)E_x(\mathbf{r},\omega)\rangle - \langle E_y^*(\mathbf{r},\omega)E_y(\mathbf{r},\omega)\rangle,$$
$$S_2(\mathbf{r},\omega) = \langle E_x^*(\mathbf{r},\omega)E_y(\mathbf{r},\omega)\rangle + \langle E_y^*(\mathbf{r},\omega)E_x(\mathbf{r};\omega)\rangle, \quad (2.112)$$
$$S_3(\mathbf{r},\omega) = i[\langle E_y^*(\mathbf{r},\omega)E_x(\mathbf{r},\omega)\rangle - \langle E_x^*(\mathbf{r},\omega)E_y(\mathbf{r},\omega)\rangle],$$

being the linear combinations of the W-matrix elements, viz.,

$$S_0(\mathbf{r},\omega) = W_{xx}(\mathbf{r},\mathbf{r},\omega) + W_{yy}(\mathbf{r},\mathbf{r},\omega),$$
$$S_1(\mathbf{r},\omega) = W_{xx}(\mathbf{r},\mathbf{r},\omega) - W_{yy}(\mathbf{r},\mathbf{r},\omega),$$
$$S_2(\mathbf{r},\omega) = W_{xy}(\mathbf{r},\mathbf{r},\omega) + W_{yx}(\mathbf{r},\mathbf{r},\omega), \quad (2.113)$$
$$S_3(\mathbf{r},\omega) = i[W_{yx}(\mathbf{r},\mathbf{r},\omega) - W_{xy}(\mathbf{r},\mathbf{r},\omega)].$$

In particular, the first Stokes parameter $S_0(\mathbf{r},\omega)$ is equal to the spectral density of the stochastic beam (see Eq. (2.97)), i.e.,

$$S_0(\mathbf{r},\omega) = S(\mathbf{r},\omega). \quad (2.114)$$

For random beams, the Stokes parameters must satisfy the inequality

$$S_1^2(\mathbf{r},\omega) + S_2^2(\mathbf{r},\omega) + S_3^2(\mathbf{r},\omega) \leq S_0^2(\mathbf{r},\omega). \quad (2.115)$$

In view of this inequality and the definition (2.116) of the degree ϱ of polarization, the possible polarization states of random beams occupy a spherical shell with radius ϱS_0 within the Poincaré sphere (see Fig. 2.1(b)).

The degree of polarization and the parameters of the polarization ellipse can be expressed via the Stokes parameters by the formulas

$$\varrho(\mathbf{r},\omega) = \frac{\sqrt{S_1^2(\mathbf{r},\omega) + S_2^2(\mathbf{r},\omega) + S_3^2(\mathbf{r},\omega)}}{S_0^2(\mathbf{r},\omega)}, \quad (2.116)$$

$$\zeta_\pm(\mathbf{r},\omega) = \left[\frac{S_0(\mathbf{r},\omega) \pm \sqrt{S_1^2(\mathbf{r},\omega) + S_2^2(\mathbf{r},\omega)}}{S_3(\mathbf{r},\omega)}\right]^{1/2}, \quad (2.117)$$

$$\psi(\mathbf{r},\omega) = \frac{1}{2}\arctan\left[\frac{S_2(\mathbf{r},\omega)}{S_1(\mathbf{r},\omega)}\right]. \quad (2.118)$$

Before concluding this section, we draw the reader's attention to the fact that at the coinciding spatial arguments, $\mathbf{r}_1 = \mathbf{r}_2 = \mathbf{r}$, the W-matrix is non-negative definite and Hermitian and, hence, its eigenvalues are real and non-negative and its eigenvectors are also real-valued and mutually orthogonal. On denoting the spectrum of this operator as

$$\mathrm{sp}[\overleftrightarrow{W}(\mathbf{r},\mathbf{r},\omega)] = \{\lambda_1(\mathbf{r},\omega), \lambda_2(\mathbf{r},\omega)\}, \qquad (2.119)$$

we find that

$$\lambda_{1,2}(\mathbf{r},\omega) = \frac{1}{2}\Big[\mathrm{Tr}\overleftrightarrow{W}(\mathbf{r},\mathbf{r},\omega) \\ \pm \sqrt{[\mathrm{Tr}\overleftrightarrow{W}(\mathbf{r},\mathbf{r},\omega)]^2 - 4\mathrm{Det}\overleftrightarrow{W}(\mathbf{r},\mathbf{r},\omega)}\Big], \qquad (2.120)$$

where we set $\lambda_1(\mathbf{r},\omega) \geq \lambda_2(\mathbf{r},\omega)$. It is evident that the *spectral strength* $ss[\overleftrightarrow{W}(\mathbf{r},\mathbf{r},\omega)]$ and the *spectral range* $sr[\overleftrightarrow{W}(\mathbf{r},\mathbf{r},\omega)]$, generally defined as the sum of all the eigenvalues of an operator and the difference between the maximum and the minimum eigenvalues, respectively, are directly related to the spectral density and the degree of polarization by the expressions

$$S(\mathbf{r},\omega) = ss[\overleftrightarrow{W}(\mathbf{r},\mathbf{r},\omega)] = \lambda_1(\mathbf{r},\omega) + \lambda_2(\mathbf{r},\omega), \qquad (2.121)$$

and

$$\varrho(\mathbf{r},\omega) = \frac{sr[\overleftrightarrow{W}(\mathbf{r},\mathbf{r},\omega)]}{ss[\overleftrightarrow{W}(\mathbf{r},\mathbf{r},\omega)]} \\ = \frac{\lambda_1(\mathbf{r},\omega) - \lambda_2(\mathbf{r},\omega)}{\lambda_1(\mathbf{r},\omega) + \lambda_2(\mathbf{r},\omega)}. \qquad (2.122)$$

2.3.3 *Two-point properties*

We recall that the degree of coherence of a scalar field is defined via the two-point field correlation function by expression (2.92), while its modulus is directly related to the visibility of interference fringes in the double-slit experiment (see Chapter 3). Since an electromagnetic beam is not characterized by a single quantity but rather by a 2×2 correlation matrix, there are a number of options for specifying its state of coherence as a single scalar quantity, but none of them are ideal. Let us now follow one of the possible approaches employing the spectral theory of operators. Recall that operator $\overleftrightarrow{W}(\mathbf{r}_1, \mathbf{r}_2, \omega)$ of a beam-like field is quasi-Hermitian on $\mathbb{R}^2$, i.e., there must exist a 2×2 operator, say $\overleftrightarrow{\Theta}(\mathbf{r}_1, \mathbf{r}_2, \omega)$, which, together with its inverse $\overleftrightarrow{\Theta}^{-1}(\mathbf{r}_1, \mathbf{r}_2, \omega)$, is bounded, positive and similar to $\overleftrightarrow{W}(\mathbf{r}_1, \mathbf{r}_2, \omega)$, meaning that

$$\overleftrightarrow{W}(\mathbf{r}_1, \mathbf{r}_2, \omega) = \overleftrightarrow{\Theta}^{-1}(\mathbf{r}_1, \mathbf{r}_2, \omega) \overleftrightarrow{W}^*(\mathbf{r}_1, \mathbf{r}_2, \omega) \overleftrightarrow{\Theta}(\mathbf{r}_1, \mathbf{r}_2, \omega). \quad (2.123)$$

Generally, the spectrum of a quasi-Hermitian operator,

$$\text{sp}[\overleftrightarrow{W}(\mathbf{r}_1, \mathbf{r}_2, \omega)] = \{\lambda_1(\mathbf{r}_1, \mathbf{r}_2, \omega), \lambda_2(\mathbf{r}_1, \mathbf{r}_2, \omega)\}, \quad (2.124)$$

is complex-valued:

$$\lambda_{1,2}(\mathbf{r}_1, \mathbf{r}_2, \omega) = \frac{1}{2}\Big[\text{Tr}\overleftrightarrow{W}(\mathbf{r}_1, \mathbf{r}_2, \omega) \\ \pm \sqrt{[\text{Tr}\overleftrightarrow{W}(\mathbf{r}_1, \mathbf{r}_2, \omega)]^2 - 4\text{Det}\overleftrightarrow{W}(\mathbf{r}_1, \mathbf{r}_2, \omega)}\Big], \\ (2.125)$$

where we set $\text{Re}[\lambda_1(\mathbf{r}_1, \mathbf{r}_2, \omega)] \geq \text{Re}[\lambda_2(\mathbf{r}_1, \mathbf{r}_2, \omega)]$, Re standing for real part. Figure 2.2(a) illustrates the complex-valued spectrum of operator $\overleftrightarrow{W}$ as well as its spectral strength and spectral range.

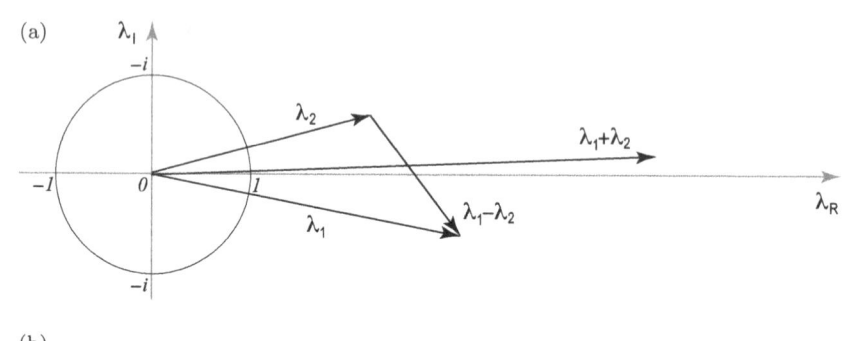

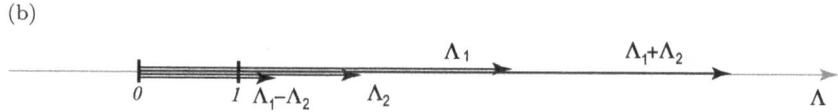

Fig. 2.2 Spectra of (a) W-operator and (b) Ω-operator.

A suitably normalized spectral strength of operator $\overleftrightarrow{W}(\mathbf{r}_1,\mathbf{r}_2,\omega)$ leads to the definition of the *spectral degree of coherence* of an electromegnetic field (see also (Wolf, 2003)):

$$\mu_W(\mathbf{r}_1,\mathbf{r}_2,\omega) = \frac{ss[\overleftrightarrow{W}(\mathbf{r}_1,\mathbf{r}_2,\omega)]}{\sqrt{ss[\overleftrightarrow{W}(\mathbf{r}_1,\mathbf{r}_1,\omega)]}\sqrt{ss[\overleftrightarrow{W}(\mathbf{r}_2,\mathbf{r}_2,\omega)]}}$$

$$= \frac{\lambda_1(\mathbf{r}_1,\mathbf{r}_2,\omega) + \lambda_2(\mathbf{r}_1,\mathbf{r}_2,\omega)}{\sqrt{\lambda_1(\mathbf{r}_1,\omega) + \lambda_2(\mathbf{r}_1,\omega)}\sqrt{\lambda_1(\mathbf{r}_2,\omega) + \lambda_2(\mathbf{r}_2,\omega)}}$$

$$= \frac{\mathrm{Tr}\overleftrightarrow{W}(\mathbf{r}_1,\mathbf{r}_2,\omega)}{\sqrt{S(\mathbf{r}_1,\omega)}\sqrt{S(\mathbf{r}_2,\omega)}}. \qquad (2.126)$$

Note that this definition only involves the diagonal elements of $\overleftrightarrow{W}$ and, hence, does not carry infomation about any possible correlations between the x and the y components.

Using the spectral strength and the spectral range of operator $\overleftrightarrow{W}(\mathbf{r}_1,\mathbf{r}_2,\omega)$, it is also possible to formally introduce the two-point complex-valued degree of cross-polarization:

$$\varrho_W(\mathbf{r}_1,\mathbf{r}_2,\omega) = \frac{sr[\overleftrightarrow{W}(\mathbf{r}_1,\mathbf{r}_2,\omega)]}{ss[\overleftrightarrow{W}(\mathbf{r}_1,\mathbf{r}_2,\omega)]}$$

$$= \frac{\lambda_1(\mathbf{r}_1,\mathbf{r}_2,\omega) - \lambda_2(\mathbf{r}_1,\mathbf{r}_2,\omega)}{\lambda_1(\mathbf{r}_1,\mathbf{r}_2,\omega) + \lambda_2(\mathbf{r}_1,\mathbf{r}_2,\omega)}, \qquad (2.127)$$

or, equivalently,

$$\varrho_W(\mathbf{r}_1,\mathbf{r}_2,\omega) = \sqrt{1 - \frac{4\mathrm{Det}[\overleftrightarrow{W}(\mathbf{r}_1,\mathbf{r}_2,\omega)]}{\mathrm{Tr}^2[\overleftrightarrow{W}(\mathbf{r}_1,\mathbf{r}_2,\omega)]}}. \qquad (2.128)$$

This measure reduces to the classic, single-point degree of polarizaton $\varrho_W(\mathbf{r},\omega)$ at the coinciding spatial arguments but provides the insight into polarization correlation as $\mathbf{r}_1$ and $\mathbf{r}_2$ become separated.

An alternative procedure for defining coherence and cross-polarization properties of the $\overleftrightarrow{W}$-operator may also be suggested via

operator
$$\overleftrightarrow{\Omega}(\mathbf{r}_1,\mathbf{r}_2,\omega) = \overleftrightarrow{W}^\dagger(\mathbf{r}_1,\mathbf{r}_2,\omega)\overleftrightarrow{W}(\mathbf{r}_1,\mathbf{r}_2,\omega). \qquad (2.129)$$

Such an operator is always Hermitian, and, hence, its spectrum
$$\mathrm{sp}[\overleftrightarrow{\Omega}] = \{\Lambda_1(\mathbf{r}_1,\mathbf{r}_2,\omega), \Lambda_2(\mathbf{r}_1,\mathbf{r}_2,\omega)\}, \qquad (2.130)$$

where $\Lambda_1(\mathbf{r}_1,\mathbf{r}_2,\omega) \geq \Lambda_2(\mathbf{r}_1,\mathbf{r}_2,\omega)$, consists of two real-valued, non-negative eigenvalues that can be found from the expression

$$\Lambda_{1,2}(\mathbf{r}_1,\mathbf{r}_2,\omega) = \frac{1}{2}\Big[\mathrm{Tr}\overleftrightarrow{\Omega}(\mathbf{r}_1,\mathbf{r}_2,\omega)$$
$$\pm \sqrt{[\mathrm{Tr}\overleftrightarrow{\Omega}(\mathbf{r}_1,\mathbf{r}_2,\omega)]^2 - 4\mathrm{Det}\overleftrightarrow{\Omega}(\mathbf{r}_1,\mathbf{r}_2,\omega)}\Big]. \qquad (2.131)$$

Figure 2.2(b) illustrates the real-valued, non-negative spectrum of operator $\overleftrightarrow{\Omega}$ as well as its spectral strength and spectral range.

In cases when operator $\overleftrightarrow{W}$ is normal, i.e., if $\overleftrightarrow{W}^\dagger\overleftrightarrow{W} = \overleftrightarrow{W}\overleftrightarrow{W}^\dagger$, there exists a simple relationship between the eigenvalues of $\overleftrightarrow{W}$ and $\overleftrightarrow{\Omega}$:

$$\Lambda_1(\mathbf{r}_1,\mathbf{r}_2,\omega) = |\lambda_1(\mathbf{r}_1,\mathbf{r}_2,\omega)|^2, \quad \Lambda_2(\mathbf{r}_1,\mathbf{r}_2,\omega) = |\lambda_2(\mathbf{r}_1,\mathbf{r}_2,\omega)|^2, \qquad (2.132)$$

otherwise the two spectra are not related. In particular, since such a Hermitian matrix is automatically normal, we get

$$\Lambda_1(\mathbf{r},\mathbf{r},\omega) = |\lambda_1(\mathbf{r},\mathbf{r},\omega)|^2, \quad \Lambda_2(\mathbf{r},\mathbf{r},\omega) = |\lambda_2(\mathbf{r},\mathbf{r},\omega)|^2. \qquad (2.133)$$

The real-valued degree of coherence can be obtained as the normalized version of the strength of operator $\overleftrightarrow{\Omega}$ as

$$\mu_\Omega(\mathbf{r}_1,\mathbf{r}_2,\omega) = \sqrt[4]{\frac{ss_\Omega^2(\mathbf{r}_1,\mathbf{r}_2,\omega)}{ss_\Omega(\mathbf{r}_1,\mathbf{r}_1,\omega)ss_\Omega(\mathbf{r}_2,\mathbf{r}_2,\omega)}}$$
$$= \sqrt[4]{\frac{\mathrm{Tr}^2[\overleftrightarrow{\Omega}(\mathbf{r}_1,\mathbf{r}_2,\omega)]}{\mathrm{Tr}[\overleftrightarrow{\Omega}(\mathbf{r}_1,\mathbf{r}_1,\omega)]\mathrm{Tr}[\overleftrightarrow{\Omega}(\mathbf{r}_2,\mathbf{r}_2,\omega)]}}, \qquad (2.134)$$

the suggested normalization not being unique. For instance, an alternative normalization was used in Tervo et al. (2003):

$$\mu_T(\mathbf{r}_1,\mathbf{r}_2) = \frac{\sqrt{ss_\Omega(\mathbf{r}_1,\mathbf{r}_2,\omega)}}{ss_W(\mathbf{r}_1,\mathbf{r}_1,\omega)ss_W(\mathbf{r}_2,\mathbf{r}_2,\omega)}$$

$$= \frac{\sqrt{\mathrm{Tr}[\Omega(\mathbf{r}_1,\mathbf{r}_2,\omega)]}}{\mathrm{Tr}[\overleftrightarrow{W}(\mathbf{r}_1,\mathbf{r}_1,\omega)]\mathrm{Tr}[\overleftrightarrow{W}(\mathbf{r}_2,\mathbf{r}_2,\omega)]}. \quad (2.135)$$

In μ_Ω and μ_T, correlations among all four elements of the cross-spectral density matrix are taken into account. However, the result is always a real number, hence, the important information about the phase (imaginary part) of the correlations is not taken into account.

There are other definitions for the degree of coherence of the electromagnetic beams, for instance, one based on maximizing the fringe visibility in the double-slit experiment by means of devices of polarization optics used at the two slits (Gori et al., 2007). If the devices are chosen to be deterministic, anisotropic, and non-absorbing, then

$$\mu_G(\mathbf{r}_1,\mathbf{r}_2,\omega) = \frac{\sum_{i=1,2} SV_i[\overleftrightarrow{W}(\mathbf{r}_1,\mathbf{r}_2,\omega)]}{\sqrt{S(\mathbf{r}_1,\omega)}\sqrt{S(\mathbf{r}_2,\omega)}}. \quad (2.136)$$

It appears that $\mu_G = (|\mu_W|)_{\max}$.

Further, the degree of cross-polarization can be defined as the range-to-strength ratio of operator $\overleftrightarrow{\Omega}$:

$$\varrho_\Omega(\mathbf{r}_1,\mathbf{r}_2,\omega) = \frac{sr_\Omega(\mathbf{r}_1,\mathbf{r}_2,\omega)}{ss_\Omega(\mathbf{r}_1,\mathbf{r}_2,\omega)} = \sqrt{1 - \frac{4\mathrm{Det}[\overleftrightarrow{\Omega}(\mathbf{r}_1,\mathbf{r}_2,\omega)]}{\mathrm{Tr}^2[\overleftrightarrow{\Omega}(\mathbf{r}_1,\mathbf{r}_2,\omega)]}}. \quad (2.137)$$

A similar quantity was introduced in Shirai and Wolf (2007) and Volkov et al. (2008) in connection with intensity fluctuations in random electromagnetic beams.

The extension of the classic Stokes parameters from one to two positions can also be made. For example, in the space-frequency formulation, the *generalized Stokes parameters* can be introduced by the expressions (Korotkova and Wolf, 2005b)

$$S_0(\mathbf{r}_1,\mathbf{r}_2,\omega) = \langle E_x^*(\mathbf{r}_1,\omega)E_x(\mathbf{r}_2,\omega)\rangle + \langle E_y^*(\mathbf{r}_1,\omega)E_y(\mathbf{r}_2,\omega)\rangle,$$
$$S_1(\mathbf{r}_1,\mathbf{r}_2,\omega) = \langle E_x^*(\mathbf{r}_1,\omega)E_x(\mathbf{r}_2,\omega)\rangle - \langle E_y^*(\mathbf{r}_1,\omega)E_y(\mathbf{r}_2,\omega)\rangle,$$
$$S_2(\mathbf{r}_1,\mathbf{r}_2,\omega) = \langle E_x^*(\mathbf{r}_1,\omega)E_y(\mathbf{r}_2,\omega)\rangle + \langle E_y^*(\mathbf{r}_1,\omega)E_x(\mathbf{r}_2,\omega)\rangle,$$
$$S_3(\mathbf{r}_1,\mathbf{r}_2,\omega) = i[\langle E_y^*(\mathbf{r}_1,\omega)E_x(\mathbf{r}_2,\omega)\rangle - \langle E_x^*(\mathbf{r}_1,\omega)E_y(\mathbf{r}_2,\omega)\rangle],$$
(2.138)

i.e., as the linear combinations of the W-matrix elements:

$$S_0(\mathbf{r}_1,\mathbf{r}_2,\omega) = W_{xx}(\mathbf{r}_1,\mathbf{r}_2,\omega) + W_{yy}(\mathbf{r}_1,\mathbf{r}_2,\omega),$$
$$S_1(\mathbf{r}_1,\mathbf{r}_2,\omega) = W_{xx}(\mathbf{r}_1,\mathbf{r}_2,\omega) - W_{yy}(\mathbf{r}_1,\mathbf{r}_2,\omega),$$
$$S_2(\mathbf{r}_1,\mathbf{r}_2,\omega) = W_{xy}(\mathbf{r}_1,\mathbf{r}_2,\omega) + W_{yx}(\mathbf{r}_1,\mathbf{r}_2,\omega),$$
$$S_3(\mathbf{r}_1,\mathbf{r}_2,\omega) = i[W_{yx}(\mathbf{r}_1,\mathbf{r}_2,\omega) - W_{xy}(\mathbf{r}_1,\mathbf{r}_2,\omega)].$$
(2.139)

2.4 Electromagnetic General Fields

Let us now consider a wide-sense statistically stationary light in which all three mutually orthogonal Cartesian components of the electric field $\vec{\mathcal{E}}(\mathbf{r},t) = [\mathcal{E}_x(\mathbf{r},t), \mathcal{E}_y(\mathbf{r},t), \mathcal{E}_z(\mathbf{r},t)]$ fluctuate in space and time. Then a 3×3 cross-spectral density matrix describing the correlations among the field components at two positions, $\mathbf{r}_1$ and $\mathbf{r}_2$, and at time delay τ, can be defined as follows (Mandel and Wolf, 1995):

$$\overleftrightarrow{W}^{(3)}(\mathbf{r}_1,\mathbf{r}_2,t) = \langle \vec{\mathcal{E}}^\dagger(\mathbf{r}_1,t) \cdot \vec{\mathcal{E}}(\mathbf{r}_2,t+\tau)\rangle_T,$$
(2.140)

where superscript (3) stands for dimension. On taking the Fourier transform of matrix $\overleftrightarrow{W}^{(3)}$ with respect to τ, we obtain the 3×3 matrix:

$$\overleftrightarrow{W}^{(3)}(\mathbf{r}_1,\mathbf{r}_2,\omega) = \langle \vec{E}^\dagger(\mathbf{r}_1,\omega) \cdot \vec{E}(\mathbf{r}_2,\omega)\rangle_\omega.$$
(2.141)

This matrix must obey the same set of properties as its beam-like counterpart: the square-integrability in space and frequency, the quasi-Hermiticity and the non-negative definiteness, with adjustment to its dimensionality.

It will be convenient to first determine the spectrum of operator $\overleftrightarrow{W}^{(3)}$ at $\mathbf{r}_1 = \mathbf{r}_2 = \mathbf{r}$ and then use it to define the second-order statistical properties. At a single position, $\overleftrightarrow{W}^{(3)}$ is Hermitian and, hence, its spectrum

$$\text{sp}[\overleftrightarrow{W}(\mathbf{r},\mathbf{r},\omega)] = \{\lambda_1(\mathbf{r},\omega), \lambda_2(\mathbf{r},\omega), \lambda_3(\mathbf{r},\omega)\} \tag{2.142}$$

is real and non-negative. The three eigenvalues can be found using the reduced Cardano formulas (Cardano, 1565):

$$\lambda_1(\mathbf{r},\omega) = [\text{Tr}[\overleftrightarrow{W}^{(3)}(\mathbf{r},\mathbf{r},\omega)] - 2\sqrt{a_1}\cos(a_4/3)]/3,$$
$$\lambda_2(\mathbf{r},\omega) = [\text{Tr}[\overleftrightarrow{W}^{(3)}(\mathbf{r},\mathbf{r},\omega)] + 2\sqrt{a_1}\cos[(a_4-\pi)/3]]/3, \tag{2.143}$$
$$\lambda_3(\mathbf{r},\omega) = [\text{Tr}[\overleftrightarrow{W}^{(3)}(\mathbf{r},\mathbf{r},\omega)] + 2\sqrt{a_1}\cos[(a_4+\pi)/3)]]/3.$$

Here,

$$a_4 = \begin{cases} \arctan\left(\frac{\sqrt{4a_1^2-a_2^2}}{a_2}\right), & \text{if } a_2 > 0, \\ \pi/2, & \text{if } a_2 = 0, \\ \arctan\left(\frac{\sqrt{4a_1^2-a_2^2}}{a_2}\right) + \pi, & \text{if } a_2 < 0, \end{cases} \tag{2.144}$$

$$a_1 = W_{xx}^2 + W_{yy}^2 + W_{zz}^2 - W_{xx}W_{yy} - W_{xx}W_{zz} - W_{yy}W_{zz}$$
$$+ 3(|W_{xy}|^2 + |W_{yz}|^2 + |W_{xz}|^2), \tag{2.145}$$

and

$$a_2 = -(2W_{xx} - W_{yy} - W_{zz})(2W_{yy} - W_{xx} - W_{zz})$$
$$\times (2W_{zz} - W_{xx} - W_{yy}) + 9[(2W_{zz} - W_{xx} - W_{yy})|W_{xy}|^2$$
$$+ (2W_{yy} - W_{xx} - W_{zz})|W_{xz}|^2 - (2W_{xx} - W_{yy} - W_{zz})|W_{yz}|^2]$$
$$- 54\text{Re}[W_{xy}W_{yz}W_{xz}^*], \tag{2.146}$$

with the argument $(\mathbf{r},\mathbf{r},\omega)$ and superscript (3) of W-matrix elements being omitted. Just like in scalar and in beam-like electromagnetic cases, the spectral density is given by the strength of operator $\overleftrightarrow{W}^{(3)}$:

$$\begin{aligned}S^{(3)}(\mathbf{r},\omega) &= ss_{W^{(3)}}(\mathbf{r},\mathbf{r},\omega) \\ &= \lambda_1(\mathbf{r},\omega) + \lambda_2(\mathbf{r},\omega) + \lambda_3(\mathbf{r},\omega) \\ &= \mathrm{Tr}[\overleftrightarrow{W}^{(3)}(\mathbf{r},\mathbf{r},\omega)].\end{aligned} \quad (2.147)$$

The insight into the polarization properties of the three-component optical fields can be gotten by first establishing conditions for their complete polarization (Ellis et al., 2004a). On recalling the condition for complete polarization for the beam-like fields expressed in Eq. (2.106), we also require that for the three-dimensional fields to be completely polarized it is necessary and sufficient that all elements $(i,j = x, y, z)$ of the 3×3 cross-spectral density matrix factorize, at the coinciding arguments, as

$$W_{ij}^{(3)}(\mathbf{r},\mathbf{r},\omega) = E_i^*(\mathbf{r},\omega)E_j(\mathbf{r},\omega). \quad (2.148)$$

In other words, all the pairs of the electric-field components must be completely correlated. Indeed, in order to show that the necessary condition holds, we first observe that in order for a pair of electric field components to be correlated, they must have the form

$$E_i(\mathbf{r},\omega) = e_i(\mathbf{r},\omega)U(\mathbf{r},\omega), \quad (2.149)$$

where $e_i(\mathbf{r},\omega)$ and $U(\mathbf{r},\omega)$ are deterministic and random functions, respectively. Hence,

$$\frac{E_i(\mathbf{r},\omega)}{E_j(\mathbf{r},\omega)} = \frac{e_i(\mathbf{r},\omega)}{e_j(\mathbf{r},\omega)}, \quad (2.150)$$

i.e., E_i and E_j may be viewed as statistically similar. Such relations can be used for associating the actual field with an equivalent monochromatic field oscillating at frequency ω:

$$E_i(\mathbf{r},\omega) = e_i(\mathbf{r},\omega)\sqrt{\langle |U(\mathbf{r},\omega)|^2 \rangle}. \quad (2.151)$$

Then to arrive at factorization in Eq. (2.148), it suffices to substitute from Eq. (2.149) into Eq. (2.141).

Conversely, to prove that factorization (2.148) implies complete polarization of the electric field, we first find the spectrum of operator $\overleftrightarrow{W}^{(3)}(\mathbf{r},\mathbf{r},\omega)$ from the characteristic equation

$$\mathrm{Det}[\overleftrightarrow{W}^{(3)}(\mathbf{r},\mathbf{r},\omega) - \lambda(\mathbf{r},\omega)\overleftrightarrow{I}^{(3)}] = 0, \quad (2.152)$$

where $\overleftrightarrow{I}^{(3)}$ is the three-dimesional identity matrix. On using Eq. (2.148) in Eq. (2.152) and some algebraic manipulations, one finds that Eq. (2.152) reduces to the form

$$\lambda^2(\mathbf{r},\omega)[\lambda(\mathbf{r},\omega) - \mathrm{Tr}[\overleftrightarrow{W}^{(3)}(\mathbf{r},\mathbf{r},\omega)]] = 0. \quad (2.153)$$

This equation implies that the only non-trivial eigenvalue $\lambda_1(\mathbf{r},\omega)$ is equal to the spectral density of the field. Further, the eigenvector, say, $\vec{q}_1(\mathbf{r},\omega)$, corresponding to $\lambda_1(\mathbf{r},\omega)$ has the same direction as the equilvalent electromagneric field (see Eq. (2.151)):

$$\vec{q}_1(\mathbf{r},\omega) = \begin{bmatrix} E_x^*(\mathbf{r},\omega) \\ E_y^*(\mathbf{r},\omega) \\ E_z^*(\mathbf{r},\omega) \end{bmatrix}. \quad (2.154)$$

One can then show that matrix $\overleftrightarrow{W}^{(3)}(\mathbf{r},\mathbf{r},\omega)$ can be transformed to a matrix $\overleftrightarrow{W}^{(3)'}(\mathbf{r},\mathbf{r},\omega)$ by applying two consecutive rotations:

$$\begin{aligned}\overleftrightarrow{W}^{(3)'}(\mathbf{r},\mathbf{r},\omega) &= R(\psi_1,\psi_2)\overleftrightarrow{W}^{(3)}(\mathbf{r},\mathbf{r},\omega)R^{-1}(\psi_1,\psi_2)\\ &= \begin{bmatrix} E_{x'}^*(\mathbf{r}_1,\omega)E_{x'}(\mathbf{r}_2,\omega) & E_{x'}^*(\mathbf{r}_1,\omega)E_{y'}(\mathbf{r}_2,\omega) & E_{x'}^*(\mathbf{r}_1,\omega)E_{z'}(\mathbf{r}_2,\omega) \\ E_{y'}^*(\mathbf{r}_1,\omega)E_{x'}(\mathbf{r}_2,\omega) & E_{y'}^*(\mathbf{r}_1,\omega)E_{y'}(\mathbf{r}_2,\omega) & E_{y'}^*(\mathbf{r}_1,\omega)E_{z'}(\mathbf{r}_2,\omega) \\ E_{z'}^*(\mathbf{r}_1,\omega)E_{x'}(\mathbf{r}_2,\omega) & E_{z'}^*(\mathbf{r}_1,\omega)E_{y'}(\mathbf{r}_2,\omega) & E_{z'}^*(\mathbf{r}_1,\omega)E_{z'}(\mathbf{r}_2,\omega) \end{bmatrix}.\end{aligned}$$
$$(2.155)$$

Here,

$$\begin{aligned}R(\psi_1,\psi_2) &= R_1(\psi_1)R_2(\psi_2)\\ &= \begin{pmatrix} \cos\psi_1 & 0 & \sin\psi_1 \\ 0 & 1 & 0 \\ -\sin\psi_1 & 0 & -\cos\psi_1 \end{pmatrix}\begin{pmatrix} \cos\psi_2 & \sin\psi_2 & 0 \\ -\sin\psi_2 & \cos\psi_2 & 0 \\ 0 & 0 & 1 \end{pmatrix},\end{aligned} \quad (2.156)$$

where

$$\psi_1(\mathbf{r},\omega) = \arctan\left[\frac{E_z(\mathbf{r},\omega)}{E_x(\mathbf{r},\omega)\cos\psi_2 + E_y(\mathbf{r},\omega)\sin\psi_2}\right], \quad (2.157)$$

and

$$\psi_2(\mathbf{r},\omega) = \arctan\left\{\frac{|E_x(\mathbf{r},\omega)|\sin[\arg[E_z(\mathbf{r},\omega)] - \arg[E_x(\mathbf{r},\omega)]]}{|E_y(\mathbf{r},\omega)|\sin[\arg[E_y(\mathbf{r},\omega)] - \arg[E_z(\mathbf{r},\omega)]]}\right\}. \quad (2.158)$$

Rotation angle ψ_1 can be readily verified to reduce to a real quantity (the imaginary part of the argument of the arctan function in Eq. (2.157) is trivial). The relations between the elements of matrices $\overleftrightarrow{W}^{(3)'}(\mathbf{r},\mathbf{r},\omega)$ and $\overleftrightarrow{W}^{(3)}(\mathbf{r},\mathbf{r},\omega)$ can be found as

$$E_{x'}(\mathbf{r},\omega) = E_x(\mathbf{r},\omega)\cos\psi_1\cos\psi_2 + E_y(\mathbf{r},\omega)\cos\psi_1\sin\psi_2$$
$$- E_z(\mathbf{r},\omega)\sin\psi_1,$$
$$E_{y'}(\mathbf{r},\omega) = -E_x(\mathbf{r},\omega)\sin\psi_2 + E_y(\mathbf{r},\omega)\cos\psi_2,$$
$$E_{z'}(\mathbf{r},\omega) = -E_x(\mathbf{r},\omega)\sin\psi_1\cos\psi_2 - E_y(\mathbf{r},\omega)\sin\psi_1\cos\psi_2$$
$$+ E_z(\mathbf{r},\omega)\cos\psi_1. \quad (2.159)$$

When rotation angles ψ_1 and ψ_2 given in Eqs. (2.157) and (2.158) are used in Eq. (2.159), then $E_{z'}(\mathbf{r},\omega)$ vanishes, yieding

$$\overleftrightarrow{W}^{(3)'}(\mathbf{r},\mathbf{r},\omega) = \begin{bmatrix} E_{x'}^*(\mathbf{r}_1,\omega)E_{x'}(\mathbf{r}_2,\omega) & E_{x'}^*(\mathbf{r}_1,\omega)E_{y'}(\mathbf{r}_2,\omega) & 0 \\ E_{y'}^*(\mathbf{r}_1,\omega)E_{x'}(\mathbf{r}_2,\omega) & E_{y'}^*(\mathbf{r}_1,\omega)E_{y'}(\mathbf{r}_2,\omega) & 0 \\ 0 & 0 & 0 \end{bmatrix}. \quad (2.160)$$

This implies that the electric field fluctuates in the (x',y') plane and its cross-spectral density matrix factorizes in that plane in the same manner as that of a beam-like field. Thus by performing the two described rotations, the plane of complete polarization has been determined and in this plane the polarization theory of the beam-like fields, based on the 2×2 cross-spectral density matrices, can be emloyed to determine the state of polarization.

The spectral degree of polarization can then be defined as the operator range normalized by the operator strength:

$$\varrho_{W^{(3)}}(\mathbf{r},\omega) = \frac{sr_{W^{(3)}}(\mathbf{r},\mathbf{r},\omega)}{ss_{W^{(3)}}(\mathbf{r},\mathbf{r},\omega)}$$

$$= \frac{\lambda_1(\mathbf{r},\omega) - \lambda_3(\mathbf{r},\omega)}{\lambda_1(\mathbf{r},\omega) + \lambda_2(\mathbf{r},\omega) + \lambda_3(\mathbf{r},\omega)}. \quad (2.161)$$

This expression can also be derived on the basis of decomposition of $\overleftrightarrow{W}^{(3)}$ matrix into submatrices describing optical fields completely polarized in one, two and three dimensions (Ellis et al., 2004b).

At a pair of spatial arguments, operator $\overleftrightarrow{W}^{(3)}$ is generally quasi-Hermitian and the individual eigenvalues are given by the most general form of the Cardano formula and might be complex-valued. However, the degree of coherence of the three-dimensional electromagnetic field may be solely expressed via the spectral strength of the $\overleftrightarrow{W}^{(3)}$-operator, which coincides with the matrix trace (see also Korotkova and Wolf, 2004)

$$\mu_{W^{(3)}}(\mathbf{r}_1,\mathbf{r}_2,\omega) = \frac{ss_{W^{(3)}}(\mathbf{r}_1,\mathbf{r}_2,\omega)}{\sqrt{ss_{W^{(3)}}(\mathbf{r}_1,\mathbf{r}_1,\omega)}\sqrt{ss_{W^{(3)}}(\mathbf{r}_2,\mathbf{r}_2,\omega)}}$$

$$= \frac{\mathrm{Tr}[\overleftrightarrow{W}^{(3)}(\mathbf{r}_1,\mathbf{r}_2,\omega)]}{\sqrt{S^{(3)}(\mathbf{r}_1,\omega)}\sqrt{S^{(3)}(\mathbf{r}_2,\omega)}}. \quad (2.162)$$

An alternative approach based on matrix

$$\Omega^{(3)}(\mathbf{r}_1,\mathbf{r}_2,\omega) = W^{(3)\dagger}(\mathbf{r}_1,\mathbf{r}_2,\omega)W^{(3)}(\mathbf{r}_1,\mathbf{r}_2,\omega), \quad (2.163)$$

having real-valued spectrum of three real-valued, non-negative eigenvalues

$$\mathrm{sp}[\Omega^{(3)}(\mathbf{r}_1,\mathbf{r}_2,\omega)] = \{\Lambda_1(\mathbf{r}_1,\mathbf{r}_2,\omega),\Lambda_2(\mathbf{r}_1,\mathbf{r}_2,\omega),\Lambda_3(\mathbf{r}_1,\mathbf{r}_2,\omega)\}$$
$$(2.164)$$

and real, mutually orthogonal eigenvectors can also be undertaken.

2.5 Model Sources

2.5.1 Schell and quasi-homogeneous models

At the end of this chapter, we will list several model families of stationary sources that will help us throughout the book to illustrate basic phenomena occuring in the evolution of random light in Chapter 4 and its interaction with scatterers and media, in Chapters 8 and 9. We will greatly augment these model sources with a large variety of other sources on considering the principles of correlation function structuring in Chapter 5. In addition, similar models will be employed for specifying spatial correlations in the refractive index of random media (Chapter 8) and temporal correlations of the non-stationary pulse trains (Chapter 10).

The first to appear and currently the most popular model for a wide-sense stationary light source and a field it radiates was introduced in a thesis (Schell, 1961). In fact, it was the only model in the scalar coherence theory used for characterization of speckled fields for several decades (Mandel and Wolf, 1995). Later, it was extended to the beam-like electromagentic case (Gori et al., 2001) and the three-dimensional electromagnetic case (Korotkova et al., 2017).

We will assume for now that all phases of the correlation functions are position-independent. We will be specifically dealing with modeling of the phase factors for a variety of sources in the following chapters. For any scalar *Schell-model* source or field, the cross-spectral density function must be of the form

$$W(\mathbf{r}_1, \mathbf{r}_2, \omega) = \sqrt{S(\mathbf{r}_1, \omega)}\sqrt{S(\mathbf{r}_2, \omega)}\mu(\mathbf{r}_2 - \mathbf{r}_1, \omega). \quad (2.165)$$

While the factorization of the cross-spectral density into three parts is a re-written definition of the degree of coherence (2.92), the assumption that μ depends on the separation between two spatial arguments makes the model of the Schell type. The Gaussian Schell-model is then obtained if the spectral density is chosen as

$$S(\mathbf{r}, \omega) = A^2 \exp\left[-\frac{r^2}{2\sigma^2}\right], \quad (2.166)$$

where A^2 is its maximum value, σ is its root-mean-square (r.m.s.) width, and the degree of coherence is

$$\mu(\mathbf{r}_2 - \mathbf{r}_1, \omega) = \exp\left[-\frac{|\mathbf{r}_2 - \mathbf{r}_1|^2}{2\delta^2}\right], \tag{2.167}$$

where δ is its r.m.s. correlation width. A source (or a field) specified by Eqs. (2.165)–(2.167) belongs to the *scalar Gaussian Schell-model* class, while $\mathbf{r}$ can be a one-, two- or three-dimansional vector, in general. In this case, source parameters A, σ and δ can be chosen independently from each other. The ratio δ/σ indicates the coherence state: if $\delta/\sigma \to 0$, light is incoherent and if $\delta/\sigma \to \infty$, light is coherent. Practically, most of the significant effects of statistical optics can be demonstrated in the regime when δ is one or several orders of magnitude smaller than σ. In addition, for beam-like fields, using the restriction that the far-zone spectral density must remain substantial only in a narrow cone around the optical axis, one can obtain the *beam condition* (Mandel and Wolf, 1995):

$$\frac{1}{4\sigma^2} + \frac{1}{\delta^2} \ll \frac{2\pi^2}{\lambda^2}. \tag{2.168}$$

The dependence of the cross-spectral density on ω can be understood in a sense that all parameters entering the model can explicitly depend on it: $A = A(\omega)$, $\sigma = \sigma(\omega)$ and $\delta = \delta(\omega)$ (Shchepakina and Korotkova, 2013).

The extension of the scalar Schell model to beam-like electromagnetic and three-dimensional electromagnetic fields involves the form

$$W_{ij}(\mathbf{r}_1, \mathbf{r}_2, \omega) = \sqrt{S_i(\mathbf{r}_1, \omega)}\sqrt{S_j(\mathbf{r}_2, \omega)}\mu_{ij}(\mathbf{r}_2 - \mathbf{r}_1, \omega), \tag{2.169}$$

where $(i, j = x, y)$ in the former case and $(i, j = x, y, z)$ in the latter. Here, S_i is the spectral density component along direction i and μ_{ij} is the correlation coeffcent between the i and the j field components. For the Gaussian Schell-model sources, one chooses

$$S_i(\mathbf{r}, \omega) = A_i^2 \exp\left[-\frac{r^2}{2\sigma_i^2}\right], \tag{2.170}$$

where A_i^2 is the maximum value of the component along direction i, σ_i is the r.m.s. width of electric field component E_i. For the electromagnetic correlations μ_{ij}, one sets

$$\mu_{ij}(\mathbf{r}_2 - \mathbf{r}_1, \omega) = B_{ij} \exp\left[-\frac{|\mathbf{r}_2 - \mathbf{r}_1|^2}{2\delta_{ij}^2}\right]. \quad (2.171)$$

In both, beam-like and three-dimensional electromagnetic cases, the source parameters are not independent from each other and must obey *realizability conditions* stemming from the quasi-Hermiticity and non-negative definiteness restrictions.

In the electromagnetic beam-like case, the sufficient condition for the source to be realizable requires that (Roychowdhury and Korotkova, 2005)

$$B_{xx} = B_{yy} = 1, \quad |B_{xy}| \leq 1, \quad |B_{yx}| \leq 1, \quad B_{xy} = B_{yx}, \quad \delta_{xy} = \delta_{yx}, \quad (2.172)$$

and

$$\max(\delta_{xx}, \delta_{yy}) \leq \delta_{xy} \leq \min\left(\frac{\delta_{xx}}{\sqrt{B_{xy}}}, \frac{\delta_{yy}}{\sqrt{B_{xy}}}\right). \quad (2.173)$$

The beam conditions of the scalar theory can be readily extended to electromagnetic beams as (Korotkova et al., 2004)

$$\frac{1}{4\sigma^2} + \frac{1}{\delta_{xx}^2} \ll \frac{2\pi^2}{\lambda^2}, \quad \frac{1}{4\sigma^2} + \frac{1}{\delta_{yy}^2} \ll \frac{2\pi^2}{\lambda^2}. \quad (2.174)$$

Since the derivation of these inequalities relies on the spectral density defined via the trace of the cross-spectral density matrix, only the r.m.s. correlation values of the diagonal components are involved.

In the case of a three-dimensional electromagnetic field, the set of the realizability conditions becomes (Korotkova et al., 2017):

$$B_{ii} = 1, \quad |B_{ij}| \leq 1, \quad B_{ij} = B_{ji}, \quad \delta_{ij} = \delta_{ji}, \quad (2.175)$$

and

$$\max(\delta_{ii}, \delta_{jj}) \leq \delta_{ij} \leq \min\left(\frac{\delta_{ii}}{\sqrt[3]{2B_{ij}}}, \frac{\delta_{jj}}{\sqrt[3]{2B_{ij}}}\right), \quad (2.176)$$

where $(i, j = x, y, z)$.

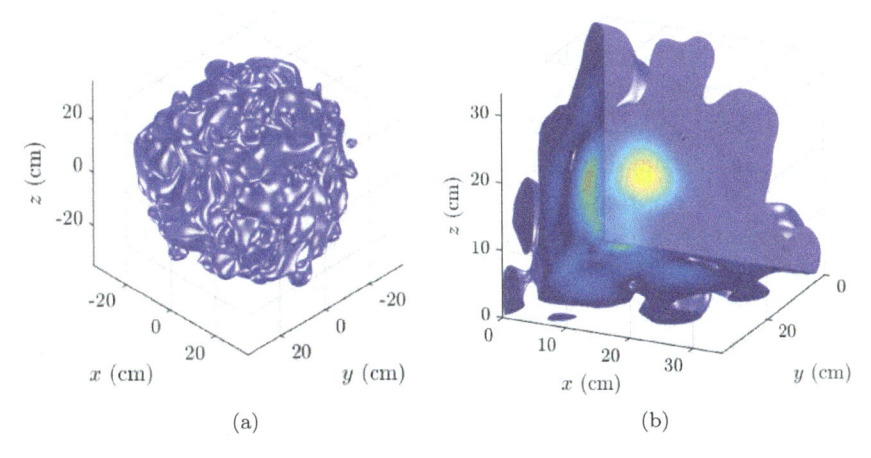

Fig. 2.3 Computer simulations of the three-dimensional Gausian Schell-model fields: surface view (a); x–y, y–z and x–z spectral density cross-sections (b). Reprinted with permission from Hyde et al. (2018) © The Optical Society.

For visualization, Fig. 2.3 shows the numerically simulated three-dimensional Gaussian Schell-model source's spectral density defined in Eq. (2.147) (Hyde et al., 2018). The realizations in lower dimensions can be obtained by projecting that in Fig. 2.3 onto a line or a plane.

The coherent mode decomposition (2.83) for one-dimensional Gaussian Schell-model sources was found in the closed form (Gori, 1980). The eigenvalues $\lambda_n(\omega)$ and the eigenfunctions $\phi_n(x,\omega)$ of the series have the forms

$$\lambda_n(\omega) = A\sqrt{\pi}\frac{a_2^n}{(a_1+a_2+a_3)^{n+1/2}}, \qquad (2.177)$$

and

$$\phi_n(x,\omega) = \sqrt{\frac{2a_3}{\pi}}\frac{1}{\sqrt{2^n n!}}H_n(x\sqrt{2a_3})\exp(-a_3 x^2), \qquad (2.178)$$

where H_n are the nth order Hermite polynomials and

$$a_1 = \frac{1}{\sigma^2}, \quad a_2 = \frac{1}{2\delta^2}, \quad a_3 = \sqrt{a_1^2 + 2a_1 a_2}. \qquad (2.179)$$

Since Gaussian functions are separable in Cartesian coordinates, for the scalar Gaussian Schell-model sources in higher dimensions, the eigenvalues and the eigenfunctions can be found as products of those for the individual coordinates.

The quasi-homogeneous model for random sources has become popular for theoretical calculations because of its intrinsic separability of spatial coordinates into sum and difference variables. In the scalar approximation, the cross-spectral density of a *quasi-homogeneous* source has the form (Carter and Wolf, 1977; Sylverman, 1958)

$$W(\mathbf{r}_1, \mathbf{r}_2, \omega) = S\left(\frac{\mathbf{r}_1 + \mathbf{r}_2}{2}\right) \mu(\mathbf{r}_2 - \mathbf{r}_1, \omega). \qquad (2.180)$$

Such an approximation is possible if spectral density S is much wider compared to the degree of coherence μ as functions of their arguments. For electromagnetic fields, this approximation can be generalized as

$$W_{ij}(\mathbf{r}_1, \mathbf{r}_2, \omega) = \sqrt{S_i\left(\frac{\mathbf{r}_1 + \mathbf{r}_2}{2}\right)} \sqrt{S_j\left(\frac{\mathbf{r}_1 + \mathbf{r}_2}{2}\right)} \mu_{ij}(\mathbf{r}_2 - \mathbf{r}_1, \omega), \qquad (2.181)$$

where i and j are Cartesian coordinates, $(i, j = x, y)$ for beam-like fields and $(i, j = x, y, z)$ for general three-dimensional fields.

Quasi-homogeneous approximation has also played an important role in scattering theory for modeling optical correlations in random media with relatively narrow refractive index correlation functions (see Chapter 8).

2.5.2 *Homogeneous model for spherical shell sources*

In a number of situations, correlations in light fields must be modeled along curves or on surfaces. An excellent illustration of this possibility is a homogeneous spherical shell source. Spherical sources are of importance to statistical optics primarily in connection with sunlight radiation. It was known since the very first discussion about sunlight correlations by Verdet (1865) that at the surface of Earth sunlight is spatially partially coherent. This was recently confirmed experimentally in Divitt and Novotny (2015) by performing the classic interference experiment with sunlight using a variable double slit (Chapter 3). In order to better comprehend the behavior of light at the Sun's surface, the mathematical models describing the optical correlations at the surface of hard-edged spherical sources have become of importance.

On following Gori and Korotkova (2009), we introduce the spherical homogeneous model source occupying a surface of a sphere with radius r and radiating outwards. Let a point on the sphere be characterized by spherical coordinates (r, θ, ϕ), where θ and ϕ are polar and azimuthal angles, respectively. Let α_{12} be the central angle between the radius vectors of two points, (r, θ_1, ϕ_1) and (r, θ_2, ϕ_2). The cross-spectral density of such a source takes the form

$$W(\theta_1, \phi_1, \theta_2, \phi_2, \omega) = S(\omega)\mu[\cos(\alpha_{12}), \omega]. \qquad (2.182)$$

The cosine-like dependence of the degree of coherence μ on α_{12} ensures proper periodicity required for statistics of light bound to a sphere. Only a selected family of functions can then be used. For example, a Gaussian function in three dimensions (2.167), that we have already used for the Gaussian Schell-model sources, becomes

$$\mu[\cos(\alpha_{12}), \omega] = \exp\left[-\frac{(1-\cos\alpha_{12})r^2}{\delta^2}\right], \qquad (2.183)$$

with the help of the relation $|\mathbf{r}_1 - \mathbf{r}_2| = r\sqrt{2(1-\cos\alpha_{12})}$. The coherent mode decomposition for these sources has the form

$$W(\theta_1, \phi_1, \theta_2, \phi_2, \omega) = \sum_{l=0}^{\infty} \lambda_{lm}(\omega) \sum_{m=-l}^{l} Y_{lm}^*(\theta_1, \phi_1) Y_{lm}(\theta_2, \phi_2), \qquad (2.184)$$

where $Y_{lm}(\theta, \phi)$ are spherical harmonic functions and

$$\lambda_{lm}(\omega) = 4\pi S(\omega) \frac{A_l(\omega)}{2l+1}, \quad (l = 0, 1, \ldots, m = -l, \ldots, l), \qquad (2.185)$$

are the eigenvalues with coefficients A_l satisfying the equation

$$A_l(\omega) = \frac{2l+1}{2} \int_{-1}^{1} \mu(\xi, \omega) P_l(\xi) d\xi, \quad (l = 0, 1, \ldots). \qquad (2.186)$$

Here, $P_l(\xi)$ is a Legendre polynomial of order l. It is of importance to note that for spherical homogeneous sources, the eigenfunctions are always spherical harmonics, regardless of the form of the degree of coherence.

For example, for the Gaussian degree of coherence (2.183), the eigenvalues of the coherent mode decomposition take the form

$$\lambda_{lm}(\omega) = \sqrt{\frac{\pi\delta}{r}} \exp\left[-\frac{r^2}{\delta^2}\right] I_{l+1/2}\left[\frac{r^2}{\delta^2}\right], \qquad (2.187)$$

where I_k is a modified Bessel function of the first kind and order k.

Chapter 3

Famous Experiments and Phenomena Relating to Random Light

The aim of this chapter is to introduce four experiments that have been fundamental for accessing the behavior of fluctuating optical fields. The first two of them, the Young and the Michelson interference experiments, provide the measurement of the spatial and the temporal degrees of coherence, respectively. The third experiment, carrying the name of Hanbury Brown and Twiss, provides the measurement of the spatial correlations in fluctuating intensity of an optical signal. Lastly, the phase conjugation experiment addresses an interesting intensity redistribution phenomenon occuring in optical signals traveling through random media in the opposite directions.

3.1 Young's Interference Experiment

The experiment was proposed by Thomas Young for illustration of the wave nature of light (Young, 1804, see also Young, 1807). It was originally set to operate with sunlight made spatially coherent using a pinhole. The procedure has been generalized to light with any coherence state by Zernike (1938) and to stationary fields with any spectral composition (see Mandel and Wolf, 1995). Much later the Young experiment has also been shown to produce similar results on interference of particles, e.g., electrons (Jönsson, 1961).

Figure 3.1 shows a schematic diagram used in the Young experiment with two pinholes. Let a realization of a stationary light field at frequency ω impinge onto an opaque screen $\mathcal{A}$ placed at $z = 0$, having

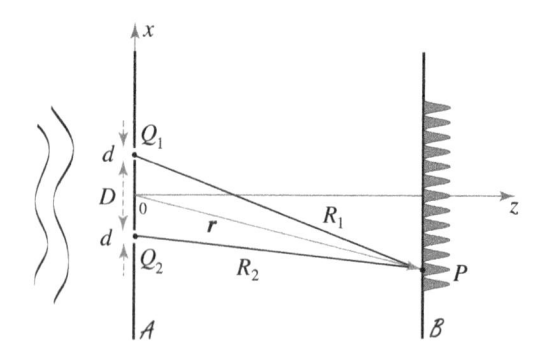

Fig. 3.1 Young's interference experiment.

two pinholes, Q_1 and Q_2, centered at coordinates $\boldsymbol{\rho}_1 = (D/2, 0, 0)$ and $\boldsymbol{\rho}_2 = (-D/2, 0, 0)$, respectively. The separation distance D between the pinholes plays a crucial part in the formation of the interference pattern. Let each pinhole have diameter d and area $A_p = \pi d^2/4$. Light originating from each pinhole propagates into the half-space $z > 0$ until it reaches the observation screen $\mathcal{B}$ placed in parallel with screen $\mathcal{A}$, where it forms an intereference pattern. The realization of the optical field at a single observation point, say P, specified by position vector $\mathbf{r}$ is the superposition of fields emerging from the two pinholes:

$$U(\mathbf{r}, \omega) = -\frac{ikA_p}{2\pi} \left[\frac{e^{ikR_1}}{R_1} U(\boldsymbol{\rho}_1, \omega) + \frac{e^{ikR_2}}{R_2} U(\boldsymbol{\rho}_2, \omega) \right], \qquad (3.1)$$

where R_α is the distance between pinhole Q_α and observation point P, $\alpha = 1, 2$. The relation between the individual fields at screens $\mathcal{A}$ and $\mathcal{B}$ is the consequence of the Huygens–Fresnel principle, which will be discussed in detail in Chapter 4. To simplify notiations, we will introduce factors

$$K_\alpha = \frac{e^{ikR_\alpha}}{R_\alpha}. \qquad (3.2)$$

The spectral density of light at point P and frequency ω can be obtained from expression (3.1) as

$$S(\mathbf{r}, \omega) = \left(\frac{kA_p}{2\pi}\right)^2 K_1^* K_2 \big[S_1(\boldsymbol{\rho}_1, \omega) + S_2(\boldsymbol{\rho}_2, \omega)$$

$$+ 2\sqrt{S_1(\boldsymbol{\rho}_1, \omega)} \sqrt{S_1(\boldsymbol{\rho}_1, \omega)} \, Re[\mu(\boldsymbol{\rho}_1, \boldsymbol{\rho}_2, \omega)] \big]. \qquad (3.3)$$

Here,

$$\mu(\boldsymbol{\rho}_1, \boldsymbol{\rho}_2, \omega) = \frac{W(\boldsymbol{\rho}_1, \boldsymbol{\rho}_2, \omega)}{\sqrt{S_1(\boldsymbol{\rho}_1, \omega)}\sqrt{S_2(\boldsymbol{\rho}_2, \omega)}} \quad (3.4)$$

is the spectral degree of coherence of the light field at pinholes Q_1 and Q_2. It can be expressed via its magnitude μ_M and phase μ_P as

$$\mu(\boldsymbol{\rho}_1, \boldsymbol{\rho}_2, \omega) = \mu_M(\boldsymbol{\rho}_1, \boldsymbol{\rho}_2, \omega) \exp[i\mu_P(\boldsymbol{\rho}_1, \boldsymbol{\rho}_2, \omega)]. \quad (3.5)$$

In a practical case when the spectral density across screen $\mathcal{A}$ is constant,

$$S_1(\boldsymbol{\rho}_1, \omega) = S_2(\boldsymbol{\rho}_2, \omega) = S(\omega), \quad (3.6)$$

we find that the interference law (3.3) reduces to the form

$$S(\mathbf{r}, \omega) = \frac{1}{2}\left(\frac{kA_p}{\pi}\right)^2 K_1^* K_2 S(\omega)\left(1 + Re[\mu(\boldsymbol{\rho}_1, \boldsymbol{\rho}_2, \omega)]\right). \quad (3.7)$$

Due to definition (3.4), μ is a complex-valued function confined to a unit circle. Hence,

$$Re[\mu] = \mu_M \cos(\mu_P), \quad (3.8)$$

giving $\mu_{\max} = \mu_M$ and $\mu_{\min} = -\mu_M$. Then on defining the contrast in the spectral density on the observation screen $\mathcal{B}$ by the expression

$$C_S(\omega) = \frac{S_{\max}(\omega) - S_{\min}(\omega)}{S_{\max}(\omega) + S_{\min}(\omega)}, \quad (3.9)$$

where $S_{\max}(\omega)$ and $S_{\min}(\omega)$ are its maximum and minimum values, we find that

$$C_S(\omega) = \mu_M(\boldsymbol{\rho}_1, \boldsymbol{\rho}_2, \omega). \quad (3.10)$$

The phase of the degree of coherence is responsible for the locations of the maxima and the minima of the fringe pattern as measured from the centerline of the setup.

A version of the Young experiment that allows measurement of the cross-spectral density matrix $\overleftrightarrow{W}$ was suggested in Roychowdhury and Wolf (2005) (see also Wolf, 2007b) where a pair of polarizers are placed across each pinhole and a phase plate is placed across one of them. By a suitable setting of the three devices in four mutually independent states, all four elements of the matrix can be sequentially obtained.

Although the Young experiment played a crucial role in optics during the last two centures, it has a number of limitations. For instance, due to the small area of the pinholes, the transmitted light intensity might be unsufficient for detection. This, in particular, happens in the case when the illumination has interacted with a random medium or scatterers prior to measurement. In such cases alternate methods, such as a folding interferometer method based on wavefront amplitude divison, are preferable (for details, see Koivurova *et al.*, 2019 and references therein). Another limitation of the classic Young experiment is the measurement of the degree of coherence at a fixed separation. In order to circumvent this, one can apply a variable slit version instead of the pinholes (Divitt *et al.*, 2014). Furthermore, with the recent development of very intricate coherence states, such as those depending not only on the separation distance but also varying with position (light with non-uniform coherence), more advanced methods based on digital measurement with the help of the spatial light modulators may be involved (Sharma *et al.*, 2016).

3.2 Michelson Interference Experiment

The Michelson interferometer is a device capable of measuring the temporal coherence properties of light by superposing two replicas of the wavefront preliminarily spatially separated and made to propagate through two paths with controllable lengths. It was first introduced in 1887 by Albert Michelson and Edward Morley in a famous experiment that was set to measure the Earth's motion through a luminiferous aether (Michelson and Morley, 1887). The result of the experiment disproved the existence of the aether and established the fact that the speed of light in vacuum is constant and does not depend on the motion of either the source or the observer. In addition, it provided with the powerful technique of wavefront self-correlation, that was later used in Fourier transform spectroscopy,

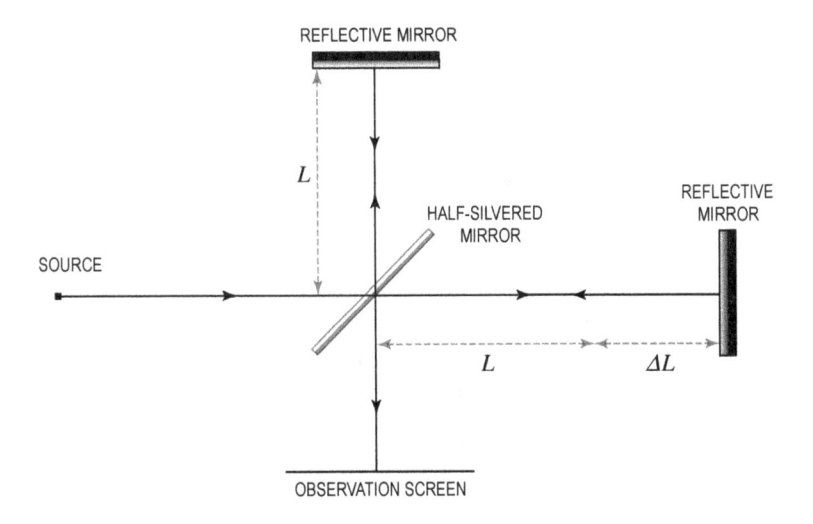

Fig. 3.2 Michelson's interferometer.

optical coherence tomography, for testing of optical components for imperfections as well as in astronomy for measurement of the diameters of the stars and, very recently, for gravitational wave detection.

Figure 3.2 illustrates the Michelson interferometer setup. A fluctuating in time but spatially uniform and quasi-monochromatic, scalar light field $\mathcal{U}_0(t)$ impinges on a half-silvered mirror which splits the wavefront 50:50 into the reflective and refractive branches. We will omit the spatial dependence of all quantities. The reflected beam propagates to a mirror placed at a fixed distance L from the beamsplitter. The refracted beam propagates to a mirror at adjustable distance $L + \Delta L$ from the beamsplitter. After reflection from the mirrors, the two beams are combined at the beamsplitter and are projected onto the observation screen. A transparent compensating plate made of the same material as the beamsplitter plate may be inserted into the reflecting branch in order to equalize the two paths. The difference ΔL between the two paths corresponds to time lag $\tau = 2\Delta L/c$. Then the combined field can be expressed as

$$\mathcal{U}(t) = \frac{1}{2}\left[\mathcal{U}_0(t) + \mathcal{U}_0(t+\tau)e^{i\Delta\varphi}\right], \quad (3.11)$$

where $\Delta\varphi$ is the phase shift due to possible transmissions and reflections at the beamsplitter. The average intensity of the combined fields then becomes

$$\mathcal{I}(\tau) = \frac{1}{2}\mathcal{I}_0 + \frac{1}{2}\mathcal{I}_0|\gamma(\tau)|\cos\left[\arg[\gamma(\tau)] + \Delta\varphi\right], \quad (3.12)$$

where $\mathcal{I}_0$ and $\gamma(\tau)$ are the average intensity and the temporal complex degree of coherence of the incident light. Due to the narrow bandwidth of the source, $\arg[\gamma(\tau)] = \alpha(\tau) - \omega_0(\tau)$ where envelope $\alpha(\tau)$ varies slowly with τ. The average intensity of the combined beams appears to be a harmonic function of time delay τ, i.e., exhibits a fringe pattern. The temporal visibility (contrast) of the interference fringes at separation τ has been defined by Michelson as

$$C_T(\tau) = \frac{\mathcal{I}_{\max}(\tau) - \mathcal{I}_{\min}(\tau)}{\mathcal{I}_{\max}(\tau) + \mathcal{I}_{\min}(\tau)}. \quad (3.13)$$

Using Eq. (3.12) in Eq. (3.13), one readily finds that

$$C_T(\tau) = |\gamma(\tau)|. \quad (3.14)$$

Thus, the fringe contrast of the combined fields in the Michelson interferometer provides information about the magnitude of the temporal degree of coherence of the incident field. For $\tau = 0$, $C_T(0) = 1$ (complete temporal coherence), while as $\tau \to \infty$ the degree of temporal coherence eventually decreases to zero (temporal incoherence), even if not monotonically.

3.3 Hanbury Brown and Twiss Interference Experiment

The method of intensity interferometry of the fluctuating signals was introduced first for radio frequencies (Brown and Twiss, 1954) and then for optical frequencies (Brown and Twiss, 1956) and is currently known as the Hanbury Brown and Twiss interferometry. The idea of the method is to employ the fourth-order correlation in the field or, equivalently, the second-order correlation in intensity of two spatially separated signals for determining the properties of their source.

Namely, the following correlation function of fluctuating intensity $\mathcal{I}(\mathbf{r},t)$ is used:

$$\langle \Delta \mathcal{I}(\mathbf{r}_1,t)\Delta \mathcal{I}(\mathbf{r}_2,t+\tau)\rangle = \langle \mathcal{I}(\mathbf{r}_1,t)\mathcal{I}(\mathbf{r}_2,t+\tau)\rangle$$
$$- \langle \mathcal{I}(\mathbf{r}_1,t)\rangle\langle \mathcal{I}(\mathbf{r}_2,t+\tau)\rangle, \quad (3.15)$$

where it is actually assumed that the average intensity is time-invariant, only depending on spatial argument $\mathbf{r}$. Since the majority of natural optical sources are governed by Gaussian statistics, in view of the Gaussian moment theorem, the first term on the right-hand side of Eq. (3.15) can be expressed as

$$\langle \mathcal{I}(\mathbf{r}_1,t)\mathcal{I}(\mathbf{r}_2,t+\tau)\rangle$$
$$= \langle \mathcal{I}(\mathbf{r}_1,t)\rangle\langle \mathcal{I}(\mathbf{r}_2,t+\tau)\rangle + |\mathcal{W}(\mathbf{r}_1,\mathbf{r}_2,\tau)|^2, \quad (3.16)$$

where $\mathcal{W}$ is the measured mutual coherence function. On suitably normalizing the intensity correlation function in Eq. (3.15), we arrive at relation

$$\gamma_I^{(4)}(\mathbf{r}_1,\mathbf{r}_2,\tau) = \frac{\langle \Delta \mathcal{I}(\mathbf{r}_1,t)\Delta \mathcal{I}(\mathbf{r}_2,t+\tau)\rangle}{\langle \mathcal{I}(\mathbf{r}_1,t)\rangle\langle \mathcal{I}(\mathbf{r}_2,t)\rangle}$$
$$= |\gamma(\mathbf{r}_1,\mathbf{r}_2,\tau)|^2. \quad (3.17)$$

Thus, the normalized intensity correlation function $\gamma_I^{(4)}$ is the square of the magnitude of the complex degree of coherence. As we will see from Chapter 4, if measured in the far field of an incoherent source, $\gamma(\mathbf{r}_1,\mathbf{r}_2,\tau)$ carries information about its size. Based on these two relations, the intensity interferometry has been successfully employed in astronomy for measuring the angular diameter of a star (Brown and Twiss, 1956). In fact, the diameters of a number of stars were measured with the help of a specially built stellar intensity interferometer in Brown et al. (1967).

3.4 BackScatter Amplification Effect

On reflection from a plane mirror the phase of a light beam changes by π and, hence, the incident and the reflected waves are always out of phase. On the contrary, if a retro-reflector is used instead of a mirror, then a wave incident on it undergoes two reflections. The total phase change is then 2π setting the two waves in phase. This outcome is independent of the form of the retro-reflector that can be a cone, a one-dimensional corner, a corner-cube, or a cat's eye. Provided that the sum of the two reflection angles is π, the returned ray always propagates in parallel to the incident ray. These ideas were first put forward in Peck (1948a,b) where the geometrical optics formulation and interferometry principles relating to the corner-cube retro-reflectors were established.

An interesting phenomenon of the BackScatter Amplification Effect (BSAE) occurs if the retro-reflector is placed in an extended random medium such as turbulent air/water or a random collection of scatterrers. Originally the BSAE mechanism was discovered in turbulent plasmas (Ruffine and de Wolfe, 1965) and later has been also shown to occur on propagation in atmopsheric (Belenkii and Mironov, 1972) and oceanic (Korotkova, 2018) turbulence (whether with mirrors or retro-reflectors). The reader is refereed to an excellent review paper Barabanenkov *et al.* (1991) for further details.

Figure 3.3 serves as an illustration of the basic idea behind the BSAE. The rays originate from a transceiver (a transmitter co-located with a receiver), pass through a random medium and, after interaction with the retro-reflector, travel back through exactly the same paths. Such double-pass systems are also known as *mono-static*, as compared with *bi-static* in which the transmitter and the

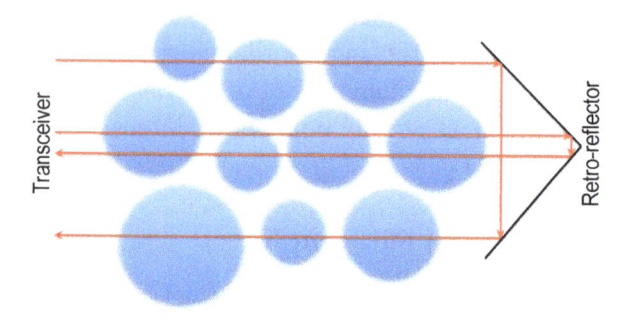

Fig. 3.3   Illustration of the BSAE formation with a retro-reflector.

receiver are spatially separated. The rays traveling far from the retro-reflector's center (optical axis) encounter different inhomogeneites, and, hence the phases accumulated on the way to the retro-reflector and back are just randomly added. On the contrary, the rays traveling sufficiently close to the optical axis pass through the same inhomogeneites on the way to and from the retro-reflector. This implies that the phase changes incurred along the two paths cancel (conjugate) each other. Hence, in a small region around the axis, roughly comparable to the size of the smallest inhomogeneity of the medium, the optical wave "bypasses" random medium, at least with a large probability. The most significant manifestation of such phase conjugation is the average intensity enhancement in the small area around the optical axis.

A practical implementation of the mono-static system may rely on the use of a beam splitter for the purpose of redirecting the incident beam to and from the retro-reflector. Consider a specific example of such a setup in which a thin phase screen mimics an extended random medium (see Fig. 3.4). The distance between the transceiver and the corner-cube can be set as several tens of meters. Physically, the random screen can be introduced with the help of a rotating ground glass diffuser, or a small volume of a turbulent air or water. The optimal placement of the screen that leads to the maximum BSAE is close to the corner cube; as it is moved towards the beamsplitter, the BSAE gradually weakens.

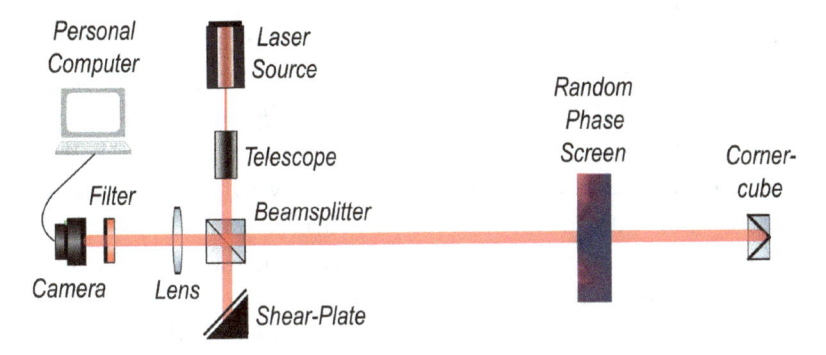

Fig. 3.4 Mono-static optical link with a beamsplitter, retro-reflector and a thin random phase screen.

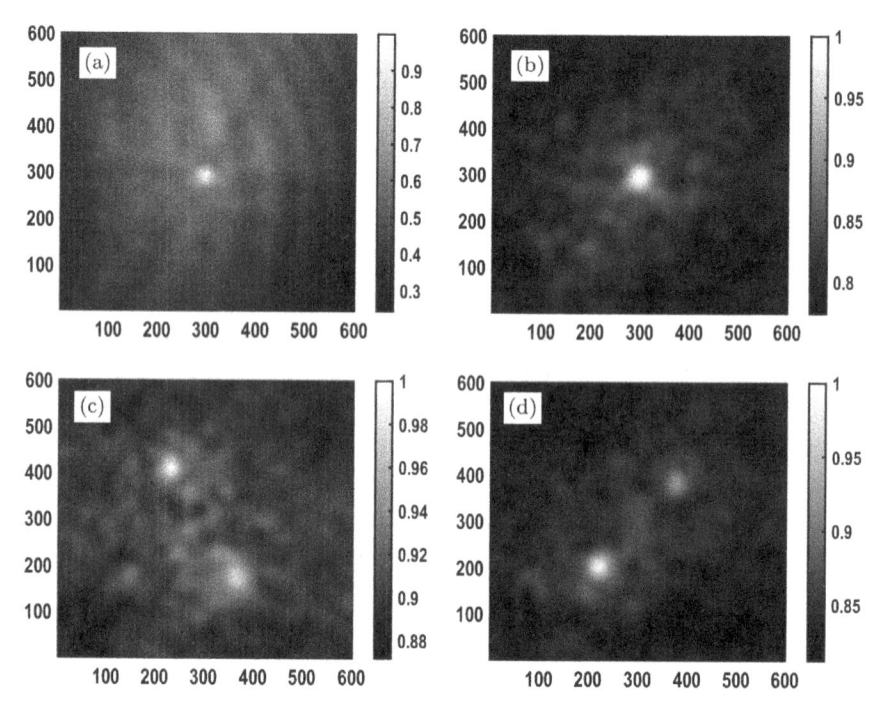

Fig. 3.5 Intensity BSAE: (a) Average intensity; (b)–(d) Intensity–intensity correlation function with: (b) central pixel (300,300); (c) pixel (233,410); (d) pixel (222,202).

Figure 3.5 shows experimental results for the intensity BSAE obtained with the help of a mono-static system in Fig. 3.5 (see also Korotkova and Soresi, 2019). In particular, Fig. 3.5(a) shows the average intensity of the beam captured by the camera. The bright intensity maximum around the optical axis is clearly visible as compared with the rest of the beam centroid. Figures 3.5(b)–(d) show the two-point intensity correlation function (see Eq. (3.15)). In particular, correlating the instantaneous intensity with the on-axis pixel in panel (b) results in appearance of a bright area around the axis being slightly wider than that for the average intensity. When the intensity is correlated with other pixels as in panels (c) and (d), two bright maxima occur, one around the chosen pixel and the other one at a point symmetric to that pixel about the center.

Chapter 4

Free-space Propagation of Stationary Light

Under some circumstances, when the stationary optical fields propagate in vacuum, their non-trivial, two-point spatial source correlations can result in changes in their single-point properties, along the path. Such changes are known as the *source correlation-induced* changes and can be shown to occur in the average beam spread, the spectral composition, the polarization properties and the OAM density flux. In this chapter, we first outline the solutions of the propagation equations for deterministic and stationary fields radiated from planar sources and then, based on the Huygens–Fresnel approximation obtained for beam-like fields, illustrate various correlation-induced changes by the numerical examples.

4.1 Deterministic Light

4.1.1 *Scalar theory*

We begin by addressing a rather general radiation problem. Let a scalar optical field $U^{(0)}(\mathbf{r}, \omega)$ produced by sources distributed in volume V propagate into half-space $z > 0$ in vacuum ($n = 1$), see Fig. 4.1. Let a source point and a field point be specified by three-dimensional vectors $\mathbf{r}' = (\boldsymbol{\rho}', z')$, $\boldsymbol{\rho}' = (x', y')$, and $\mathbf{r} = (\boldsymbol{\rho}, z)$, $\boldsymbol{\rho} = (x, y)$, $z > 0$, respectively. Wave equation (2.11) governing this field's propagation can be solved by the method of Green's functions

(Debnath and Bhatta, 2015), on writing it as surface integral

$$U(\mathbf{r},\omega) = \iint_{\partial V} [U^{(0)}(\mathbf{r}',\omega)\nabla_{r_s} G(\mathbf{r},\mathbf{r}',\omega) \\ -G(\mathbf{r},\mathbf{r}',\omega)\nabla U^{(0)}(\mathbf{r}',\omega)]d\mathbf{s}(\mathbf{r}') \qquad (4.1)$$

if $\mathbf{r} \in V$, and trivial right-hand side if $\mathbf{r} \notin V$, V being half-space $z > 0$. Here, Green's function $G(\mathbf{r},\mathbf{r}_s,\omega)$ must satisfy the equation

$$[\nabla_r^2 + k^2]G(\mathbf{r},\mathbf{r}',\omega) = \delta^{(3)}(\mathbf{r}-\mathbf{r}'), \quad \mathbf{r} \in V. \qquad (4.2)$$

If V is a large semisphere and $G(\mathbf{r},\mathbf{r}',\omega)$ obeys radiation conditions, vanishing for large $\mathbf{r}$, then Eq. (4.1) becomes, for $z > 0$,

$$U(\mathbf{r},\omega) = \int_{-\infty}^{\infty}\int_{-\infty}^{\infty} \left[G(\mathbf{r}',\mathbf{r},\omega)U_{z'}^{(0)}(\mathbf{r}')\Big|_{z'=0} \right. \\ \left. -U^{(0)}(\mathbf{r}')G_{z'}(\mathbf{r}',\mathbf{r},\omega)\Big|_{z'=0}\right]d^2r', \qquad (4.3)$$

where $G(\mathbf{r}',\mathbf{r},\omega)$ satisfies the equation

$$\left[\nabla_{r'}^2 + \frac{\partial^2}{\partial z'^2} + k^2\right]G(\mathbf{r},\mathbf{r}',\omega) = \delta^{(2)}(\boldsymbol{\rho}-\boldsymbol{\rho}')\delta(z-z'), \quad z' > 0. \qquad (4.4)$$

When sources are distributed in the plane $z' = 0$, one can, in addition to the condition above, impose the Dirichlet boundary condition

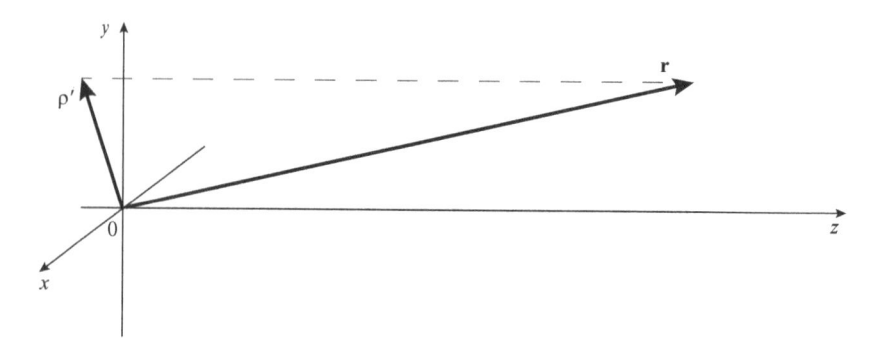

Fig. 4.1 Illustration of the notation relating to electric field propagation from plane $z = 0$ into half-space $z > 0$.

on the $z' = 0$ plane: $G^{(D)}(\mathbf{r}', \mathbf{r}, \omega)\big|_{z'=0} = 0$. Then Eq. (4.3) becomes

$$U(\mathbf{r}, \omega) = -\int_{-\infty}^{\infty}\int_{-\infty}^{\infty} U^{(0)}(\mathbf{r}', \omega) G_{z'}^{(D)}(\mathbf{r}', \mathbf{r}, \omega)\big|_{z'=0} d^2 r', \quad (4.5)$$

where $z > 0$. The Green function in Eq. (4.5) can be found by the method of images. Indeed, it must satisfy modification of Eq. (4.4):

$$\left[\nabla_{\mathbf{r}'}^2 + \frac{\partial^2}{\partial z'^2} + k^2\right] G^{(D)}(\mathbf{r}, \boldsymbol{\rho}', |z'|, \omega)$$
$$= \delta^{(2)}(\boldsymbol{\rho} - \boldsymbol{\rho}')[\delta(z - z') - \delta(z + z')], \quad (4.6)$$

valid for $z > 0$ and all z'. In vacuum, the solution of Eq. (4.6) becomes

$$G^{(D)}(\mathbf{r}, \mathbf{r}', \omega) = \frac{1}{4\pi}[G^{(S)}(\mathbf{r}', \boldsymbol{\rho}, -z, \omega) - G^{(S)}(\mathbf{r}', \boldsymbol{\rho}, z, \omega)], \quad (4.7)$$

where, with $\mathbf{R} = \mathbf{r} - \mathbf{r}'$, $R = |\mathbf{R}|$,

$$G^{(S)}(\mathbf{r}', \mathbf{r}, \omega) = \frac{\exp(ikR)}{R}, \quad (4.8)$$

is the Green function of the outgoing spherical wave. Substitution of $G^{(D)}$ into Eq. (4.5) yields

$$U(\mathbf{r}, \omega) = \frac{z}{2\pi} \int_{-\infty}^{\infty}\int_{-\infty}^{\infty} \left[\frac{1}{\sqrt{(\boldsymbol{\rho} - \boldsymbol{\rho}')^2 + z^2}} - ik\right]$$
$$\times \frac{U^{(0)}(\mathbf{r}', \omega) \exp[ik\sqrt{(\boldsymbol{\rho} - \boldsymbol{\rho}')^2 + z^2}]}{(\boldsymbol{\rho} - \boldsymbol{\rho}')^2 + z^2} d^2 \rho'. \quad (4.9)$$

For propagation distances much larger/smaller than the wavelength, the first/second term in the square brackets is neglected.

4.1.2 *Electromagnetic theory*

The analysis of the previous section can be generalized to the case of the electric vector field radiating from a deterministic source situated in the plane $z = 0$ into the positive half-space, $z > 0$ (see Fig. 4.1).

In the plane of the source, it is sufficient for the electric field to be specified by two mutually orthogonal Cartesian components

$$\vec{E}^{(0)}(\boldsymbol{\rho}',\omega) = [E_x^{(0)}(\boldsymbol{\rho}',\omega), E_y^{(0)}(\boldsymbol{\rho}',\omega)], \tag{4.10}$$

where $\boldsymbol{\rho}' = (x', y')$ is a source point, and $z' = 0$ (omitted from now on). According to relations of diffraction theory (see Luneberg, 1964 for details) the three Cartesian components of the electric field at a point $\mathbf{r} = (x, y, z)$ in the half-space $z > 0$ are given by the following formulas:

$$E_x(\mathbf{r},\omega) = -\frac{1}{2\pi} \int_{-\infty}^{\infty} \int_{-\infty}^{\infty} E_x^{(0)}(\boldsymbol{\rho}',\omega) G_z^{(S)}(\mathbf{r},\boldsymbol{\rho}',\omega) d^2\rho', \tag{4.11}$$

$$E_y(\mathbf{r},\omega) = -\frac{1}{2\pi} \int_{-\infty}^{\infty} \int_{-\infty}^{\infty} E_y^{(0)}(\boldsymbol{\rho}',\omega) G_z^{(S)}(\mathbf{r},\boldsymbol{\rho}',\omega) d^2\rho', \tag{4.12}$$

$$E_z(\mathbf{r},\omega) = \frac{1}{2\pi} \int_{-\infty}^{\infty} \int_{-\infty}^{\infty} \sum_{j=x,y} E_j^{(0)}(\boldsymbol{\rho}',\omega) G_j^{(S)}(\mathbf{r},\boldsymbol{\rho}',\omega) d^2\rho'. \tag{4.13}$$

A single matrix transformtaion can also be conveniently used to express these relations (Alonso et al., 2006) as

$$\vec{E}(\mathbf{r},\omega) = \int_{-\infty}^{\infty} \int_{-\infty}^{\infty} \vec{E}^{(0)}(\boldsymbol{\rho}',\omega) \cdot \overleftrightarrow{K}(\boldsymbol{\rho}',\mathbf{r},\omega) d^2\rho', \tag{4.14}$$

where $\cdot$ denotes product of matrices and $\overleftrightarrow{K}$ is the 3 × 2 matrix propagator

$$\overleftrightarrow{K}(\boldsymbol{\rho}',\mathbf{r},\omega) = \frac{1}{2\pi} \begin{pmatrix} -G_z^{(S)}(\mathbf{r},\boldsymbol{\rho}',\omega) & 0 & G_x^{(S)}(\mathbf{r},\boldsymbol{\rho}',\omega) \\ 0 & -G_z^{(S)}(\mathbf{r},\boldsymbol{\rho}',\omega) & G_y^{(S)}(\mathbf{r},\boldsymbol{\rho}',\omega) \end{pmatrix}$$

$$= \frac{L(R,\omega)}{2\pi} \begin{pmatrix} -z & 0 & x - x' \\ 0 & -z & y - y' \end{pmatrix}, \tag{4.15}$$

with

$$L(R,\omega) = \frac{(ikR - 1)\exp(ikR)}{R^3}, \quad R = |\mathbf{r} - \boldsymbol{\rho}'|. \tag{4.16}$$

Just like in the scalar theory, one distinguishes the far zone and the near zone that can be determined by retaining and/or approximating the first or the second term in Eq. (4.16), respectively.

In the Fraunhoffer zone of the source ($kr \to \infty$) approximation

$$R \approx r - \mathbf{u} \cdot \boldsymbol{\rho}', \qquad (4.17)$$

is used, where $\mathbf{u} = (u_x, u_y, u_z) = \mathbf{r}/r$ is a unit vector and $r = |\mathbf{r}|$ (Mandel and Wolf, 1995) and, hence, propagator $\overleftrightarrow{K}$ can be also approximated as

$$\overleftrightarrow{K}^{(\infty)}(\boldsymbol{\rho}', \mathbf{r}, \omega) = \frac{k}{2\pi} \begin{pmatrix} -u_z & 0 & u_x \\ 0 & -u_z & u_y \end{pmatrix} \frac{\exp(ikr)}{r} \exp(-ik\boldsymbol{\rho}' \cdot \mathbf{u}). \qquad (4.18)$$

Alternatively, the Fresnel-zone approximation can be applied under the assumption that $u_x^2 + u_y^2 \ll u_z^2$ and then under the approximation of R in the denominator of the resulting expression using binomial theorem:

$$R = \sqrt{(x-x')^2 + (y-y')^2 + z^2}$$
$$\approx z \left[1 + \frac{(x-x')^2 + (y-y')^2}{2z^2} + \ldots \right] \qquad (4.19)$$

Such an approximation stipulates that the radiated field occupies a narrow cone around the axis z. The resulting expression for the propagator becomes the same for the x and y components:

$$K(\boldsymbol{\rho}', \mathbf{r}, \omega) = \frac{-ik}{2\pi z} \exp\left[ik \frac{(x-x')^2 + (y-y')^2}{2z} \right] \exp[-ikz]. \qquad (4.20)$$

It agrees in form with propagator of Eq. (4.9) of a scalar field, if only the second term is retained, the narrow angle approximation is applied:

$$\frac{z}{\sqrt{(\boldsymbol{\rho} - \boldsymbol{\rho}')^2 + z^2}} = 1, \qquad (4.21)$$

and the binomial approximation (4.19) is then used.

4.2 Stationary Light

4.2.1 3×3 cross-spectral density tensor propagation

Suppose that in the source plane $z = 0$ the electric vector field is stationary and is specified by the 2×2 cross-spectral density matrix (see Eq. (2.94)):

$$\overleftrightarrow{W}^{(0)}(\boldsymbol{\rho}'_1, \boldsymbol{\rho}'_2, \omega) = [\langle E_i^{(0)*}(\boldsymbol{\rho}'_1, \omega) E_j^{(0)}(\boldsymbol{\rho}'_2, \omega)\rangle], \quad (i,j = x, y). \quad (4.22)$$

Further, let the field propagating into the half-space $z > 0$ be characterized by the 3×3 matrix

$$\overleftrightarrow{W}^{(3)}(\mathbf{r}_1, \mathbf{r}_2, \omega) = [\langle E_i^*(\mathbf{r}_1, \omega) E_j(\mathbf{r}_2, \omega)\rangle], \quad (i,j = x, y, z). \quad (4.23)$$

Then using Eqs. (4.11)–(4.13) in Eq. (4.14) we find, with the help of matrix identity $(\overleftrightarrow{B} \cdot \overleftrightarrow{A})^\dagger = \overleftrightarrow{A}^\dagger \cdot \overleftrightarrow{B}^\dagger$, that

$$\overleftrightarrow{W}^{(3)}(\mathbf{r}_1, \mathbf{r}_2, \omega)$$
$$= \iint \overleftrightarrow{K}^\dagger(\boldsymbol{\rho}'_1, \mathbf{r}_1, \omega) \cdot \overleftrightarrow{W}^{(0)}(\boldsymbol{\rho}'_1, \boldsymbol{\rho}'_2, \omega) \cdot \overleftrightarrow{K}(\boldsymbol{\rho}'_2, \mathbf{r}_2, \omega) d^2\rho'_1 d^2\rho'_2.$$
(4.24)

Equation (4.24) is the most general propagation law for an electromagnetic field generated by a planar source and propagating anywhere in the positive half-space. It is valid in all zones and is applicable to either a full, three-dimensional electromagentic case or in scalar or vectorial beam-like approximations. We now write this equation more explicitly as

$$\begin{pmatrix} W_{xx}(\mathbf{r}_1, \mathbf{r}_2, \omega) & W_{xy}(\mathbf{r}_1, \mathbf{r}_2, \omega) & W_{xz}(\mathbf{r}_1, \mathbf{r}_2, \omega) \\ W_{yx}(\mathbf{r}_1, \mathbf{r}_2, \omega) & W_{yy}(\mathbf{r}_1, \mathbf{r}_2, \omega) & W_{yz}(\mathbf{r}_1, \mathbf{r}_2, \omega) \\ W_{zx}(\mathbf{r}_1, \mathbf{r}_2, \omega) & W_{zy}(\mathbf{r}_1, \mathbf{r}_2, \omega) & W_{zz}(\mathbf{r}_1, \mathbf{r}_2, \omega) \end{pmatrix}$$
$$= \frac{1}{4\pi^2} \iint \frac{(-ikR_1 - 1)\exp(-ikR_1)}{R_1^3} \frac{(ikR_2 - 1)\exp(ikR_2)}{R_2^3}$$
$$\times \begin{pmatrix} -z_1 & 0 \\ 0 & -z_1 \\ x_1 - x'_1 & y_1 - y'_1 \end{pmatrix} \cdot \begin{pmatrix} W_{xx}^{(0)}(\boldsymbol{\rho}'_1, \boldsymbol{\rho}'_2, \omega) & W_{xy}^{(0)}(\boldsymbol{\rho}'_1, \boldsymbol{\rho}'_2, \omega) \\ W_{yx}^{(0)}(\boldsymbol{\rho}'_1, \boldsymbol{\rho}'_2, \omega) & W_{yy}^{(0)}(\boldsymbol{\rho}'_1, \boldsymbol{\rho}'_2, \omega) \end{pmatrix}$$
$$\cdot \begin{pmatrix} -z_2 & 0 & x_2 - x'_2 \\ 0 & -z_2 & y_2 - y'_2 \end{pmatrix} d^2\rho'_1 d^2\rho'_2, \quad (4.25)$$

where $R_\alpha = |\mathbf{r}_\alpha - \boldsymbol{\rho}'_\alpha|$, $\alpha = 1, 2$. It is evident from Eq. (4.25) that while W_{xx}, W_{yy}, W_{xy} and W_{yx} components of the 3 × 3 matrix in the half-space $z > 0$ solely depend on the corresponding components of the 2 × 2 matrix in the source plane (and are independent from each other), the five components of the field matrix involving index z depend on the source components with the x and y indices.

In particular, with the help of Eq. (4.18) we find at once that in the far field of the source the 3 × 3 cross-spectral density tensor takes the form

$$\begin{pmatrix} W_{xx}^{(\infty)}(r\mathbf{u}_1, r\mathbf{u}_2, \omega) & W_{xy}^{(\infty)}(r\mathbf{u}_1, r\mathbf{u}_2, \omega) & W_{xz}^{(\infty)}(r\mathbf{u}_1, r\mathbf{u}_2, \omega) \\ W_{yx}^{(\infty)}(r\mathbf{u}_1, r\mathbf{u}_2, \omega) & W_{yy}^{(\infty)}(r\mathbf{u}_1, r\mathbf{u}_2, \omega) & W_{yz}^{(\infty)}(r\mathbf{u}_1, r\mathbf{u}_2, \omega) \\ W_{zx}^{(\infty)}(r\mathbf{u}_1, r\mathbf{u}_2, \omega) & W_{zy}^{(\infty)}(r\mathbf{u}_1, r\mathbf{u}_2, \omega) & W_{zz}^{(\infty)}(r\mathbf{u}_1, r\mathbf{u}_2, \omega) \end{pmatrix}$$

$$= \frac{k^2}{4\pi^2} \frac{\exp[k(r_2 - r_1)]}{r_1 r_2} \begin{pmatrix} -u_{1z} & 0 \\ 0 & -u_{1z} \\ u_{1x} & u_{1y} \end{pmatrix}$$

$$\cdot \iint \exp[-ik(\boldsymbol{\rho}'_2 \cdot \mathbf{u}_2 - \boldsymbol{\rho}'_1 \cdot \mathbf{u}_1)]$$

$$\times \begin{pmatrix} W_{xx}^{(0)}(\boldsymbol{\rho}'_1, \boldsymbol{\rho}'_2, \omega) & W_{xy}^{(0)}(\boldsymbol{\rho}'_1, \boldsymbol{\rho}'_2, \omega) \\ W_{yx}^{(0)}(\boldsymbol{\rho}'_1, \boldsymbol{\rho}'_2, \omega) & W_{yy}^{(0)}(\boldsymbol{\rho}'_1, \boldsymbol{\rho}'_2, \omega) \end{pmatrix} d^2\boldsymbol{\rho}'_1 d^2\boldsymbol{\rho}'_2$$

$$\cdot \begin{pmatrix} -u_{2z} & 0 & u_{2x} \\ 0 & -u_{2z} & u_{2y} \end{pmatrix}, \tag{4.26}$$

where $\mathbf{u}_\alpha = \mathbf{r}_\alpha / r_\alpha$ are unit vectors, and $r_\alpha = |\mathbf{r}_\alpha|$, ($\alpha = 1, 2$). Moreover, in the paraxial regime where $u_x^2 + u_y^2 \ll u_z^2$, the 3 × 3 matrix elements involving index z become negligible and the transformation law reduces to that between 2 × 2 matrices:

$$\begin{pmatrix} W_{xx}^{(\infty)}(r\mathbf{u}_1, r\mathbf{u}_2, \omega) & W_{xy}^{(\infty)}(r\mathbf{u}_1, r\mathbf{u}_2, \omega) \\ W_{yx}^{(\infty)}(r\mathbf{u}_1, r\mathbf{u}_2, \omega) & W_{yy}^{(\infty)}(r\mathbf{u}_1, r\mathbf{u}_2, \omega) \end{pmatrix}$$

$$= \frac{k^2}{4\pi^2} \frac{\exp[ik(r_2 - r_1)]u_{1z}u_{2z}}{r_1 r_2}$$

$$\times \iint \begin{pmatrix} W_{xx}^{(0)}(\boldsymbol{\rho}_1', \boldsymbol{\rho}_2', \omega) & W_{xy}^{(0)}(\boldsymbol{\rho}_1', \boldsymbol{\rho}_2', \omega) \\ W_{yx}^{(0)}(\boldsymbol{\rho}_1', \boldsymbol{\rho}_2', \omega) & W_{yy}^{(0)}(\boldsymbol{\rho}_1', \boldsymbol{\rho}_2', \omega) \end{pmatrix}$$

$$\times \exp[-ik(\boldsymbol{\rho}_2' \cdot \mathbf{u}_2 - \boldsymbol{\rho}_1' \cdot \mathbf{u}_1)] d^2\rho_1' d^2\rho_2'. \tag{4.27}$$

Further, in the scalar approximation the propagation law reduces to the form

$$W^{(\infty)}(r\mathbf{u}_1, r\mathbf{u}_2, \omega) = \frac{k^2}{4\pi^2} \frac{\exp[ik(r_2 - r_1)] u_{1z} u_{2z}}{r_1 r_2}$$

$$\times \iint W^{(0)}(\boldsymbol{\rho}_1', \boldsymbol{\rho}_2', \omega) \exp[-ik(\boldsymbol{\rho}_2' \cdot \mathbf{u}_2 - \boldsymbol{\rho}_1' \cdot \mathbf{u}_1)]$$

$$\times d^2\rho_1' d^2\rho_2'. \tag{4.28}$$

Here, ratios $u_{\alpha z}/r_\alpha = \cos\phi_\alpha$, $\alpha = 1, 2$, where ϕ_α is the angle made by vector $\mathbf{r}_\alpha$ with the optical axis.

4.2.2 The van Cittert–Zernike theorem

A significant special case follows from Eq. (4.28) under the assumption that the source is spatially incoherent,

$$W^{(0)}(\boldsymbol{\rho}_1', \boldsymbol{\rho}_2', \omega) = S^{(0)}(\boldsymbol{\rho}_1', \omega)\delta^{(2)}(\boldsymbol{\rho}_2' - \boldsymbol{\rho}_1'). \tag{4.29}$$

with $\delta^{(2)}$ being the two-dimensional Dirac delta-function. On using Eq. (4.29) in Eq. (4.28) and applying the sifting property of the delta function, we arrive at the expression

$$W^{(\infty)}(r\mathbf{u}_1, r\mathbf{u}_2, \omega) = \frac{k^2}{4\pi^2} \frac{\exp[ik(r_2 - r_1)] u_{1z} u_{2z}}{r_1 r_2}$$

$$\times \int S^{(0)}(\boldsymbol{\rho}', \omega) \exp[-ik(\boldsymbol{\rho}' \cdot (\mathbf{u}_2 - \mathbf{u}_1))] d^2\rho'$$

$$= \frac{k^2}{4\pi^2} \frac{\exp[ik(r_2 - r_1)] u_{1z} u_{2z}}{r_1 r_2}$$

$$\times FT[S^{(0)}](k(\mathbf{u}_2 - \mathbf{u}_1), \omega), \tag{4.30}$$

where the two-dimensional Fourier transform of the source spectral density was recognized. Calculation of the modulus of the degree of coherence from Eq. (2.92) then yields

$$|\mu^{(\infty)}(r\mathbf{u}_1, r\mathbf{u}_2, \omega)| = \frac{FT[S^{(0)}](k(\mathbf{u}_2 - \mathbf{u}_1), \omega)}{FT[S^{(0)}](0, \omega)}. \tag{4.31}$$

The derived expression is known as the *van Cittert–Zernike theorem* (van Cittert, 1934) formulated in the space–frequency domain for the far zone of the source and can be stated as follows. *The modulus of the spectral degree of coherence at two points in the far field of a planar, incoherent source is equal to the two-dimensional Fourier transform of the source spectral density, calculated at vector $k(\mathbf{u}_2 - \mathbf{u}_1)$, normalized by its maximum value.* It immediately follows from Eq. (4.31) that radiation produced by an incoherent source develops non-trivial coherence state on free-space propagation.

As an example, we apply the van Cittert–Zernike theorem to a circular planar source of radius a and constant spectral density $S^{(0)}(\omega)$. On using polar coordinates, we readily find that

$$|\mu^{(\infty)}(r\mathbf{u}_1, r\mathbf{u}_2, \omega)| = \frac{2J_1(q)}{q}, \tag{4.32}$$

where

$$q = \frac{ka}{r}\sqrt{(x_1 - x_2)^2 + (y_1 - y_2)^2} \tag{4.33}$$

is a dimensionless parameter and J_1 is the first-order Bessel function of the first kind. Figure 4.2 shows the degree of coherence (4.32) as a function of q. The first zero of this function occurs at $q = 3.83$, corresponding to separation distance $0.61 r\lambda/a$.

The van Cittert–Zernike theorem can also be written in the intermediate zone and can be readily generalized to beam-like electromagentic fields and complete three-dimensional electromagnetic fields, on using the delta-correlation assumption (4.29) in the corresponding equations of this section (for detailed discussion of these cases, refer to Gori *et al.* (1998) and Alonso *et al.* (2006), respectively).

The importance of the van Cittert–Zernike theorem is in the possibility of determining the extent of the planar sources (or sources modeled as such) from the far-field measumenets of the degree of coherence, the connection being very useful in astronomy (see Chapter 3).

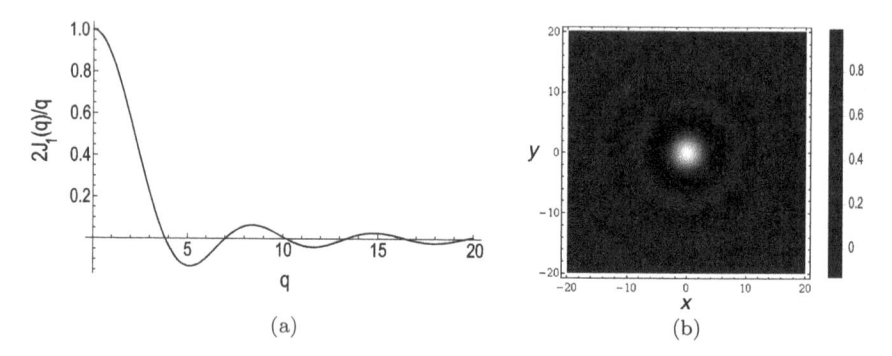

Fig. 4.2 Illustration of the van Cittert–Zernike theorem for a planar circular aperture. The far-zone degree of coherence (a) as a function of q and (b) as density plot depending on x and y.

4.3 Source Correlation-induced Changes

4.3.1 *The Huygens–Fresnel integral*

We will now apply the Fresnel approximation to propagator K introduced in Eq. (4.20) to fields radiated by random sources. On assuming that the source is scalar, on substituting from Eq. (4.20) into Eq. (4.14) and then taking the second-order correlations on both sides, we can relate the cross-spectral density functions (2.77) in the source and in the field as

$$W(\mathbf{r}_1, \mathbf{r}_2, \omega) = \frac{k^2 \exp[ik(z_2 - z_1)]}{4\pi^2 z^2} \iiiint_{-\infty}^{\infty} W^{(0)}(\boldsymbol{\rho}_1', \boldsymbol{\rho}_2', \omega) \\ \times \exp\left[-\frac{ik}{2z_1}(\boldsymbol{\rho}_1 - \boldsymbol{\rho}_1')^2\right] \exp\left[\frac{ik}{2z_2}(\boldsymbol{\rho}_2 - \boldsymbol{\rho}_2')^2\right] d\boldsymbol{\rho}_1' d\boldsymbol{\rho}_2'. \tag{4.34}$$

Equation (4.34) is the propagation law for the cross-spectral density of a light field radiated from any scalar, stationary source to the Fresnel zone. Furthermore, Eq. (2.92) implies that the cross-spectral density and all the statistical properties of the beam derived from it, such as the spectral density, the spectral degree of coherence, the OAM flux density, the spectrum, etc., also quantitatively depend on two factors: the source spectral density, and the source spectral degree of coherence. For electromagnetic beam-like fields, Eq. (4.34) is the same in form, but with W, $W^{(0)}$ being replaced by W_{ij}, $W_{ij}^{(0)}$ and can be used for illustration of the polarization changes.

4.3.2 Examples

The Gaussian Schell-model source (see Chapter 2) used in Eq. (4.34) yields

$$W(\mathbf{r}_1, \mathbf{r}_2, \omega) = \frac{S^{(0)}(\omega)}{\Delta_s^2} \exp\left[-\frac{(\boldsymbol{\rho}_1 - \boldsymbol{\rho}_2)^2}{2\Delta_s^2 \alpha_s^2}\right] \exp\left[-\frac{ik(\rho_1^2 - \rho_2^2)}{2\beta_s}\right]$$
$$\times \exp\left[-\left[\frac{1}{8\sigma^2} + i\frac{z_2 - z_1}{8k\sigma^2 \alpha_s^2}\right] \frac{(\boldsymbol{\rho}_1 + \boldsymbol{\rho}_2)^2}{\Delta_s^2}\right],$$
(4.35)

where the generalized beam spread coefficient Δ_s^2 and the radius of curvature coefficient β_s are given as

$$\Delta_s^2 = 1 + \frac{z_1 z_2}{k^2 \sigma^2 \alpha_s^2} + i\frac{z_2 - z_1}{k\alpha_s^2}, \quad \beta_s = \sqrt{z_1 z_2}\left[1 + \frac{k^2 \sigma^2}{z_1 z_2 \alpha_s^2}\right],$$
$$\frac{1}{\alpha_s^2} = \frac{1}{4\sigma^2} + \frac{1}{\delta^2}.$$
(4.36)

The very first source correlation-induced effect was revealed in Collett and Wolf (1978): the source degree of coherence was shown to control the free-space diffraction rate of a generated random beam. This effect was illustrated by constructing a family of Collett-Wolf sources, having different spectral densities and degrees of coherence that produce the same far-field spectral density distributions. Moreover, this family was also shown to include a coherent source with a well-chosen spectral density. Figure 4.3 illustrates the spectral densities $I^{(0)}$ and the degrees of coherence $g^{(0)}$ ($S^{(0)}$ and $\mu^{(0)}$ in our notations) of two sources: a laser source (with subscript L) and a quasi-homogeneous source (with subscript Q) that produce the same far-zone spectral density. For sufficiently small δ, the quasi-homogeneous and the Schell-model source produce the same spectral densities.

An even more surprising phenomenon of the correlation-induced spectral changes was later discovered (Wolf, 1986). Provided that a random beam is generated by a source with a sufficiently narrow

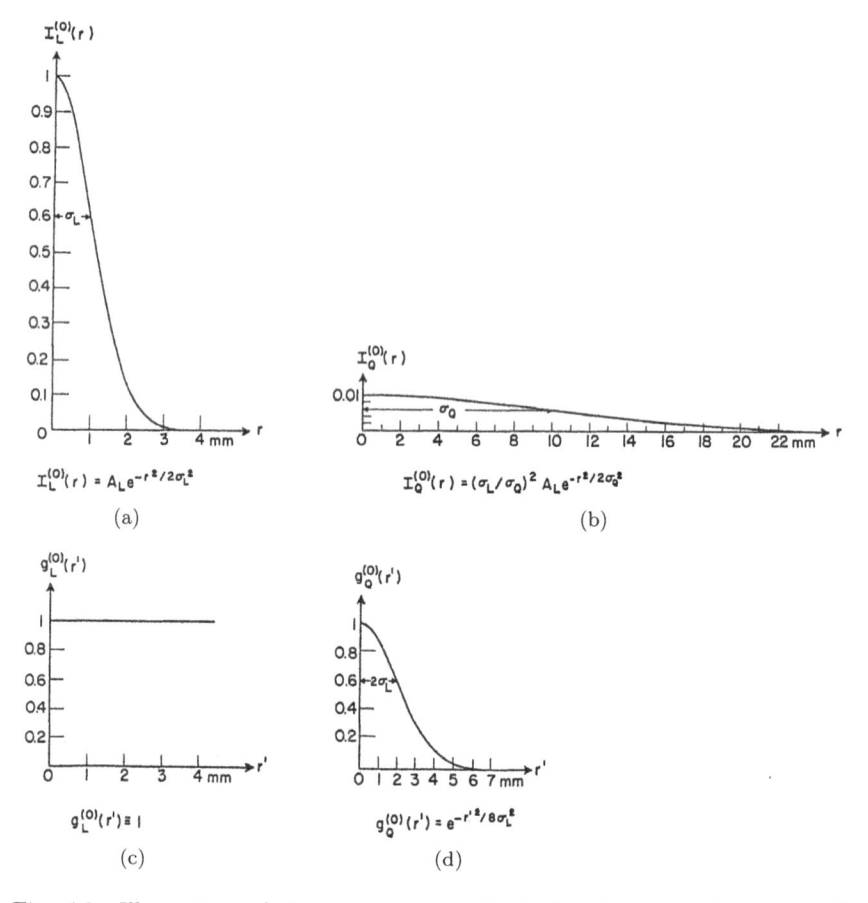

Fig. 4.3 Illustration of the source correlation-induced average beam spread. Reprinted with permission from Collett and Wolf (1978). © The Optical Society.

spectral composition, it is possible to observe the redistribution of the spectrum in the transverse cross-section as the beam propagates in free space. Figure 4.4 shows the changes in the spectral composition of a beam radiated by a scalar quasi-homogeneous source having the initial spectrum

$$S^{(0)}(\omega) = \exp\left[-\frac{(\omega - \omega_0)^2}{2\sigma_\omega^2}\right], \quad (4.37)$$

where σ_ω is the effective width of the source spectral line and ω_0 is the central frequency (Dačić and Wolf, 1988). As compared with the

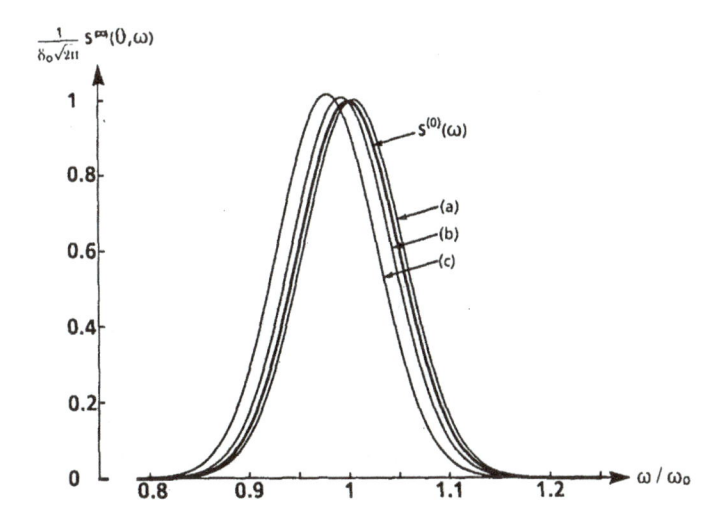

Fig. 4.4 Illustration of the source correlation-induced spectral changes. Reprinted with permission from Dačić and Wolf (1988). © The Optical Society.

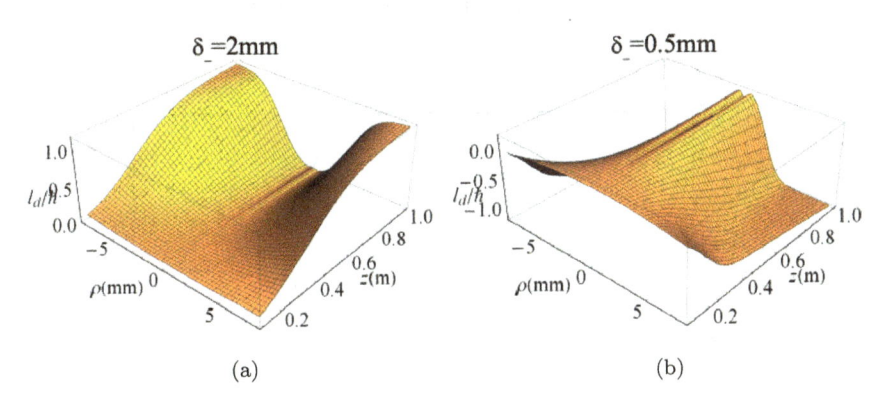

Fig. 4.5 Illustration of the source correlation-induced changes in the OAM flux density. Reprinted with permission from Zhang et al. (2020b).

source spectrum $S_0(\omega)$, the on-axis spectrum propagated to the far field of the source (a) exhibits a blue shift while the far-field spectra (b) and (c) measured at small angles from the axis are red-shifted.

Figure 4.5 shows the source correlation-induced changes in the OAM flux density $l_d^{(\mathrm{orb})}(\mathbf{r}, \omega)$ (see Eq. (2.90)), recently illustrated in Zhang et al. (2020b) for a beam radiated by a source being an incoherent superposition of two Laguerre–Gaussian modes and

Gaussian degrees of coherence with different root-mean-square correlation widths:

$$W^{(0)}(\boldsymbol{\rho}_1,\boldsymbol{\rho}_2,\omega) = \sum_{l=-1,1} U_l^*(\boldsymbol{\rho}_1,\omega)U_l(\boldsymbol{\rho}_2,\omega)\exp\left[-\frac{|\boldsymbol{\rho}_1-\boldsymbol{\rho}_2|^2}{2\delta_l^2}\right], \tag{4.38}$$

where

$$U_l(\boldsymbol{\rho},\omega) = \sqrt{\frac{2}{\pi\sigma^2|l|!}}\left(\frac{\sqrt{2}}{\sigma}\right)^{|l|}\rho^{|l|}\exp\left[-\frac{r^2}{\sigma^2}\right]\exp[il\theta]. \tag{4.39}$$

Here, the position vector is expressed in polar coordinates, $\boldsymbol{\rho} = (\rho,\theta)$ and σ is the source width. On substituting from Eqs. (4.38) and (4.39), first into Eq. (4.34) and then using the result in Eq. (2.89), one can trace the changes in the OAM flux density $l_d^{(\mathrm{orb})}$ given by Eq. (2.90). In the case of a beam radiated by a deterministic source ($\delta_l \to \infty$, $l = \pm 1$), $l_d^{(\mathrm{orb})}$ remains trivial everywhere in the transverse beam cross-section, since for the individual Laguerre–Gaussian

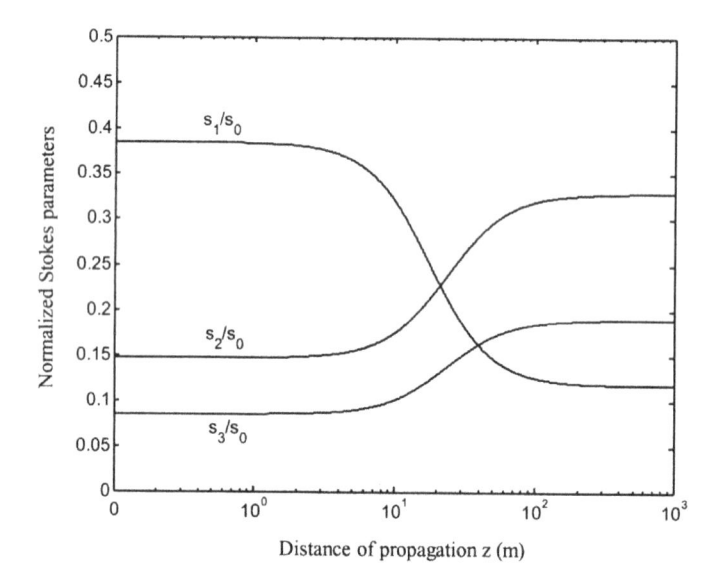

Fig. 4.6 Illustration of the source correlation-induced changes in the normalized Stokes parameters. Reprinted with permission from Korotkova and Wolf (2005b). © The Optical Society.

modes it is also a constant, $l_d^{(\mathrm{orb})} = lh$. However, for beams generated by random sources ($\delta_l < \infty$, $l = \pm 1$) the OAM flux density becomes spatially distributed as the beam propagates from the source plane.

Figure 4.6 demonstrates the source correlation-induced changes in the polarization properties of an electromagnetic Gaussian Schell-model beam generated by a source given in Eqs. (2.169)–(2.171). The normalized Stokes parameters $s_j = S_j/S_0$, ($j = 1, 2, 3$) defined in Eq. (2.113) exhibit a single change at the intermediate distances from the source and saturate to certain values in the far field.

Chapter 5

Structured Light Coherence

Since the spatial correlation functions of random optical sources and fields must obey a number of constraints, the set of legitimate mathematical models describing them is quite limited. However, elaborate structuring of source correlations was shown to result in re-shaping of the spectral density of the radiated beams on propagation, and leading, for example, to beam self-focusing, splitting, twisting, radial acceleration, etc., making it very useful for remote beam shaping applications. In this chapter, we discuss the basic approaches for structuring of the magnitude and the phase of the source correlation functions starting from one-dimensional scalar optical fields and then generalizing the analysis to two dimensions. At the end of the chapter, we briefly overview the possibilities of structuring of source coherence resulting in the polarization and OAM control in radiated beams.

5.1 One-dimensional Sources

5.1.1 *Bochner's theorem method*

We begin by gaining insight into the possible structure of the source coherence functions in a one-dimensional, planar case, i.e., when the cross-spectral density has the form

$$W^{(0)}(x_1, x_2, \omega) = \langle U^{(0)*}(x_1, \omega) U^{(0)}(x_2, \omega) \rangle, \qquad (5.1)$$

where x_1 and x_2 are the Cartesian coordinates of one-dimensional position vectors in the source plane $z = 0$. Alternatively, the cross-spectral density can be represented by the integral (Gori and Santarsiero, 2007)

$$W^{(0)}(x_1, x_2, \omega) = \int_{-\infty}^{\infty} H_0^*(x_1, v, \omega) p(v, \omega) H_0(x_2, v, \omega) dv, \quad (5.2)$$

for some scalar variable v. Such a representation is known as the Bochner theorem, being a consequence of the fact that the cross-spectral density is a Hermitian, non-negative definite operator defined in a Hilbert space. In representation (5.2), $H_0(x, v, \omega)$ is an arbitrary function, generally complex-valued, corresponding to the *class* of correlations and $p(v, \omega)$ is a non-negative *profile* function characterizing the angular distribution of the spectral density in the far field. The classification of the one-dimensional scalar random sources with respect to class H_0 involves two possibilities: *uniform correlations* and *non-uniform correlations*. In the former case, the coherence state is the same throughout the source plane, while in the latter case it depends on the choice of the source points. Since in this chapter our interest will be confined to spatial correlations of stationary fields, we will imply but omit the angular frequency dependence of all quantities.

To obtain sources with uniform correlations, the correlation class is set in Fourier-like form:

$$H_0(x, v) = a(x) \exp[-2\pi i b(x) v], \quad (5.3)$$

where $a(x)$ is an amplitude function, generally complex-valued, relating to the source spectral density as $|a(x)|^2 = S(x)$, and $b(x)$ is a real-valued function. On substitution from Eq. (5.3) into Eq. (5.2), we arrive at the following form for the source cross-spectral density function:

$$W^{(0)}(x_1, x_2) = a^*(x_1) a(x_2) FT^{-1} \left[p[b(x_2) - b(x_1)] \right], \quad (5.4)$$

In particular, for $b(x) = \beta x$ one obtains scaled, one-dimensional Schell-model sources, with β being a scaling factor:

$$W^{(0)}(x_1, x_2) = a^*(x_1) a(x_2) FT^{-1} \left[p(\beta x_d) \right], \quad (5.5)$$

where x_d is the separation between two points,

$$x_d = x_2 - x_1. \quad (5.6)$$

On setting $\beta = 1$ and recognizing that

$$\mu(x_d) = FT^{-1}[p(v)], \tag{5.7}$$

i.e., the inverse Fourier transform of $p(v)$ is the source degree of coherence, we arrive at the one-dimensional Schell-model source. It was assumed here that the choice of $p(v)$ is such that $\mu(0) = 1$. Otherwise, the normalized version can be used:

$$\mu(x_d) = \frac{FT^{-1}[p(v)]}{FT^{-1}[p(0)]}. \tag{5.8}$$

If, in addition, a Gaussian profile for amplitude $a(x)$ is chosen, then one-dimensional Gaussian Schell-model sources are obtained. Indeed, if

$$a(x) = \exp\left(-\frac{x^2}{2\sigma_0^2}\right), \quad p(v) = \delta \exp\left(-\frac{\delta^2 v^2}{2}\right), \tag{5.9}$$

then

$$W^{(0)}(x_1, x_2) = \exp\left(-\frac{x_1^2 + x_2^2}{2\sigma^2}\right)\mu(x_1, x_2), \tag{5.10}$$

where the degree of coherence becomes

$$\mu(x_1, x_2) = \exp\left[-\frac{(x_1 - x_2)^2}{2\delta^2}\right]. \tag{5.11}$$

The beams radiated by the Schell-like sources re-shape the spectral density from the source profile $|a|^2$ to a profile proportional to p on passing to the far zone. More complex fields can be radiated if in Eq. (5.5) $a(x)$ is chosen to be different from Gaussian, for example, a higher-order Laguerre–Gaussian function (Mei, 2017).

An example of a class of modes leading to sources with non-uniform correlations has the form (Lajunen and Saastamoinen, 2011; Wu et al., 2018)

$$H_0(x, v) = a(x) \exp\left[-ik(x - \beta_u)^n v\right]. \tag{5.12}$$

Here, $n \geq 2$ and β_u specifies a position within the source where correlation attains its maximum value. For instance, let us set $a(x)$

and $p(v)$ as in Eq. (5.9). On substituting from Eqs. (5.9) and (5.12) into Eq. (5.2), one finds that

$$W^{(0)}(x_1, x_2) = \exp\left(-\frac{x_1^2 + x_2^2}{2\sigma^2}\right)\mu(x_1, x_2), \qquad (5.13)$$

where the degree of coherence assumes the form

$$\mu(x_1, x_2) = \exp\left[-\frac{[(x_1 - \beta_u)^2 - (x_2 - \beta_u)^2]^2}{\delta_c^{2n}}\right], \qquad (5.14)$$

with $\delta_c = \sqrt{2\delta/k}$. We notice at once that such degree of coherence does not solely depend on separation distance x_d, but rather is a more complex function of source points x_1 and x_2. The beams generated by such sources self-focus at transverse position β_u (whether on or off-axis) within the beam at some intermediate propagation distance. The beam then diffracts further while the transverse position of the maximum remains the same. Such a phenomenon can be intuitively understood as follows. In the neigborhood of position β_u, the source degree of coherence is locally higher compared to that in other source portions. Hence, as the beam originating from such a source propagates, its spread from β_u-neigborhood is smaller than its spread from other portions into β_u-neigborhood. Thus, the region in the beam corresponding to β_u-neigborhood of the source acquires more power compared to other regions. But once the far field is reached, such power redistribution is settled.

The propagation law for the one-dimensional cross-spectral density from source points x_1' and x_2' to field points x_1 and x_2 takes the form

$$W(x_1, x_2) = \int W^{(0)}(x_1', x_2') K^*(x_1', x_1) K(x_2', x_2) dx_1' dx_2'. \qquad (5.15)$$

On substituting from Eq. (5.2) into Eq. (5.15), we express $W(x_1, x_2)$ as

$$W(x_1, x_2) = \int p(v) H_z^*(x_1, v) H_z(x_2, v) dv, \qquad (5.16)$$

where

$$H_z(x,v) = \int H_0(x',v)K(x',v)dx'. \tag{5.17}$$

In particular, the spectral density reduces to the expression

$$S(x) = \int p(v)|H_z(x,v)|^2 dv. \tag{5.18}$$

5.1.2 *Examples*

Let us now introduce several examples of uniformly correlated sources and the corresponding far fields they radiate. Apart from the Gaussian Fourier transform pair that we have already discussed,

$$p(v) = \delta \exp\left[-\frac{\delta^2 v^2}{2}\right], \quad \mu(x_d) = \exp\left[-\frac{x_d^2}{2\delta^2}\right], \tag{5.19}$$

other standard Fourier transform pairs, such as sinc and flat profiles, sinc^2 and triangle profiles, can also be used (Voelz et al., 2015). The following pairs have also been proposed that are of particular interest in optics. If a Gaussian profile function is modulated by a cosh function, then the following pair is obtained (Liang et al., 2014):

$$p(v) = \delta \exp\left[-\frac{\delta^2 v^2}{2}\right]\cosh(av), \quad \mu(x_d) = \exp\left[-\frac{x_d^2}{2\delta^2}\right]\cos(ax_d). \tag{5.20}$$

The one-dimensional multi-Gaussian family of functions can be obtained from the distribution

$$p(v) = 1 - \left[1 - \exp\left(-\frac{\delta^2 v^2}{2}\right)^M\right], \tag{5.21}$$

where $M > 0$. For $M \geq 1$ integer values, flat-top profiles with Gaussian edges are formed (Korotkova, 2014), and for $M \leq 1$, rational values in the form $M = 1/N$, where N is an integer, cusped shapes are generated (Li et al., 2016). For $M = 1$, the classic Gaussian Schell-model source is deduced from either family. In both cases ($M \geq 1$

and $M \leq 1$), Eq. (5.21) can be expressed as the series of Gaussian functions:

$$p(v) = \delta \sum_{m=1}^{M} \binom{M}{m} (-1)^{m-1} \exp\left[-\frac{m\delta^2 v^2}{2}\right], \quad M \geq 1, \qquad (5.22)$$

$$p(v) = \sum_{m=1}^{\infty} \frac{(-1)^m M^m}{m!} \prod_{p=1}^{m} \left[1 - (p-1)\frac{1}{M}\right] \exp(-2m\pi^2 \delta^2 v^2),$$
$$M \leq 1, \qquad (5.23)$$

where the Taylor series is employed in the second line. Such representations are particularly useful for analytical calculations: not only do they help producing distributions with smooth edges but also they allow to express the results via easily tractable Gaussian functions. The corresponding formulas for the degrees of coherence have the forms

$$\mu(x_d) = \frac{1}{C_0} \sum_{m=1}^{M} \binom{M}{m} \frac{(-1)^{m-1}}{\sqrt{m}} \exp\left[-\frac{x_d^2}{2m\delta^2}\right], \qquad (5.24)$$

where $C_0 = \sum_{m=1}^{M} \binom{M}{m} (-1)^{m-1}/\sqrt{m}$ and

$$\mu(x_d) = \frac{1}{C_0} \sum_{m=1}^{\infty} \frac{(-1)^{m+1} M^m}{\sqrt{m} m!} \prod_{p=1}^{m} \left[1 - (p-1)\frac{1}{M}\right] \exp\left(-\frac{x_d^2}{2m\delta^2}\right),$$
$$(5.25)$$

with $C_0 = \sum_{m=1}^{\infty} (-1)^{m+1} \prod_{p=1}^{m} [1-(p-1)/M] M^m/(\sqrt{m} m!)$. We note that all the listed models exhibit symmetry. That is, each profile function $p(v)$ is centered about zero. This naturally results in the real-valued $\mu(x_d)$.

5.1.3 Sliding function method

We will now outline another technique for modeling of the source coherence states in the case of Schell-like (uniform) correlations. As Eq. (5.7) indicates, any (non-negative) far-field intensity distribution along a line can be associated with a complex coherence state of a one-dimensional source (via Fourier transform relation). For arbitrary intensity distributions, the analytic form of the coherence state might not exist and in such cases the discrete Fourier transform can be applied (Voelz et al., 2015). However, it is useful to establish simple mathematical models for source coherence states leading to specific shapes of the far-field spectral densities. It is typically not difficult to show that a certain $p(v)$ profile is non-negative. However, the direct design of a profile for μ that must carry certain desired features may prove to be a daunting task. In what follows, we outline a procedure for simplifying such modeling.

Since $p(v) \geq 0$, it can be written as product (Korotkova and Chen, 2018)

$$p(v) = \sqrt{p(v)}\sqrt{p(v)} = h_p(v)h_p(v), \quad (5.26)$$

for some real-valued function $h_p(v) = \sqrt{p(v)}$. Let us also set $\int_{-\infty}^{\infty} |h_p(v)| dv < \infty$, i.e., we require that $h_p(v)$ belongs to $L^1(\mathbb{R})$ space. Then, by the one-dimensional convolution theorem, $\mu(x_d)$ can be expressed as

$$\mu(x_d) = FT^{-1}[h_p(v)] \circledast FT^{-1}[h_p(v)] = g(x_d) \circledast g(x_d), \quad (5.27)$$

where $\circledast$ denotes convolution and $g(x_d) = FT^{-1}[h_p(v)]$ is a *sliding function*. We will also require that it is a member of $L^1(\mathbb{R})$ space, i.e., $\int_{-\infty}^{\infty} |g(x_d)| dx_d < \infty$.

From the properties of the Fourier transform, we can readily establish that the sliding function is Hermitian:

$$g(-x_d) = g^*(x_d). \quad (5.28)$$

This can be shown by expressing $h_p(v)$ via the real and the imaginary parts of $g(x_d)$, say $g_R(x_d)$ and $g_I(x_d)$:

$$\begin{aligned}
h_p(v) &= \int_{-\infty}^{\infty} g(x_d) \exp[2\pi i x_d v] dx_d \\
&= \int_{-\infty}^{\infty} [g_R(x_d) + i g_I(x_d)][\cos(2\pi x_d v) + i \sin(2\pi x_d v)] dx_d \\
&= \int_{-\infty}^{\infty} [g_R(x_d) \cos(2\pi x_d v) - g_I(x_d) \sin(2\pi x_d v)] dx_d \\
&\quad + i \int_{-\infty}^{\infty} [g_R(x_d) \sin(2\pi x_d v) + g_I(x_d) \cos(2\pi x_d v)] dx_d.
\end{aligned}$$

$$(5.29)$$

Since $h_p(v)$ must be real, its imaginary part must vanish, implying that $g_R(x_d)$ and $g_I(x_d)$ must be an even and and odd functions, respectively. Thus, we can write $g(-x_d) = g^*(x_d)$, i.e., $g(x_d)$ must be a Hermitian function.

On the other hand, on representing the sliding function via its magnitude $g_M(x_d)$ and phase $g_P(x_d)$, i.e.,

$$g(x_d) = g_M(x_d) \exp[i g_P(x_d)], \tag{5.30}$$

we immediately establish that

$$g_M(-x_d) = g_M(x_d), \quad g_P(-x_d) = -g_P(x_d). \tag{5.31}$$

This implies that the magnitude and the phase of $g(x_d)$ must also be an even and an odd function, respectively. We now make two important observations. First, while the integrability condition on $g(x_d)$ directly affects $g_M(x_d)$, i.e., $\int_{-\infty}^{\infty} g_M(x_d) dx_d < \infty$, its phase, $g_P(x_d)$, is free from such a restriction. Second, the real and the imaginary parts of $g(x_d)$ can be chosen independently, they are not bound by any fundamental relations.

Thus, we have illustrated a very simple but general procedure for modeling sources with novel complex coherence states: any Hermitian sliding function $g(x_d)$ (even magnitude/real part and odd phase/imaginary part) with absolutely integrable magnitude leads to a new coherence state: complex degree of coherence $\mu(x_d)$ is just the self-convolution of sliding function $g(x_d)$. The fundamental difference between $\mu(x_d)$ and $g(x_d)$ is that the former function is necessarily non-negative definite (by construction), but the latter is generally not. However, due to the fact that both are Hermitian, and $\mu(x_d)$ is self-convolution of $g(x_d)$, some of the features of $g(x_d)$ can be inherited by $\mu(x_d)$. The examples that we consider in the next section illustrate this statement at best.

One-dimensional Schell-like coherence states enjoy a convenient geometrical representation, a *coherence curve* (Korotkova and Chen, 2018). Let x_d be a parameter varying in the interval $[0, \infty)$ and let the real and the imaginary parts of $\mu(x_d)$ (or, equivalently, its magnitude and phase) trace a curve. Since $\mu(x_d)$ can take on values everywhere within the unit circle of the complex plane (including the boundary),

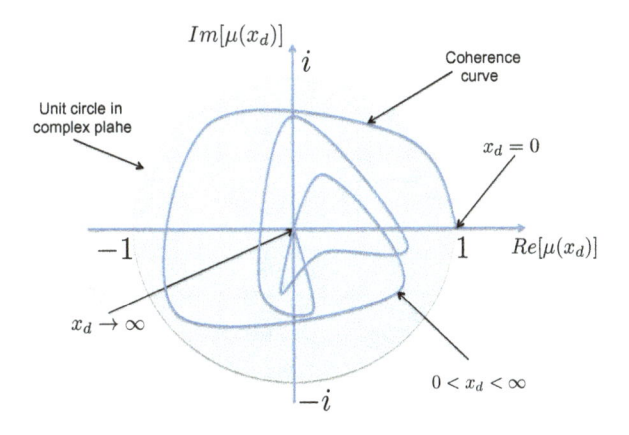

Fig. 5.1 Illustration of a coherence curve.

and since its value at $x_d = 0$ is unity, the typical scenario for the coherence curve is to start from real number 1, trace a curve within the unit circle and approach origin as $x_d \to \infty$. Since for $x_d < 0$ the curve is just the reflection of that for $x_d > 0$ with respect to the real line, its presence is not necessary and might overcomplicate the presentation. In cases when $g_P(x_d) = 0$, coherence curve is confined to the real line. Figure 5.1 presents a generic coherence curve in the complex plane.

5.1.4 *Examples*

We will now explore the effects of the phase of the source degree of coherence on the radiated light beams, which will be particularly convenient to do with the help of the sliding function method. Hence, we will set the magnitude of the sliding function as Gaussian:

$$g_M(x_d) = \frac{1}{\sqrt{\delta}\pi^{1/4}} \exp\left(-\frac{x_d^2}{2\delta^2}\right), \tag{5.32}$$

while focusing entirely on the effect of its phase $g_P(x_d)$ (Korotkova and Chen, 2018). Of course, in the trivial case $g_P(x_d) = 0$, the source degree of coherence remains Gaussian.

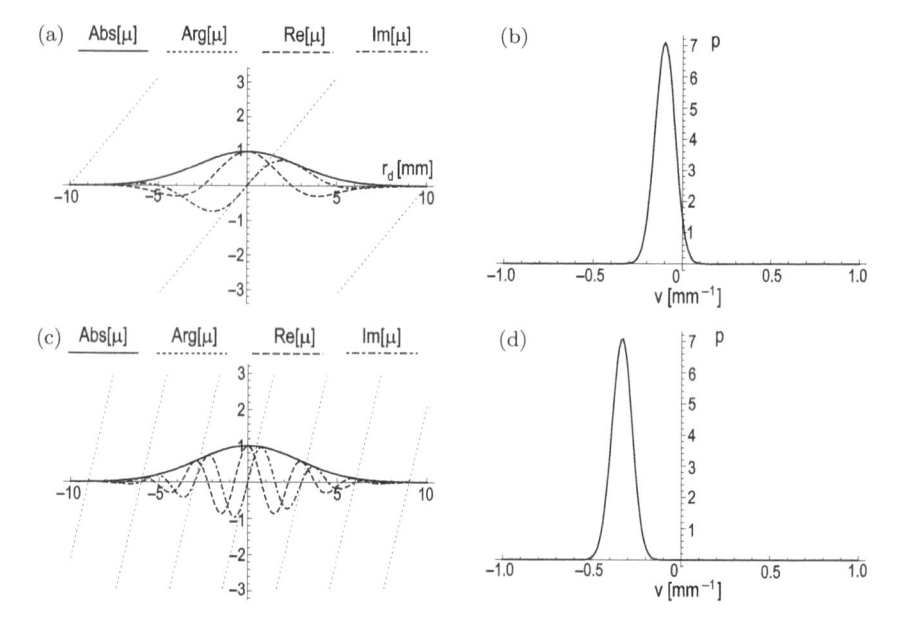

Fig. 5.2 Source degree of coherence $\mu(x_d)$ and far-field profile function $p(v)$ for sliding function with a linear phase with smaller (a), (c) and larger (b), (d) values of a. Reprinted with permission from Korotkova and Chen (2018). © The Optical Society.

For the linear phase of the sliding function:

$$g_P(x_d) = ax_d. \tag{5.33}$$

We find, on substituting from Eqs. (5.32) and (5.33) into Eq. (5.27), that the source degree of coherence becomes

$$\mu(x_d) = \exp\left(-\frac{x_d^2}{4\delta^2} + iax_d\right). \tag{5.34}$$

The corresponding far-field profile function, being the Fourier transform of $\mu(x_d)$, then takes the form

$$p(v) = 2\delta\sqrt{\pi}\exp\left[-\delta^2(2\pi v + a)^2\right]. \tag{5.35}$$

Figure 5.2 illustrates source degree of coherence $\mu(x_d)$ and far-field profile $p(v)$. Thus, the linear phase of the source coherence function acts as a shift for the far-field spectral density. One-dimensional cos-Gaussian correlation (Liang et al., 2014) can also be considered as a linear combination of two tilts symmetric about the origin.

Another simple odd function that one can choose for g_P is signum:

$$g_P(x_d) = a * \text{sign}(x_d), \tag{5.36}$$

i.e., $g_P(x_d) = -a, 0, +a$ for $x_d < 0, = 0, > 0$, respectively. For a Gaussian magnitude and a signum phase of $g(x_d)$, the degree of coherence can be expressed via the error function Erf as

$$\mu(x_d) = \begin{cases} \exp\left(-\dfrac{x_d^2}{4\delta^2}\right)\left[1 + [1 - \exp(-2ai)]\,\text{Erf}\left(\dfrac{x_d}{2\delta}\right)\right], & x_d \leq 0, \\ \exp\left(-\dfrac{x_d^2}{4\delta^2}\right)\left[1 + [\exp(2ai) - 1]\,\text{Erf}\left(\dfrac{x_d}{2\delta}\right)\right], & x_d > 0. \end{cases} \tag{5.37}$$

Then the corresponding far-field spectral density can be expressed in terms of the imaginary error function Erfi as

$$p(v) = 2\delta\sqrt{\pi}\exp[-4\pi^2\delta^2 v^2](\cos a - \text{Erfi}[\sqrt{2}\pi\delta v]\sin a)^2. \tag{5.38}$$

The effect of the signum-based phase is in redistribution of the spectral density about the axis: one side is enhanced and the other is suppressed.

Of particular interest are the effects of a nonlinear phase. Let

$$g_P(x_d) = ax_d^n, \quad n = 2m+1, \quad m = 1, 2, \ldots \tag{5.39}$$

For example, for $n = 3$, the degree of coherence has the analytic expression

$$\mu(x_d) = \exp\left(-\dfrac{x_d^2}{4\delta^2}\right)\exp\left(\dfrac{iax_d^3}{4}\right)(1 - i3a\delta x_d)^{-1/2}, \tag{5.40}$$

but it is not so for higher values of n. For $n = 3$, $p(v)$ does not have a simple closed form expression. Figure 5.3 illustrates the degree of coherence $\mu(x_d)$ and the numerical $p(v)$ curve, which resembles a coherent Airy beam, famous for its radial acceleration (Siviloglou and Christodoulides, 2007). Indeed, in the limiting case $\delta = \infty$, it reduces to $p(v) = |\text{Ai}(v)|^2$, where Ai is the Airy function of the first kind. Hence, just like in the case of a deterministic source with a

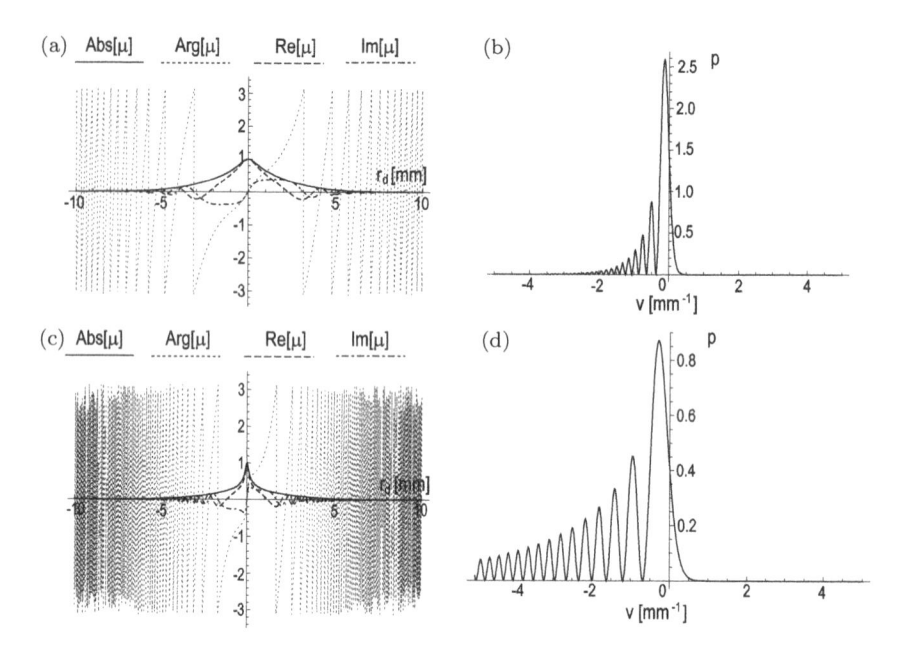

Fig. 5.3 Source degree of coherence $\mu(x_d)$ and far-field profile function $p(v)$ for sliding function with a cubic phase with smaller (a), (c) and larger (b), (d) values of a. Reprinted with permission from Korotkova and Chen (2018). © The Optical Society.

cubic phase, the random sources constructed from a sliding function with the cubic phase radiate radially accelerating spectral densities. Higher-order monomials produce effects similar to cubic monomials and the sign of a determines the side to which the spectral density accelerates.

The sine phase of the sliding function also provides an interesting example. The Maclaurin expansion of this function, $\sin(\xi) = \xi - \xi^3/3! + \xi^5/5! - \ldots$, indicates that it is an infinite linear superposition of all odd monomials with coefficients rapidly decreasing to zero. Hence, on setting

$$g_P(x_d) = 2\pi \sin(ax_d), \tag{5.41}$$

we see that depending on the values of a the far-field spectral density can transition through a number of profiles: for very small a, only the linear term is present producing a small tilt, then for larger a, the cubic term becomes dominant and produces an Airy-like pattern, then for even larger a, the cubic term and the fifth-power terms compete, producing complex distributions on both sides of the optical axis.

In order to provide a deeper insight into coherence structuring, we include in Fig. 5.4 the source coherence curves of all the examples considered so far in this section.

Until now we assumed that the magnitude and the phase of the sliding function are independent of each other. However, they might be bound by certain relations expressed via planar curves in the complex plane (Chen and Korotkova, 2018). Of course, among all such curves only those would apply which have the following properties: (I) $g(0) = 1$, (II) $g(0) \geq g(x_d)$ for any $x_d \neq 0$; (III) $\text{Re}[g(-x_d)] = \text{Re}[g(x_d)]$, $\text{Im}[g(-x_d)] = -\text{Im}[g(x_d)]$ and (IV) $\int_{-\infty}^{\infty} |g(x_d)| dx_d < \infty$. As long as $g(x_d)$ satisfies these four constraints, $\mu(x_d)$ is a legitimate choice, regardless of relationships that might exist between its magnitude and phase.

For instance, a spiral-like *cochleoid* defined as $r(\theta) = \sin(\theta)/\theta$, where r and θ are polar coordinates, meets restrictions (I)–(IV). On choosing x_d as a parameter for the polar curve, we can set $g(x_d)$ in the form

$$g(x_d) = \frac{\sin(x_d)}{x_d}[\cos(x_d) + i \sin x_d] \exp(-x_d^2/\delta^2). \tag{5.42}$$

Here, the first two factors represent the cochleoid parametrized by x_d and the third factor (optional) is a Gaussian function assuring the sufficiently rapid decrease of $g(x_d)$ to zero for large values of x_d.

Another eligible sliding function is based on Bernoulli's *lemniscate* that is described as $r(\theta) = \pm\sqrt{\cos(2\theta)}$ in polar coordinates. Using parametrization by x_d, we now may set $g(x_d)$ as

$$g(x_d) = \frac{\cos x_d}{1 + \sin^2 x_d}[1 + i \sin x_d] \exp(-x_d^2/\delta^2), \tag{5.43}$$

where the truncating Gaussian function is necessary for ensuring the absolute integrability of $g(x_d)$. Figure 5.5 shows the source coherence state formation and the far-field spectral density for the sliding function case formed with the lemniscate.

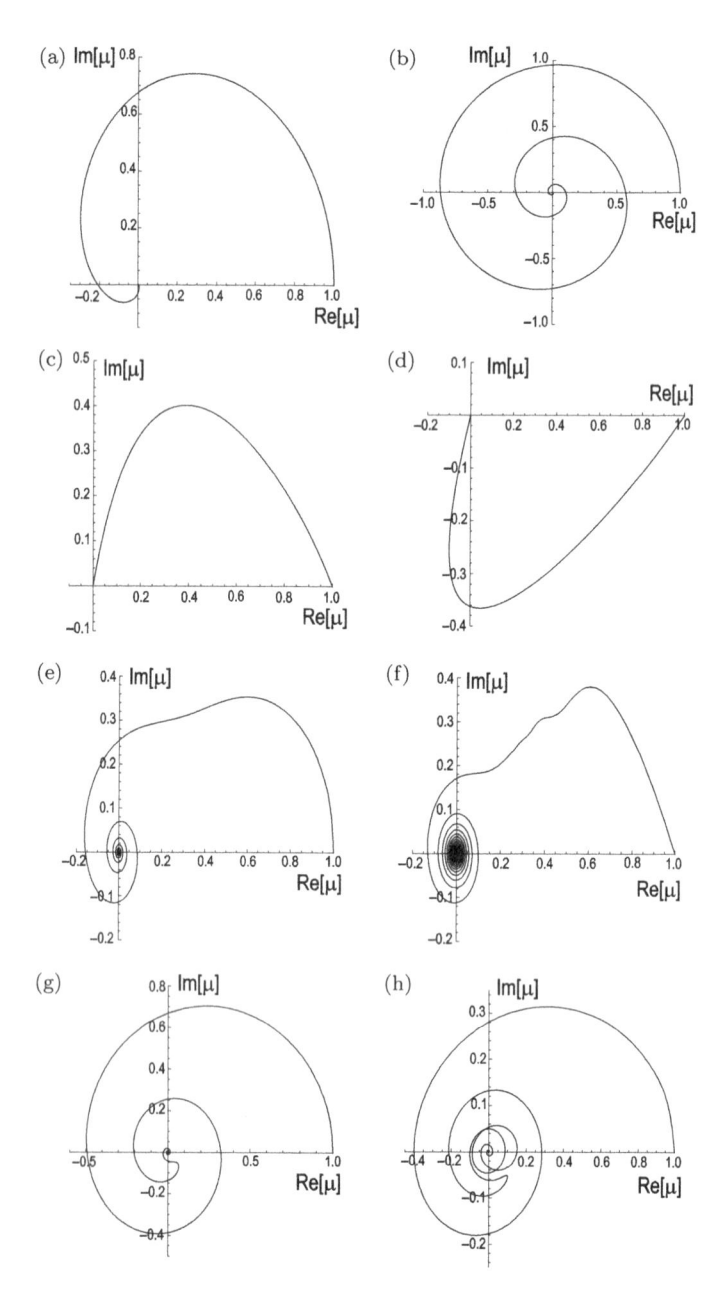

Fig. 5.4 Coherence curves obtained for the examples considered in this section: (a), (b) linear phase; (b), (c) signum phase; (e), (f) cubic phase; (g), (h) sine phase. Each case is given for a smaller and a larger value of parameter a. Reprinted with permission from Korotkova and Chen (2018). © The Optical Society.

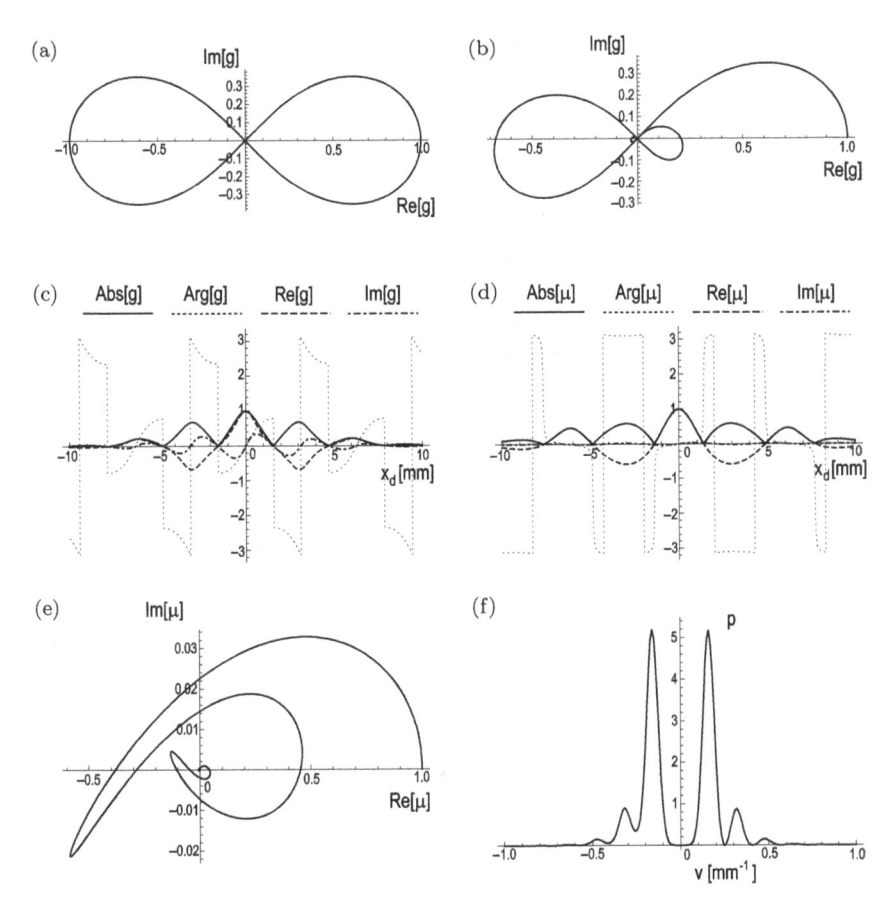

Fig. 5.5 Structuring of complex coherence state with lemniscate. (a) untruncated lemniscate; (b) truncated lemniscate; (c) magnitude, phase, real and imaginaty parts of sliding function; (d) magnitude, phase, real and imaginaty parts of degree of coherence; (e) coherence curve; (f) corresponding $p(v)$. Reprinted with permission from Chen and Korotkova (2018). © The Optical Society.

Yet another planar curve termed *teardrop* due to its shape, given in Cartesian coordinates as $y(x) = \pm(1-x^2)[(1-x)/2]^m$, where m is a positive integer controlling the teardrop base profile, can also be employed. Using its parametric representation with x_d and truncating its absolute value by a Gaussian function, we set $g(x_d)$ in the form

$$g(x_d) = [\cos x_d + i \sin x_d \sin^m(x_d/2)] \exp(-x_d^2/\delta^2). \quad (5.44)$$

A broad family of curves known as *centered trochoids* can be formed on tracing a curve by a point attached to a circle with radius b, at distance d from its center, as the circle rolls without slipping along a fixed circle of radius a. The sliding functions corresponding to this family become

$$g(x_d) = r_1 \exp[ix_d] + r_2 \exp[i(\omega_2/\omega_1)x_d], \quad (5.45)$$

where $r_1 = a \pm b$, $r_2 = d$ and $\omega_2/\omega_1 = 1 \pm a/b$. If $\omega_1\omega_2 > 0$, then the curve is *epitrochoid* (the rolling circle moves outside of the fixed circle) and if $\omega_1\omega_2 < 0$, then it is *hypotrochoid* (the rolling circle moves inside of the fixed circle). There are also three other subgroups: *roses* or *rhodoneas* if $r_1 = r_2$; *centered cycloids* if $r_1\omega_1 = r_2\omega_2$ (equal angular speeds); and *trochoids with a meplat* if $r_1\omega_1^2 = r_2\omega_2^2$ (equal angular accelerations). In particular, a *rhodonea* curve and a *cardioid* can be, respectively, employed as sliding functions

$$g(x_d) = \cos(kx_d)\left[\cos x_d + i \sin x_d\right]\exp(-x_d^2/\delta^2), \quad (5.46)$$

$$g(x_d) = 0.5(1 + \cos x_d)\left[\cos x_d + i \sin x_d\right]\exp(-x_d^2/\delta^2). \quad (5.47)$$

5.2 Two-dimensional, Scalar Sources and Beams

5.2.1 Bochner's theorem method

The cross-spectral density function of a scalar two-dimensional field at a pair of points $\boldsymbol{\rho}_1$ and $\boldsymbol{\rho}_2$ in the source plane and frequency ω can be represented as superposition (Gori and Santarsiero, 2007)

$$W^{(0)}(\boldsymbol{\rho}_1, \boldsymbol{\rho}_2) = \int H_0^*(\boldsymbol{\rho}_1, \mathbf{v}) p(\mathbf{v}) H_0(\boldsymbol{\rho}_2, \mathbf{v}) d^2v, \quad (5.48)$$

with $\mathbf{v} = v_x\hat{x} + v_y\hat{y}$ being a two-dimensional vector. It is a version of the Bochner theorem for two dimensions.

Just like in the one-dimensional case, the Fourier-like modes

$$H_0(\boldsymbol{\rho}, \mathbf{v}) = a(\boldsymbol{\rho})\exp[-2\pi i b(\boldsymbol{\rho}) \cdot \mathbf{v}], \quad (5.49)$$

where $a(\boldsymbol{\rho})$ is a complex-valued amplitude function and $b(\boldsymbol{\rho})$ is a real-valued function, lead to sources with uniform correlations.

On substitution from Eq. (5.49) into Eq. (5.48), we arrive at the cross-spectral density function

$$W^{(0)}(\boldsymbol{\rho}_1, \boldsymbol{\rho}_2) = a^*(\boldsymbol{\rho}_1)a(\boldsymbol{\rho}_2)FT^{-1}\left[p[b(\boldsymbol{\rho}_2) - b(\boldsymbol{\rho}_1)]\right]. \quad (5.50)$$

In particular, for $b(\boldsymbol{\rho}) = \beta\boldsymbol{\rho}$, one obtains the Schell-model sources, with β being the scaling factor

$$W^{(0)}(\boldsymbol{\rho}_1, \boldsymbol{\rho}_2) = a^*(\boldsymbol{\rho}_1)a(\boldsymbol{\rho}_2)FT^{-1}\left[p[\beta(\boldsymbol{\rho}_2 - \boldsymbol{\rho}_1)]\right]. \quad (5.51)$$

On setting $\beta = 1$ and choosing a Gaussian profile for $a(\boldsymbol{\rho})$, we arrive at the Gaussian Schell-model source. The main feature of the beams these sources radiate is re-shaping of the spectral density from the source a profile $|a(\boldsymbol{\rho})|^2$ to a profile proportional to p, on passing to the far field.

For two-dimensional sources with non-uniform correlations, one can set (see Lajunen and Saastamoinen, 2011; Wu et al., 2018)

$$H_0(\boldsymbol{\rho}, \mathbf{v}) = a(\boldsymbol{\rho})\exp\left[-ik(\boldsymbol{\rho} - \boldsymbol{\beta}_u)^n v\right]. \quad (5.52)$$

Here, $n \geq 2$ and the two-dimensional vector $\boldsymbol{\beta}_u = (\beta_x, \beta_y)$ specifies a position within the source where correlation attains the maximum value. The beams generated by such sources self-focus at transverse position $\boldsymbol{\beta}_u$ whether on or off-axis of the beam, at some intermediate propagation distance. The beam then diffracts further, while the transverse position of the maximum remains the same.

It was known for a long time that certain stationary beams may be furnished by a so-called twist phase (Simon and Mukunda, 1993). If, in addition, the spectral density of the source is not rotationally symmetric, the twist phase results in the rotation of the beam through the angle of $\pi/2$ as it propagates from the source to the far field (Mei and Korotkova, 2017). Such a possibility arrises from setting

$$H_0(\boldsymbol{\rho}, \mathbf{v}) = a(\boldsymbol{\rho})\exp\left[-[(ay + ix)v_x - (bx - iy)v_y]\right], \quad (5.53)$$

where a and b are positive real constants.

The propagation law for the two-dimensional, scalar cross-spectral density from source points $\boldsymbol{\rho}'_1$ and $\boldsymbol{\rho}'_2$ to field points $\mathbf{r}_1$ and $\mathbf{r}_2$ is

$$W(\mathbf{r}_1, \mathbf{r}_2) = \int_{-\infty}^{\infty} \int_{-\infty}^{\infty} W^{(0)}(\boldsymbol{\rho}'_1, \boldsymbol{\rho}'_2) K^*(\boldsymbol{\rho}'_1, \mathbf{r}_1) K(\boldsymbol{\rho}'_2, \mathbf{r}_2) d^2\rho'_1 d^2\rho'_2, \tag{5.54}$$

which can also be expressed, on substituting from Eq. (5.48) into Eq. (5.54) as

$$W(\mathbf{r}_1, \mathbf{r}_2) = \int_{-\infty}^{\infty} \int_{-\infty}^{\infty} p(\mathbf{v}) H_z^*(\mathbf{r}_1, \mathbf{v}) H_z(\mathbf{r}_2, \mathbf{v}) d^2v. \tag{5.55}$$

Here,

$$H_z(\mathbf{r}, \mathbf{v}) = \int_{-\infty}^{\infty} \int_{-\infty}^{\infty} H_0(\boldsymbol{\rho}', \mathbf{v}) K(\boldsymbol{\rho}', \mathbf{v}) d^2\rho' \tag{5.56}$$

are the propagating modes at point $\mathbf{r}$. Further, the spectral density reduces to the expression

$$S(\mathbf{r}) = \int_{-\infty}^{\infty} \int_{-\infty}^{\infty} p(\mathbf{v}) |H_z(\mathbf{r}, \mathbf{v})|^2 d^2v. \tag{5.57}$$

In the next several sections, we will narrow our attention down to propagation modes H_0 in the form of Eq. (5.49) with $b(\boldsymbol{\rho}) = 1$ and discuss the expressions for the source degree of coherence in three situations, when the radiated spectral densities of the far field must be radially symmetric in the polar system, have Cartesian symmetry and be separable in the polar coordinate system into the radial and the polar variables.

5.2.2 Examples: Uniform correlations, radial symmetry

We first consider a simple case when the profile function does not have azimuthal dependence, i.e.,

$$p(\mathbf{v}) = p(v), \tag{5.58}$$

where $v = |\mathbf{v}|$. The source's cross-spectral density then reduces to the expression

$$W^{(0)}(\boldsymbol{\rho}_1, \boldsymbol{\rho}_2) = 2\pi a^*(\boldsymbol{\rho}_1) a(\boldsymbol{\rho}_2) \int_0^\infty p(v) J_0(2\pi v \rho_d) v \, dv, \qquad (5.59)$$

where J_0 is the 0-order Bessel function of the first kind, the integral being the inverse Hankel transform (denoted here by HT) of the profile function and $\rho_d = |\boldsymbol{\rho}_2 - \boldsymbol{\rho}_1| = \sqrt{x_d^2 + y_d^2}$. Hence, the degree of coherence becomes

$$\mu(\rho_d) = \frac{HT^{-1}[p(v)]}{HT^{-1}[p(0)]}. \qquad (5.60)$$

Of course, for the basic example, we set the two-dimensional Gaussian pair of functions:

$$p(v) = \delta^2 \exp\left[-\frac{\delta^2 v^2}{2}\right], \quad \mu(\rho_d) = \exp\left[-\frac{\rho_d^2}{2\delta^2}\right]. \qquad (5.61)$$

Several correlation families can be obtained by assigning the profile function in the form of a Gaussian function multipled by another rotationally symmetric distribution. Two examples of this kind can be obtained by use of Bessel functions or monomials (Mei and Korotkova, 2013). In the former case, we set

$$p(v) = 2\pi \delta^2 I_0(2\pi \beta \delta v) \exp\left(-\beta^2/2 - 2\pi^2 \delta^2 v^2\right), \qquad (5.62)$$

where I_0 is the zero-order modified Bessel function, to obtain the source degree of coherence

$$\mu(\rho_d) = J_0\left(\frac{\beta}{\delta} \rho_d\right) \exp\left(-\frac{\rho_d^2}{2\delta^2}\right), \qquad (5.63)$$

J_0 being the zero-order Bessel function. In the latter case, on setting

$$p(v) = (\pi^{2n+1} \delta^{2n+2} 2^{n+1}/n!) v^{2n} \exp\left(-2\pi^2 \delta^2 v^2\right), \qquad (5.64)$$

where n is an integer, we obtain source degree of coherence in the form

$$\mu(\rho_d) = L_n\left(\frac{\rho_d^2}{2\delta^2}\right) \exp\left(-\frac{\rho_d^2}{2\delta^2}\right), \qquad (5.65)$$

where L_n is the Laguerre polynomial of order n. Both families radiate to ring-shaped spectral density profiles with controllable size and thickness.

Finite and infinite alternating series of Gaussian source correlation functions with well-suited widths and amplitudes can serve for generating beams with flat and cusped spectral density profiles. Of course, other functional forms can be used to describe such distributions, but the series of Gaussian functions prove to be very convenient for analytical calculations, for instance. Let us set the following profile function with rotational symmetry (Korotkova et al., 2012; Sahin and Korotkova, 2012):

$$p(v) = 1 - \left[1 - \exp\left(-\frac{\delta^2 v^2}{2}\right)\right]^M, \quad (5.66)$$

where, as before, $v = |\mathbf{v}|$. For $M = 1$, the profile reduces to Gaussian, while for the integer values of M, $p(v)$ represents profile with a flat area around the beam center and a Gaussian-like edge. We can write (5.66) as a finite series

$$p(v) = \sum_{m=1}^{M} \binom{M}{m} (-1)^{m+1} \exp\left[-\frac{m\delta^2 v^2}{2}\right]. \quad (5.67)$$

On taking the two-dimensional Fourier transform of $p(v)$, we find that

$$\mu(\rho_d) = \frac{1}{C_0} \sum_{m=1}^{M} \binom{M}{m} \frac{(-1)^m}{m} \exp\left[-\frac{\rho_d^2}{2m\delta^2}\right], \quad (5.68)$$

with normalization factor

$$C_0 = \sum_{m=1}^{M} \binom{M}{m} \frac{(-1)^m}{m}. \quad (5.69)$$

On the other hand, when M is a rational number in the form $M = 1/N$, where N is an integer, $p(v)$ represents a profile cusped around the beam center (Wang and Korotkova, 2017). In this case, function $[1 - \exp(-\delta^2 v^2/2)]^M$ in Eq. (5.66) can be expanded in the

Maclaurin series, and the profile function becomes

$$p(v) = \sum_{n=1}^{\infty} \frac{(-1)^{n+1}M^n}{n!} \prod_{m=1}^{n} \left[1 - (m-1)\frac{1}{M}\right] \exp\left(-\frac{n\delta^2 v^2}{2}\right). \tag{5.70}$$

The corresponding source degree of coherence then has the form

$$\mu(\rho_d) = \frac{1}{C_0} \sum_{n=1}^{\infty} \frac{(-1)^{n+1}M^n}{n!n} \prod_{m=1}^{n} \left[1 - (m-1)\frac{1}{M}\right] \exp\left[-\frac{\rho_d^2}{2n\delta^2}\right], \tag{5.71}$$

where we have used the normalization factor

$$C_0 = \sum_{n=1}^{\infty} \frac{(-1)^{n+1}M^n}{n!n} \prod_{m=1}^{n} \left[1 - (m-1)\frac{1}{M}\right]. \tag{5.72}$$

5.2.3 Examples: Uniform correlations, Cartesian symmetry

If the profile function has rectangular symmetry, i.e., it factorizes as

$$p(\mathbf{v}) = p_x(v_x)p_y(v_y), \tag{5.73}$$

where v_x and v_y are the Cartesian coordinates of vector $\mathbf{v}$, then the source cross-spectral density takes the form

$$W^{(0)}(\boldsymbol{\rho}_1,\boldsymbol{\rho}_2) = a^*(\boldsymbol{\rho}_1)a(\boldsymbol{\rho}_2) \int_{-\infty}^{\infty} p_x(v_x)\exp[-2\pi i x_d v_x]dv_x$$
$$\times \int_{-\infty}^{\infty} p_y(v_y)\exp[-2\pi i y_d v_y]dv_y, \tag{5.74}$$

where we recognize the one-dimensional Fourier transforms taken along the x and y directions. Hence, the degree of source coherence becomes

$$\mu^{(0)}(\rho_d) = \frac{FT^{-1}[p_x(x_d)]}{FT^{-1}[p_x(0)]} \frac{FT^{-1}[p_y(y_d)]}{FT^{-1}[p_y(0)]}. \tag{5.75}$$

Recall that the sources and the beams stemming from formula (5.66) possess rotational symmetry. It is also possible to obtain the

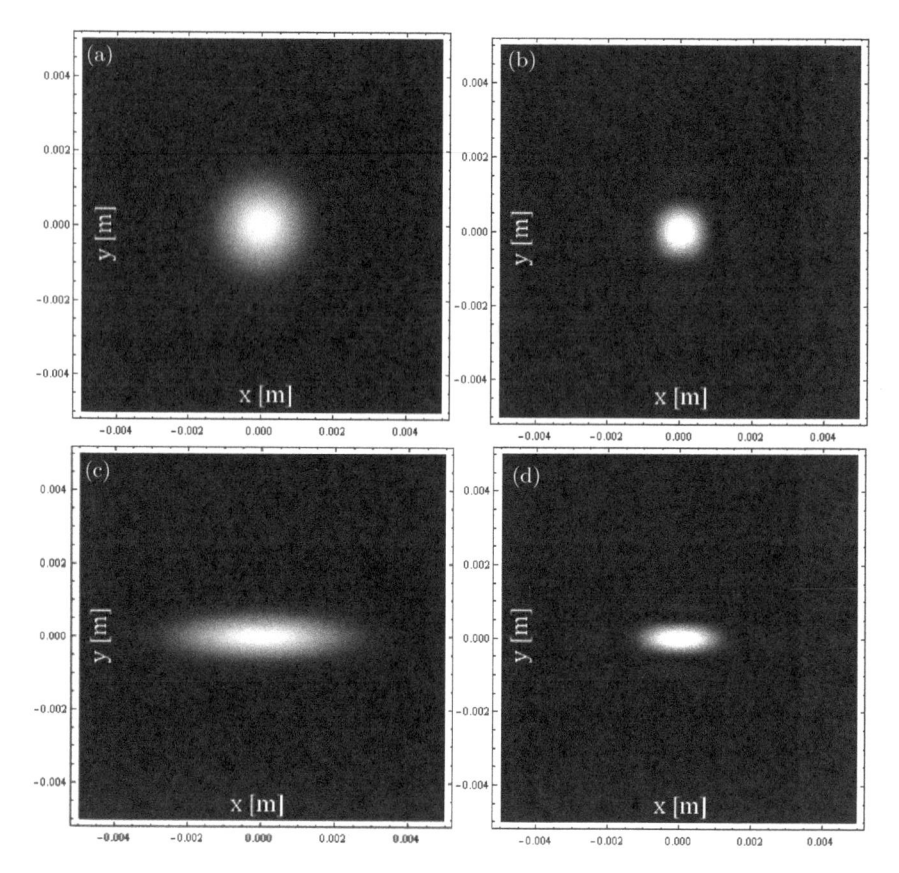

Fig. 5.6 The source degree of coherence for forming rectangular flat-top profiles in the far field. (a) and (c) $M = 1$; (b) and (d) $M = 40$; (a) and (b) $\delta_x = \delta_y = 0.8$ mm; (c) and (d) $\delta_x = 1.6$ mm, $\delta_y = 0.56$ mm. Reprinted with permission from Korotkova (2014). © The Optical Society.

multi-Gaussian analog with the Cartesian symmetry. This can be done starting from the profile function:

$$p(\mathbf{v}) = \{1 - [1 - \exp(-\delta_x^2 v_x^2/2)]^M\}\{1 - [1 - \exp(-\delta_y^2 v_y^2/2)]^M\}. \quad (5.76)$$

Then for $M > 1$, rectangle-like flat shapes are formed (Korotkova, 2014), for $M < 1$, rhomb-like cusped shapes are generated (Li et al., 2016), and, if $M = 1$, then the Gaussian function is deduced. Figures 5.6 and 5.7 illustrate the source degree of coherence used

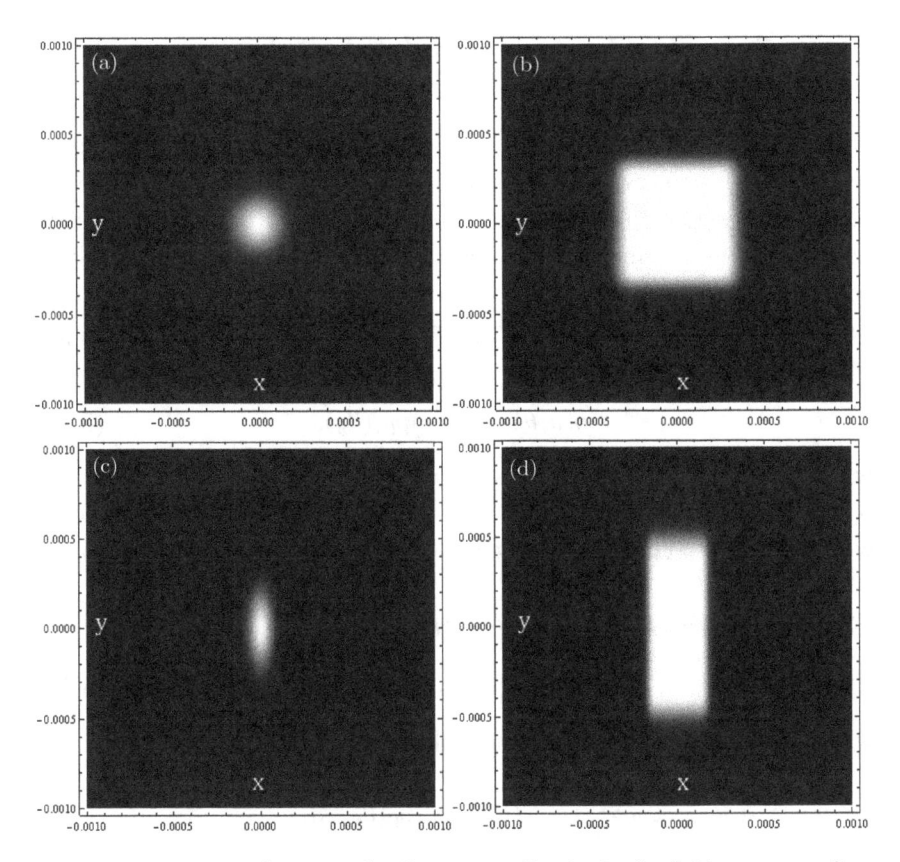

Fig. 5.7 Formation of rectangular flat-top profiles in the far field, corresponding to degrees of coherence in Fig. 5.6. Reprinted with permission from Korotkova (2014). © The Optical Society.

in the formation of the flat-top rectangular beams and the formed far-field spectral densities, respectively.

A number of models can be established by modulation of a Gaussian profile by product $p_x(x)p_y(y)$. For instance, cosh-Gaussian p-function with Cartesian symmetry may be obtained if (Liang *et al.*, 2014)

$$p(\mathbf{v}) = \exp\left(-\delta_x^2 v_x^2/2\right) \exp\left(-\delta_y^2 v_y^2/2\right) \cosh(n\delta_x) \cosh(n\delta_y). \quad (5.77)$$

The corresponding source degree of coherence is a product of a Gaussian function modulated by the cosine functions in x and y directions.

5.2.4 Examples: Uniform correlations, no symmetry

For generation of a beam having a spectral density with radial or Cartesian symmetry, it is sufficient to use the real-valued degree of source coherence function. However, if the control of the spectral density in the polar variable must be achieved, the complex-valued source degree of coherence is required (Wang and Korotkova, 2016a). To see this, we must start with prescribing profile $p(v, \theta_v)$ as a separable function in polar coordinates, i.e.,

$$p(v, \theta) = p_v(v) p_{\theta_v}(\theta_v), \tag{5.78}$$

where v and θ_v are the radius and the polar angle. Both functions $p_v(v)$ and $p_\theta(\theta_v)$ must be non-negative and $p_\theta(\theta_v)$ must be periodic with period 2π. The cross-spectral density of the source then becomes

$$W^{(0)}(\boldsymbol{\rho}_1, \boldsymbol{\rho}_2) = \frac{1}{C_0} a^*(\boldsymbol{\rho}_1) a(\boldsymbol{\rho}_2) \int_0^{2\pi} \int_0^{\infty} p_v(v) p_\theta(\theta_v)$$
$$\times \exp[-2\pi i \rho_d v \cos(\theta_v - \theta_d)] v \, dv \, d\theta_v, \tag{5.79}$$

where $\theta_d = \arctan(y_d/x_d)$ is the polar angle of vector $\boldsymbol{\rho}_d$ and

$$C_0 = \int_0^{2\pi} \int_0^{\infty} p_v(v) p_\theta(\theta_v) v \, dv \, d\theta_v. \tag{5.80}$$

The exponent in the integrand of Eq. (5.79) can be represented as

$$\exp[-2\pi i \rho_d v \cos(\theta_v - \theta_d)]$$
$$= \sum_{m=-\infty}^{\infty} (-i)^m J_m(2\pi \rho_d v) \exp[im(\theta_v - \theta_d)], \tag{5.81}$$

where J_m is the mth order Bessel function of the first kind. Hence, on substituting from Eq. (5.81) into Eq. (5.79), we find that

$$W^{(0)}(\boldsymbol{\rho}_1, \boldsymbol{\rho}_2) = \frac{1}{C_0} \sum_{m=-\infty}^{\infty} (-i)^m \exp[-im\theta_d]$$
$$\times \int_0^{\infty} p_v(v) J_m(2\pi r_d v) v \, dv \int_0^{2\pi} p_\theta(\theta_v) \exp[im\theta_v] d\theta_v. \tag{5.82}$$

Since $p_\theta(\theta_v)$ is 2π-periodic, it can be represented by the Fourier series:

$$p_\theta(\theta_v) = \frac{a_0}{2} + \sum_{m=1}^{\infty}[a_m \cos(m\theta_v) + b_m \sin(m\theta_v)], \qquad (5.83)$$

$$a_m = \frac{1}{\pi}\int_{-\pi}^{\pi} p_\theta(\theta_v)\cos(m\theta_v)d\theta_v, \quad m = 0,1,2,\ldots$$

$$b_m = \frac{1}{\pi}\int_{-\pi}^{\pi} p_\theta(\theta_v)\sin(m\theta_v)d\theta_v, \quad m = 1,2,\ldots \qquad (5.84)$$

On substituting from Eqs. (5.83) and (5.84) into Eq. (5.82) and using the definition of the degree of coherence, we arrive at the relation

$$\mu(\rho_d, \theta_d) = \frac{FB[p(v,\theta_v)]}{FB[p(0)]}, \qquad (5.85)$$

where FB denotes the Fourier–Bessel transform (Goodman, 2000). Explicitly, the degree of coherence can be expressed as

$$\mu(\rho_d, \theta_d) = \frac{1}{C_0}\Bigg[\pi a_0 \int_0^\infty p_v(v) J_0(2\pi r_d v) v\, dv$$

$$+ 2\pi \sum_{m=1}^{\infty} a_m \cos(m\theta_d)(-i)^m \int_0^\infty p_v(v) J_m(2\pi r_d v) v\, dv$$

$$+ 2\pi \sum_{m=1}^{\infty} b_m \sin(m\theta_d)(-i)^m \int_0^\infty p_v(v) J_m(2\pi r_d v) v\, dv\Bigg]. \qquad (5.86)$$

The presense of the imaginary terms in Eq. (5.86) indicates the fact that the full angular control of the far-field spectral density can be acheived only if both the magnitude and the phase of the source spectral density are spatially modulated. We will now outline several model sources of this kind. They will all have Gaussian distribution in the radial part, i.e.,

$$p_v(v) = \exp\left(-\frac{\delta^2 v^2}{2}\right), \qquad (5.87)$$

but will differ by the azimuthal part $p_\theta(\theta_v)$.

It is evident that assigning a harmonic variation for function $p_\theta(\theta_v)$ can reduce infinite series (5.86) to one or two terms. However, since function $p_\theta(\theta_v)$ must be non-negative, perhaps the best setting is

$$p_\theta(\theta_v) = \cos^2\left(\frac{n\theta_v}{2}\right) = \frac{1}{2} + \frac{1}{2}\cos(n\theta_v). \tag{5.88}$$

The coefficients in Fourier series (5.83) are $a_0 = 1$, $a_m = 1/2$ only for $m = n$, otherwise $a_m = 0$ and $b_m = 0$. Substituting them into Eq. (5.86) leads to the source degree of coherence represented only by two terms:

$$\mu(\rho_d, \theta_d) = \frac{\pi}{C_0}\int_0^\infty p_v(v)[J_0(2\pi\rho_d v) + \cos(n\theta_d)(-i)^n J_n(2\pi\rho_d v)]v\,dv. \tag{5.89}$$

Further, after substituting from Eq. (5.87), the integrals over variable v can be evaluated in closed form, and, hence,

$$\mu(\rho_d, \theta_d) = \frac{1}{C_0}\exp\left(-\frac{\rho_d^2}{2\delta^2}\right) + \frac{(-1)^n\sqrt{\pi}\rho_d}{2\sqrt{2}C_0\delta}\exp\left(-\frac{\rho_d^2}{4\delta^2}\right)$$
$$\times\left[I_{(n-1)/2}\left(\frac{\rho_d^2}{4\delta^2}\right) - I_{(n+1)/2}\left(\frac{\rho_d^2}{4\delta^2}\right)\right]\cos(n\theta_d), \tag{5.90}$$

reducing to that for Gaussian Schell-model source if $n = 0$. Here, I_n stands for the modified Bessel function of the first kind and order n. When n is even, the profile function has Cartesian symmetry, while for odd values of n, such symmetry is broken. Indeed, Eq. (5.90) implies that in the former case the source degree of coherence must be real, and in the latter case, it must have non-zero imaginary terms.

Only a very limited number of choices for p_θ lead to reduction of the infinite series in Eq. (5.86) to a finite number of terms all expressed in closed form. Indeed, even for relatively simple p_θ, the corresponding degree of coherence may retain a rather complicated form. As an example, consider p_θ linearly increasing with the angle:

$$p_\theta(\theta_v) = l\theta_v - 2l_I\pi. \tag{5.91}$$

Here, l_I is the integer part of $l\theta_v/2\pi$, ensuring that the value of p_θ is in the interval $[0, 2\pi]$, while parameter l indicates the number of times p_θ increases from 0 to π on the interval $[0, 2\pi]$. Such a distribution resembles a vortex in the spectral density of the beam. By representing this profile function via the Fourier series as

$$p_\theta(\theta_v) = \pi - \sum_{m=1}^{\infty} \frac{2}{m} \sin(ml\theta_v), \qquad (5.92)$$

we find that in comparison with Eq. (5.83) the Fourier coefficients are $a_0 = 2\pi$, $a_m = 0$ and $b_m = 2/m$. On substituting these values into Eq. (5.90) and using Eq. (5.87), we finally find that

$$\mu^{(0)}(\rho_d, \theta_d) = \frac{1}{C_0} \exp\left(-\frac{\rho_d^2}{2\delta^2}\right) - \frac{\rho_d}{C_0 \delta \sqrt{2\pi}} \exp\left(-\frac{\rho_d^2}{4\delta^2}\right)$$
$$\times \sum_{m=1}^{\infty} \frac{(-i)^{ml}}{m} \sin(ml\theta_d) \left[I_{(ml-1)/2}\left(\frac{\rho_d^2}{4\delta^2}\right) \right.$$
$$\left. - I_{(ml+1)/2}\left(\frac{\rho_d^2}{4\delta^2}\right) \right]. \qquad (5.93)$$

Figures 5.8 and 5.9 illustrate the source degree of coherence used in the formation of the intensity vortices and the formed far fields, respectively.

It is also possible to model the uniformly correlated sources that radiate beams that are radially accelerating on propagation, in the sense that the maximum of their spectral desnity accelerates away from the axis with the increasing distance from the source. One distribution of this kind is a beam resembling a crescent (Wang et al., 2017). Suppose that the angular profile function has the form

$$p(\theta_v) = \cos^2(\theta_v/2), \qquad (5.94)$$

being a single trigonometric mode, while the radial part,

$$p(v) = \frac{\delta^2 (\delta v)^{2n}}{2^n n! \pi} \exp\left(-\frac{\delta^2 v^2}{2}\right), \qquad (5.95)$$

gives rise to a ring with value of n determining its thickness and size. The product of these two functions produces a crescent. The

140   *Theoretical Statistical Optics*

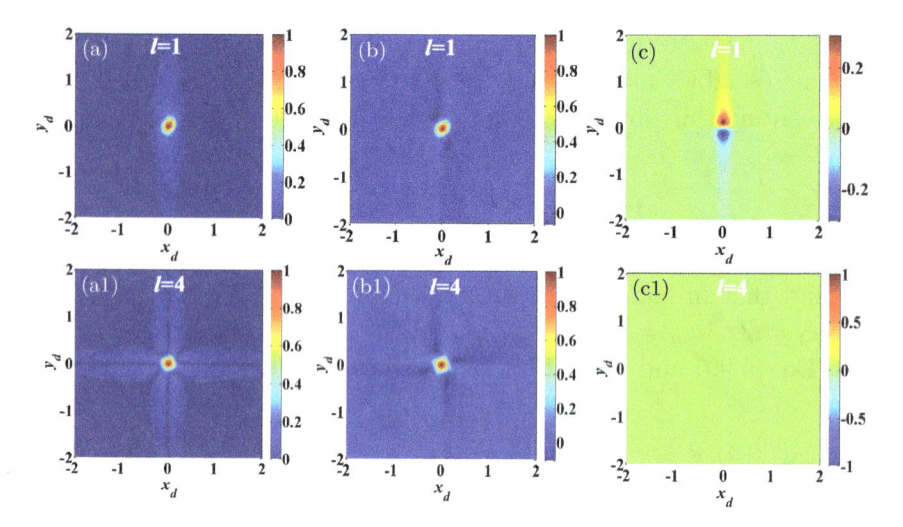

Fig. 5.8 The source degree of coherence for forming intensity vortices in the far field. Reprinted with permission from Wang and Korotkova (2016a). © The Optical Society.

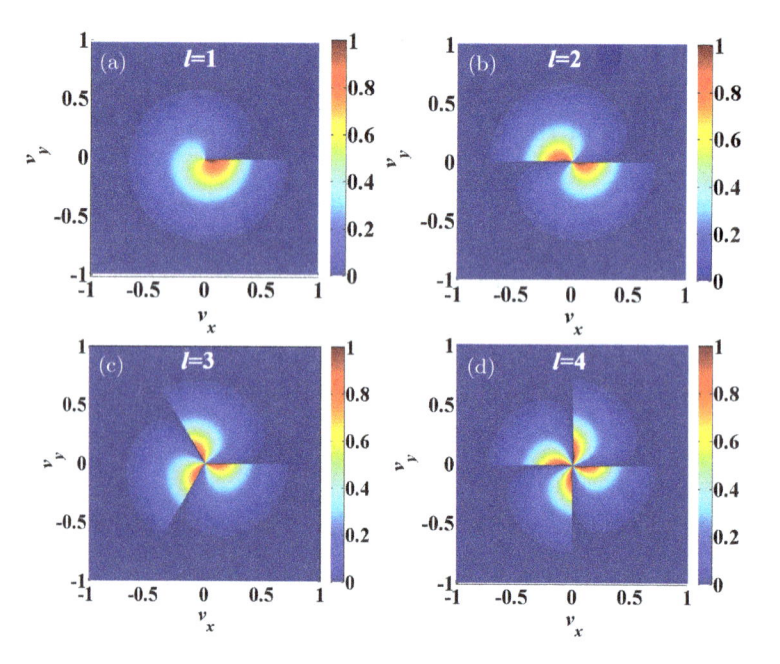

Fig. 5.9 Intensity vortices in the far field. Reprinted with permission from Wang and Korotkova (2016a). © The Optical Society.

corresponding source degree of coherence is complex:

$$\mu^{(0)}(\rho_d, \theta_d) = L_n^0 \left(-\frac{\rho_d^2}{2\delta^2}\right) \exp\left(-\frac{\rho_d^2}{2\delta^2}\right)$$

$$- \frac{i\sqrt{2}(n-1/2)!\rho_d}{2n!\delta} L_{n-1/2}^1 \left(-\frac{\rho_d^2}{2\delta^2}\right)$$

$$\times \exp\left(-\frac{\rho_d^2}{2\delta^2}\right) \cos(\theta_d), \quad (5.96)$$

where m_{n0} is the generalized Laguerre polynomial. The imaginary part of the degree of coherence is non-trivial for all integer values of n.

5.2.5 Examples: Twisted correlations

An interesting possibility for source coherence structuring is presented by the twisted correlation class given in Eq. (5.53). Originally, Twisted Gaussian Schell-Model (TGSM) sources and beams were introduced in Simon and Mukunda (1993) but were intrinsically constrained to the Gaussian spectral density and Gaussian degree of coherence. However, the Bochner theorem approach can be efficiently used for modeling of a variety of other twisted optical fields. Indeed, on choosing in Eq. (5.53) the Gaussian profile for amplitude function $a(\rho)$, we obtain for the modes the expression

$$H_0(\boldsymbol{\rho}, \mathbf{v}) = \exp\left[-\frac{\rho^2}{4\sigma^2}\right] \exp\left[-[(a_xy + ix)v_x - (a_y - iy)v_y]\right], \quad (5.97)$$

where σ, a_x and a_y are positive real constants. Also, on setting the profile function in an elliptic Gaussian form,

$$p(\mathbf{v}) = \frac{\sqrt{b_xb_y}}{\pi} \exp\left[-b_xv_x^2 - b_yv_y^2\right], \quad (5.98)$$

and substituting modes (5.97) and profile function (5.98) into the two-dimensional Bochner integral in Eq. (5.48), we find that

$$W^{(0)}(\boldsymbol{\rho}_1, \boldsymbol{\rho}_2) = \exp\left(-\frac{x_1^2 + x_2^2}{4\sigma_x^2}\right) \exp\left(-\frac{y_1^2 + y_2^2}{4\sigma_y^2}\right)$$

$$\times \exp[-i\theta_t(x_1y_2 - x_2y_1)] \exp\left[-\frac{(x_1 - x_2)^2}{2\delta_x^2}\right]$$

$$\times \exp\left[-\frac{(y_1 - y_2)^2}{2\delta_y^2}\right]. \quad (5.99)$$

Here, parameters are found to be

$$\frac{1}{4\sigma_x^2} = \frac{1}{4\sigma^2} - \frac{b_y \theta_t^2}{2}, \quad \frac{1}{4\sigma_y^2} = \frac{1}{4\sigma^2} - \frac{b_x \theta_t^2}{2},$$
$$\frac{1}{2\delta_x^2} = \frac{1}{4b_x^2} + \frac{b_y \theta_t^2}{4}, \quad \frac{1}{2\delta_y^2} = \frac{1}{4b_y^2} + \frac{b_x \theta_t^2}{4}, \quad (5.100)$$

and

$$\theta_t = a_x/b_x = a_y/b_y, \quad (5.101)$$

is the *twist factor* that can be shown to satisfy inequalities

$$\theta_t \delta_x^2 \leq 1, \quad \theta_t \delta_y^2 \leq 1. \quad (5.102)$$

This implies that parameters b_x, b_y and θ_t are constrained as

$$\theta_t \geq \frac{1}{b_y} + \sqrt{\frac{b_x - b_y}{b_x b_y^2}}, \quad \text{or} \quad \theta_t \leq \frac{1}{b_y} - \sqrt{\frac{b_x - b_y}{b_x b_y^2}}, \quad \text{if } b_x > b_y,$$

$$\theta_t \geq \frac{1}{b_x} + \sqrt{\frac{b_y - b_x}{b_x^2 b_y}}, \quad \text{or} \quad \theta_t \leq \frac{1}{b_x} - \sqrt{\frac{b_y - b_x}{b_x^2 b_y}}, \quad \text{if } b_y > b_x.$$

(5.103)

In the particular case when $\delta_x = \delta_y = \delta$ and $\sigma_x = \sigma_y = \sigma$, the anisotropic TGSM source reduces to the classic isotropic TGSM source of Simon and Mukunda (1993). The interesting outcome of the twist phase is in its ability to rotate the spectral density distribution through an angle of $\pi/2$ as the beam propagates from the source plane to the far zone. This can be established by evaluating the propagating modes H_z and using them in Eq. (5.57) for the propagating spectral density. This will not be visually revealed in the isotropic TGSM source but will manifest in the rotation of the elliptic TGSM spectral density.

One can also furnish the two-dimensional multi-Gaussian Schell-model (TMGSM) source with the twist phase. To do so, it suffices to set

$$p(\mathbf{v}) = \frac{1}{\sqrt{\pi}} \prod_{j=x,y} \frac{\sqrt{b_j}}{C_j} \sum_{m_j=1}^{M_j} \binom{M_j}{m_j} (-1)^{m_j-1} \exp[-b_j m_j v_j^2] \quad (5.104)$$

with normaliztion factors

$$C_j = \sum_{m_j=1}^{M_j} \binom{M_j}{m_j} \frac{(-1)^{m_j-1}}{\sqrt{m_j}} \exp[-b_j m_j v_j^2], \quad (j=x,y). \quad (5.105)$$

This leads to the cross-spectral density

$$W^{(0)}(\boldsymbol{\rho}_1, \boldsymbol{\rho}_2) = \frac{1}{C_x} \sum_{m_x=1}^{M_x} \binom{M_x}{m_x} \frac{(-1)^{m_x-1}}{\sqrt{m_x}}$$

$$\times \exp\left[-\left(\frac{1}{4\sigma^2} - \frac{b_x \theta_t^2}{2m_x}\right)(y_1^2 + y_2^2)\right]$$

$$\times \exp\left[-\frac{(x_1-x_2)^2}{4m_x b_x} - \frac{b_x \theta_t^2 (y_1-y_2)^2}{4m_x}\right.$$

$$\left. - \frac{i\theta_t (x_1-x_2)(y_1+y_2)}{2m_x}\right] \frac{1}{C_y} \sum_{m_y=1}^{M_y} \binom{M_y}{m_y} \frac{(-1)^{m_y-1}}{\sqrt{m_y}}$$

$$\times \exp\left[-\left(\frac{1}{4\sigma^2} - \frac{b_y \theta_t^2}{2m_y}\right)(x_1^2 + x_2^2)\right]$$

$$\times \exp\left[-\frac{(y_1-y_2)^2}{4m_y b_y} - \frac{b_y \theta_t^2 (x_1-x_2)^2}{4m_y}\right.$$

$$\left. - \frac{i\theta_t(x_1+x_2)(y_2-y_1)}{2m_y}\right]. \quad (5.106)$$

Figure 5.10 illustrates rotation of the spectral density of a beam generated by the source specified in Eq. (5.106) about the optical axis. The actual evolution of the TMGSM beam is more complex than that of the anisotropic TGSM: the spectral density simultaneously converts from initial Gaussian to the smooth-edged square shape and twists through angle $\pi/2$.

Twisted beams are known to carry the OAM. For instance, for the classic isotropic TGSM beam, the normalized OAM flux density $l_d^{(\text{orb})}$ defined in Eq. (2.90) and the total OAM per photon $l_{\text{tot}}^{(\text{orb})}$ defined in Eq. (2.91) are

$$l_d^{(\text{orb})}(\boldsymbol{\rho}) = \hbar \theta_t \rho^2, \quad l_{\text{tot}}^{(\text{orb})} = \frac{\sigma^2 \theta_t}{kc}. \quad (5.107)$$

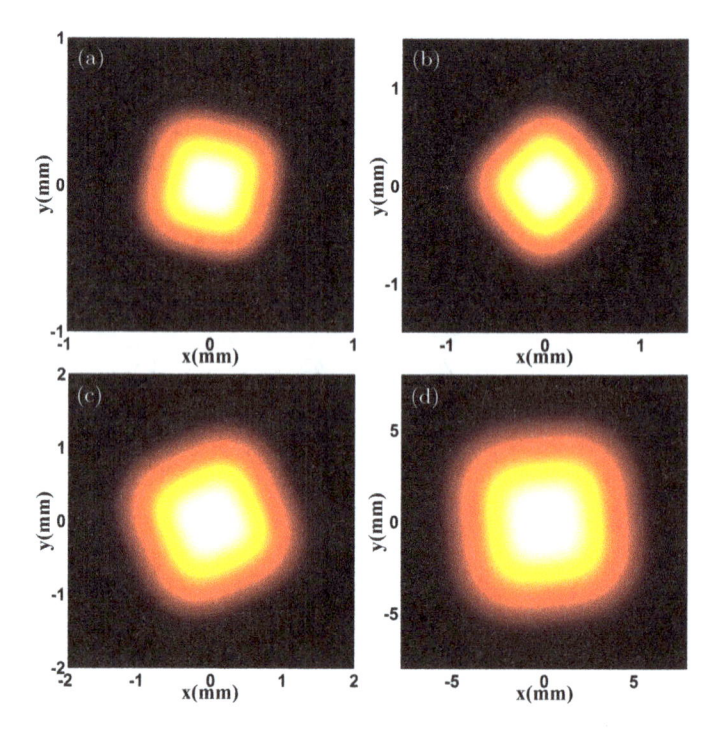

Fig. 5.10 Rotation of the TMGSM beam on propagation at increasing distances z from the source plane.

5.2.6 Examples: Separable phases

An entirely different possibility arises if function $a(\boldsymbol{\rho})$ in the description of the correlation class H_0 becomes complex-valued. For example, in the two-dimensional beam case, instead of a Gaussian mode we may set a zero-order Laguerre–Gaussian mode

$$a(\boldsymbol{\rho}) = C_l \rho^{|l|} \exp\left[-\frac{\rho^2}{\sigma^2}\right] \exp[il\theta], \tag{5.108}$$

where index l represents the *topological charge*, and

$$C_l = \sqrt{\frac{2}{\pi\sigma^2 |l|!}} \left(\frac{\sqrt{2}}{\sigma}\right)^{|l|}. \tag{5.109}$$

Substituting from Eq. (5.108) into the uniform correlation class H_0 given by Eq. (5.49) and then into Eq. (5.48) yields (Mei and

Korotkova, 2018)

$$W_0(\boldsymbol{\rho}_1,\boldsymbol{\rho}_2) = \left(\frac{\rho_1\rho_2}{\sigma^2}\right)^l \exp\left[-\frac{\rho_1^2+\rho_2^2}{\sigma^2}\right]\mu(\boldsymbol{\rho}_2-\boldsymbol{\rho}_1)\exp[il(\theta_1-\theta_2)]. \tag{5.110}$$

Depending on the choice of μ, the optical beam can be shaped in a variety of ways. In particular, if $\mu = \mu(\boldsymbol{\rho}_2-\boldsymbol{\rho}_1)$, i.e., phase-independent, then the source has *separable phase*. The sources with separable phases have the OAM flux density and the total OAM per photon as follows:

$$l_d^{(\text{orb})}(\boldsymbol{\rho}) = \hbar l, \quad l_{\text{tot}}^{(\text{orb})} = \hbar l. \tag{5.111}$$

We note that the source correlation-induced OAM changes that we have discussed in Chapter 4 use the incoherent superposition of several Laguerre–Gaussian beams with different values of l and different profiles of the source degrees of coherence. In such a case, the OAM flux density does not necessarily remain constant as the beam propagates.

5.2.7 *Sliding function method*

Just like in the one-dimensional case, we begin from condition $p(\mathbf{v}) \geq 0$ that must be valid for any $\mathbf{v}$ and write it as the product

$$p(\mathbf{v}) = \sqrt{p(\mathbf{v})}\sqrt{p(\mathbf{v})} = h_p(\mathbf{v})h_p(\mathbf{v}), \tag{5.112}$$

for some uniquely defined, real-valued function $h_p(\mathbf{v}) = \sqrt{p(\mathbf{v})}$, such that $\iint_{-\infty}^{\infty}|h_p(\mathbf{v})|d^2v < \infty$, i.e., belonging to the $L^1(\mathbb{R}^2)$ space. Then by the convolution theorem $\mu(\mathbf{r}_d)$ can be expressed as

$$\mu(\mathbf{r}_d) = FT^{-1}[h_p(\mathbf{v})] \circledast FT^{-1}[h_p(\mathbf{v})] = g(\mathbf{r}_d) \circledast g(\mathbf{r}_d), \tag{5.113}$$

where $\circledast$ stands for two-dimensional convolution and $g(\mathbf{r}_d) = FT^{-1}[h_p(\mathbf{v})] = FT^{-1}[\sqrt{p(\mathbf{v})}]$. We also assume that $g(\mathbf{r}_d)$ belongs to the $L^1(\mathbb{R}^2)$ space, i.e., $\iint_{-\infty}^{\infty}|g(\mathbf{r}_d)|d^2r_d < \infty$.

The conditions set on function $g(\mathbf{r}_d)$ that are sufficient for obtaining the real-valued function $h_p(\mathbf{v})$ can be derived as follows. Representing $h_p(\mathbf{v})$ as the two-dimensional Fourier transform of the real and imaginary parts of $g(\mathbf{r}_d)$, say, $g_R(\mathbf{r}_d)$ and $g_I(\mathbf{r}_d)$, yields

$$h_p(v_x, v_y) = \int_{-\infty}^{\infty}\int_{-\infty}^{\infty} g(x_d, y_d)\exp[2\pi i(x_d v_x + y_d v_y)]dx_d dy_d$$

$$= \int_{-\infty}^{\infty}\int_{-\infty}^{\infty}[g_R(x_d, y_d) + ig_I(x_d, y_d)]$$

$$\times [\cos[2\pi(x_d v_x + y_d v_y)] + i\sin[2\pi(x_d v_x + y_d v_y)]]dx_d dy_d. \tag{5.114}$$

This implies that the imaginary part of $h_p(v_x, v_y)$ has the form

$$Im[h_p(v_x, v_y)] = \int_{-\infty}^{\infty}\int_{-\infty}^{\infty}[g_R(x_d, y_d)\sin[2\pi(x_d v_x + y_d v_y)]$$

$$+ g_I(x_d, y_d)\cos[2\pi(x_d v_x + y_d v_y)]]dx_d dy_d. \tag{5.115}$$

Using the summation formulas for sine and cosine, we get

$$Im[h_p(v_x, v_y)] = \int_{-\infty}^{\infty}\int_{-\infty}^{\infty} g_R(x_d, y_d)[\sin(2\pi x_d v_x)\cos(2\pi y_d v_y)$$

$$+ \cos(2\pi x_d v_x)\sin(2\pi y_d v_y)]dx_d dy_d$$

$$+ \int_{-\infty}^{\infty}\int_{-\infty}^{\infty} g_I(x_d, y_d)[\cos(2\pi x_d v_x)\cos(2\pi y_d v_y)$$

$$- \sin(2\pi x_d v_x)\sin(2\pi y_d v_y)]dx_d dy_d. \tag{5.116}$$

Thus, in order to have trivial $Im[h_p(v_x, v_y)]$, $g_R(x_d, y_d)$ must be either even with respect to x_d and y_d, simultaneously, or odd with respect to both of these variables, simultaneously. Either of these conditions will make the first integral in Eq. (5.116) vanish. Alternatively, this is equivalent to condition $g_R(-x_d, -y_d) = g_R(x_d, y_d)$. At the same time, $g_I(x_d, y_d)$ must be either even with respect to x_d and odd with respect to y_d or even with respect to y_d and odd with respect to x_d. Either of these conditions will make the second integral in Eq. (5.116) vanish. This leads to relation $g_I(-x_d, -y_d) = -g_I(x_d, y_d)$.

Thus, the complex degree of coherence of a two-dimensional, scalar, Schell-model source may be expressed as an auto-convolution of a two-dimensional sliding function whose real and imaginary parts are proper combinations of even and odd functions, viz.,

$$g_R(-x_d, -y_d) = g_R(x_d, y_d), \quad g_I(-x_d, -y_d) = -g_I(x_d, y_d). \quad (5.117)$$

Hence, $g(x_d, y_d)$ must be Hermitian in two dimensions, i.e.,

$$g^*(-x_d, -y_d) = g(x_d, y_d). \quad (5.118)$$

Alternatively, if the sliding function is expressed by its magnitude $g_M(x_d, y_d)$ and phase $g_P(x_d, y_d)$:

$$g(x_d, y_d) = g_M(x_d, y_d) \exp[i g_P(x_d, y_d)], \quad (5.119)$$

then (5.118) implies that

$$g_M(-x_d, -y_d) = g_M(x_d, y_d), \quad g_P(-x_d, -y_d) = -g_P(x_d, y_d), \quad (5.120)$$

i.e., the magnitude and the phase of $g(x_d, y_d)$ must be an even and an odd function, respectively.

5.2.8 *Example*

Just like in the one-dimensional case, to focus on the phase effects of the degree of coherence, we set g_M as a Gaussian function:

$$g_M(x_d, y_d) = \frac{1}{\delta \sqrt{\pi}} \exp\left(-\frac{x_d^2 + y_d^2}{2\delta^2}\right). \quad (5.121)$$

Let us now choose the cubic phase of the sliding function, leading to the two-dimensinal Airy-like, far-field spectral density distributions. In two dimensions, the phase of $g(x_d, y_d)$ may be separable or non-separable:

$$g_P(x_d, y_d) = a[x_d^3 + y_d^3], \quad (5.122)$$

$$g_P(x_d, y_d) = a[x_d^2 y_d + y_d^2 x_d]. \quad (5.123)$$

Figures 5.11(a), 5.11(c) and 5.11(e) show the example of the separable-phase case in Eq. (5.122), and Figs. 5.11(b), 5.11(d) and 5.11(f) correspond to the non-separable-phase case in Eq. (5.123). The spectral density produced with the separable phase is just the two-dimensional version of the Airy-like beam. The non-separable phase leads to a more complex spectral density profile.

148 *Theoretical Statistical Optics*

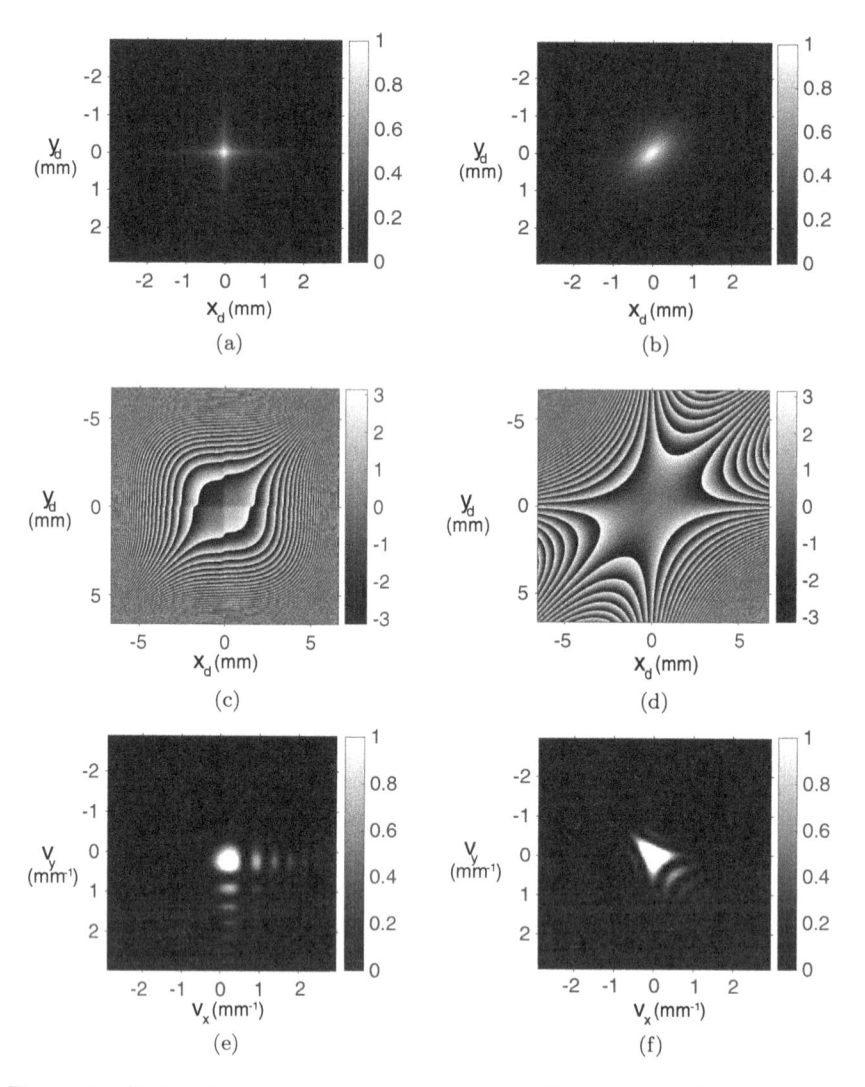

Fig. 5.11 Sliding function with cubic phase in 2D: separable (left) and non-separable (right). Source degree of coherence: (a) and (b) magnitude; (c) and (d) phase; (e) and (f) far-field spectral density. Reprinted with permission from Chen and Korotkova (2019). © The Optical Society.

5.3 Other Methods, Models, Statistics

We have described two methods for structuring of the source degree of coherence to be used for beam shaping applications. The first method is based on the Bochner theorem and is useful for construction of the source coherence states suitable for radiation with the required far-field spectral densities. The second method is based on self-convolution of a "sliding" function, being helpful for modeling of entirely novel beams "from the source plane". The two methods do not exhaust all the possibilities. For example, one can generate new source coherence states and, hence, new beam families, by performing some operations on already known states. So far, some legitimate linear combinations (Gori and Santarsiero, 2014; Mei and Korotkova, 2015; Santarsiero et al., 2014), products, powers (Mei et al., 2014),

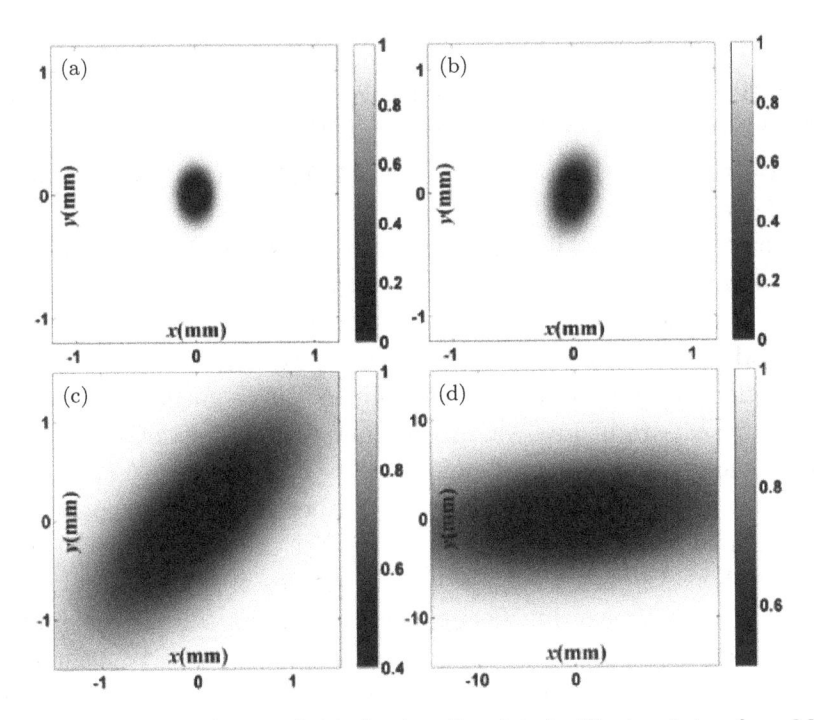

Fig. 5.12 Rotating degree of polarization. Reprinted with permission from Mei and Korotkova (2018). © The Optical Society.

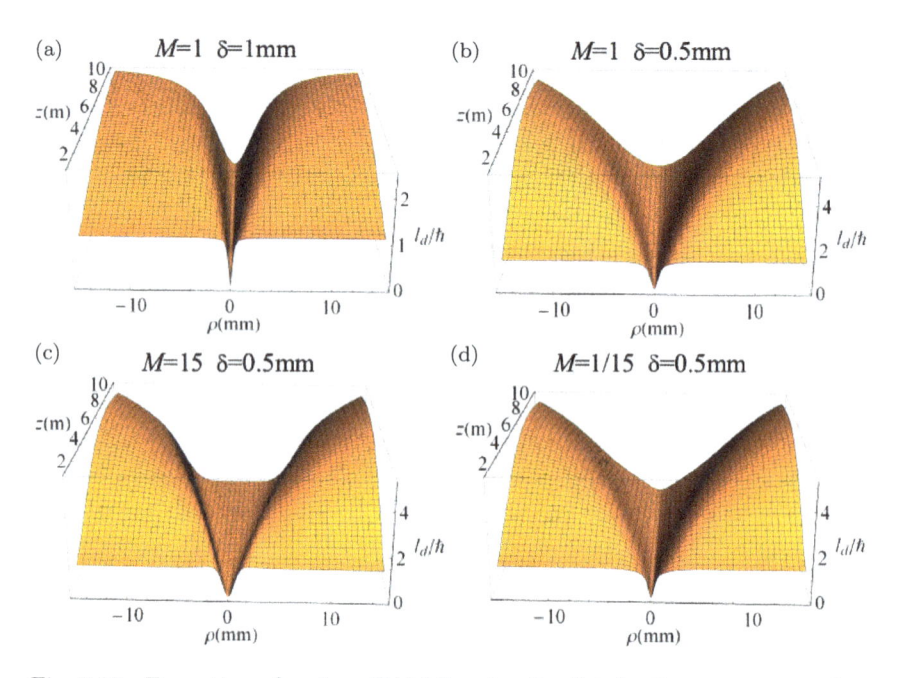

Fig. 5.13 Formation of various OAM flux density distributions on propagation. Reprinted with permission from Zhang *et al.* (2020b)

and convolutions (Korotkova and Mei, 2015) were shown to achieve these goals.

Although we have illustrated a variety of source models, an enormous number of them could not be overviewed. Excellent examples of this are one- and two-dimensional beam arrays with different symmetries (Ma and Ponomarenko, 2014; Mei *et al.*, 2015; Wan and Zhao, 2018) that can be produced from a single source. Such multi-replica distributions are of interest in communications, particle manupilation, imaging and other technologies.

In addition to the spectral density, other statistical properties of optical fields can be spatially structured with the help of source correlations. For example, the spatial distributions of polarimetric properties of the propagating beams can be controlled from the source (Mei *et al.*, 2013; Tong and Korotkova, 2012). Figure 5.12 illustrates the possibility of rotating the degree of polarization of an electromagnetic twisted Gaussian Schell-model beam on propagation (Mei and Korotkova, 2018). Another example is the profiling of the OAM density flux of a beam generated by a stationary source. Figure 5.13 shows the OAM flux density of the Gaussian Schell-model source with different source coherence widths, (A) and (B), controlling the size of the vortex, as well as of the rotationally symmetric multi-Gaussian Schell-model source with (C) $M > 1$ leading to the flat area in the vortex center, and with (D) $M < 1$ producing a cone-like vortex (Zhang et al., 2020b).

Chapter 6

Light Interaction with Devices of Polarization Optics

As we established in Chapter 4, the polarization properties of stationary light beams can change on propagation in free space if they are produced by electromagnetic sources with self- and cross-correlations being different for two electric field components. A more practical way to change polarization locally (point by point) is by passing the beam through a device of polarization optics, or their combination. These devices can be of deterministic or random nature. Polarizers, absorbers, rotators and compensators are four basic devices that change the state of light uniformly across its cross-section. Some devices, such as a depolarizer can modulate the polarization state deterministically but differently at different points. Other devices can modulate polarization of the incident beam randomly with certain prescribed statistics. Two theories, Jones and Stokes–Mueller calculi, account for deterministic and random polarization changes, respectively. They are capable of solving the direct problem of characterizing the change in light polarization with a known device of polarization optics and the inverse problem of characterizing the device from the knowledge of the incident and the transmitted beams' polarization states.

6.1 Jones Calculus

We begin by outlining the basic approach for characterizing a local (point by point) transformation of a deterministic electromagnetic beam by a deterministic device of polarization optics (Jones, 1941). According to the Jones calculus, the space-dependent part of the electric field vector $\vec{\mathcal{E}}$ at position $\mathbf{r}$ within the beam and oscillating at frequency ω is represented as a two-dimensional vector (see Section 2.4.1 for row-vector characterization)

$$\begin{bmatrix} E_x(\mathbf{r},\omega) \\ E_y(\mathbf{r},\omega) \end{bmatrix} = \begin{bmatrix} a_x(\mathbf{r},\omega)\exp[i\varphi_x(\mathbf{r},\omega)] \\ a_y(\mathbf{r},\omega)\exp[i\varphi_y(\mathbf{r},\omega)] \end{bmatrix}. \tag{6.1}$$

Alternatively, we can write

$$\begin{bmatrix} E_x(\mathbf{r},\omega) \\ E_y(\mathbf{r},\omega) \end{bmatrix} = \exp[i\varphi_x(\mathbf{r},\omega)] \begin{bmatrix} a_x(\mathbf{r},\omega) \\ a_y(\mathbf{r},\omega)\exp[i\varphi(\mathbf{r},\omega)] \end{bmatrix}, \tag{6.2}$$

where $\varphi(\mathbf{r},\omega) = \varphi_y(\mathbf{r},\omega) - \varphi_x(\mathbf{r},\omega)$. Further simplification can then be made

$$\begin{bmatrix} E_x(\mathbf{r},\omega) \\ E_y(\mathbf{r},\omega) \end{bmatrix} = \begin{bmatrix} \cos\phi(\mathbf{r},\omega) \\ \sin\phi(\mathbf{r},\omega)\exp[i\varphi(\mathbf{r},\omega)] \end{bmatrix}, \tag{6.3}$$

where $\exp[i\varphi_x(\mathbf{r},\omega)]$ is dropped and angle ϕ is defined as

$$\phi(\mathbf{r},\omega) = \arctan\left[a_y(\mathbf{r},\omega)/a_x(\mathbf{r},\omega)\right]. \tag{6.4}$$

In terms of Jones vectors, the important special cases of Eq. (6.3) are linear and circular polarization states

$$\begin{bmatrix} E_x(\mathbf{r},\omega) \\ E_y(\mathbf{r},\omega) \end{bmatrix}_{(L)} = \begin{bmatrix} \cos\phi(\mathbf{r},\omega) \\ \sin\phi(\mathbf{r},\omega) \end{bmatrix}, \quad \begin{bmatrix} E_x(\mathbf{r},\omega) \\ E_y(\mathbf{r},\omega) \end{bmatrix}_{(C)} = \frac{1}{\sqrt{2}} \begin{bmatrix} 1 \\ \pm i \end{bmatrix}, \tag{6.5}$$

where positive and negative signs stand for left-hand and right-hand circular polarizations.

The most general elliptical polarization states are described by vectors

$$\begin{bmatrix} E_x(\mathbf{r},\omega) \\ E_y(\mathbf{r},\omega) \end{bmatrix}_{(E)} = \frac{1}{\sqrt{q_1(\mathbf{r},\omega)^2 + q_2(\mathbf{r},\omega)^2 + q_3(\mathbf{r},\omega)^2}}$$

$$\times \begin{bmatrix} q_1(\mathbf{r},\omega) \\ q_2(\mathbf{r},\omega) \pm iq_3(\mathbf{r},\omega) \end{bmatrix}, \tag{6.6}$$

where positive and negative signs stand for left-hand and right-hand elliptical states and

$$q_1(\mathbf{r},\omega) = a_x(\mathbf{r},\omega),$$
$$\sqrt{q_2(\mathbf{r},\omega)^2 + q_3(\mathbf{r},\omega)^2} = a_y(\mathbf{r},\omega), \quad (6.7)$$
$$\mp \tan^{-1}\left(\frac{q_3(\mathbf{r},\omega)}{q_2(\mathbf{r},\omega)}\right) = \phi(\mathbf{r},\omega).$$

When a beam of light characterized by Jones vector $\vec{E}^{(i)}$ is incident on a device and another beam of light with Jones vector $\vec{E}^{(t)}$ emerges from it (see Fig. 6.1), the linear transformation characterizing the interaction can be expressed as a 2×2 matrix, known as *Jones matrix*:

$$\vec{E}^{(t)}(\mathbf{r},\omega) = \overleftrightarrow{J}(\mathbf{r},\omega)\vec{E}^{(i)}(\mathbf{r},\omega), \quad (6.8)$$

or, more explicitly,

$$\begin{bmatrix} E_x^{(t)}(\mathbf{r},\omega) \\ E_y^{(t)}(\mathbf{r},\omega) \end{bmatrix} = \begin{bmatrix} J_{xx}(\mathbf{r},\omega) & J_{xy}(\mathbf{r},\omega) \\ J_{yx}(\mathbf{r},\omega) & J_{yy}(\mathbf{r},\omega) \end{bmatrix} \begin{bmatrix} E_x^{(i)}(\mathbf{r},\omega) \\ E_y^{(i)}(\mathbf{r},\omega) \end{bmatrix}. \quad (6.9)$$

There are four fundamental devices of polarization optics which introduce spatially uniform modulation on the incident light:

(I) *Rotators* induce the change in the axis of the optical field oscillation without changes in intensity. A rotation of the field

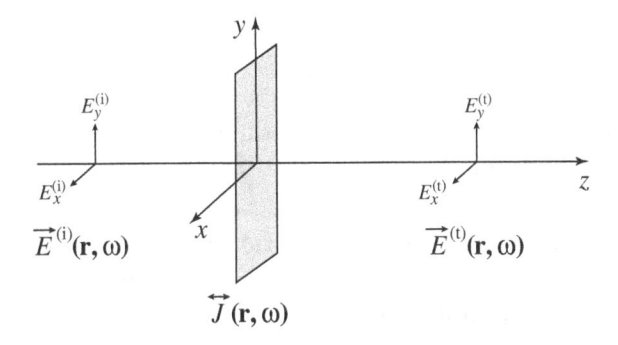

Fig. 6.1 Illustarting notation used in Jones matrix formulation.

through angle φ_r is represented by transformation

$$\overleftrightarrow{\mathcal{J}}_{(O)} = \begin{bmatrix} \cos\varphi_r & -\sin\varphi_r \\ \sin\varphi_r & \cos\varphi_r \end{bmatrix}. \tag{6.10}$$

(II) *Linear polarizers* filter out linearly polarized light. For example, a linear polarizer placed at angle φ_p with respect to the positive x-direction has Jones matrix

$$\overleftrightarrow{\mathcal{J}}_{(P)} = \begin{bmatrix} \cos^2\varphi_p & \cos\varphi_p\sin\varphi_p \\ \cos\varphi_p\sin\varphi_p & \sin^2\varphi_p \end{bmatrix}. \tag{6.11}$$

The effect of the polarizer is in changing the direction of oscillations in the incident electric field by angle φ_p and in reducing its intensity by a factor of $\cos^2(\varphi - \varphi_p)$.

(III) *Phase retarders* (phase plates) introduce a phase shift for one or both electric field components and are characterized by Jones matrix

$$\overleftrightarrow{\mathcal{J}}_{(R)} = \begin{bmatrix} \exp(i\varphi_{Rx}) & 0 \\ 0 & \exp(i\varphi_{Ry}) \end{bmatrix}. \tag{6.12}$$

Two important cases are Half-Wave Plates (HWPs) ($\varphi_R = \varphi_{Rx} - \varphi_{Ry} = \pm\pi$) and the Quarter-Wave Plates (QWPs) ($\varphi_R = \varphi_{Rx} - \varphi_{Ry} = \pm\pi/2$), the latter being useful for converting between linear and circular states.

(IV) *Absorbers* reduce the amplitudes of one or both components of the electric field vector and are given by diagonal Jones matrices

$$\overleftrightarrow{\mathcal{J}}_{(A)} = \begin{bmatrix} \exp(-\alpha_{Ax}) & 0 \\ 0 & \exp(-\alpha_{Ay}) \end{bmatrix}. \tag{6.13}$$

The four devices listed above modulate the polarization of the incident beam uniformly across the source. However, the Jones matrix calculus is also capable of characterizing devices producing modulation depending on spatial position **r**. An excellent example of this is the corner-cube retro-reflector made of a barefringent crystal. After a light beam enters the crystal, it undergoes two total internal reflections from the cube's facets but re-emerges along the same direction as that of the incident beam. The polarization modulation of the incident polarized beam depends on the two crystal's facets that it interacts with and on the order of interaction. Therefore, the polarization of the reflected beam is structured as six 60° sectors with uniform but different states (Liu and Azzam, 1997).

A cascaded system of N aligned devices characterized by Jones matrices $\overleftrightarrow{J}^{(n)}$, $n = 1, \ldots, N$ can be found as the product of the individual matrices,

$$\overleftrightarrow{J}^{(t)}(\mathbf{r}, \omega) = \prod_{n=1}^{N} \overleftrightarrow{J}^{(n)}(\mathbf{r}, \omega), \tag{6.14}$$

where matrix multiplication is performed from right to left.

6.2 Stokes–Mueller Calculus

6.2.1 *Mueller matrix*

The idea of characterization of a linear optical system, whether deterministic or random, in terms of its ability to modify the Stokes vector (Stokes, 1852) of a light beam incident on it belongs to (Mueller, 1948). The Mueller matrix is a 4×4 linear operator that relates the Stokes vectors $\vec{S}^{(i)}$ and $\vec{S}^{(t)}$ of the input and the output beams as

$$\vec{S}^{(t)}(\mathbf{r}, \omega) = \overleftrightarrow{M}(\mathbf{r}, \omega)\vec{S}^{(i)}(\mathbf{r}, \omega), \tag{6.15}$$

where $\overleftrightarrow{M} = \{m_{kl}\}$, $(k, l = 0, 1, 2, 3)$, or, more explicitly,

$$\begin{bmatrix} S_0^{(t)}(\mathbf{r}, \omega) \\ S_1^{(t)}(\mathbf{r}, \omega) \\ S_2^{(t)}(\mathbf{r}, \omega) \\ S_3^{(t)}(\mathbf{r}, \omega) \end{bmatrix} = \begin{bmatrix} m_{00}(\mathbf{r}, \omega) & m_{01}(\mathbf{r}, \omega) & m_{02}(\mathbf{r}, \omega) & m_{03}(\mathbf{r}, \omega) \\ m_{10}(\mathbf{r}, \omega) & m_{11}(\mathbf{r}, \omega) & m_{12}(\mathbf{r}, \omega) & m_{13}(\mathbf{r}, \omega) \\ m_{20}(\mathbf{r}, \omega) & m_{21}(\mathbf{r}, \omega) & m_{22}(\mathbf{r}, \omega) & m_{23}(\mathbf{r}, \omega) \\ m_{30}(\mathbf{r}, \omega) & m_{31}(\mathbf{r}, \omega) & m_{32}(\mathbf{r}, \omega) & m_{33}(\mathbf{r}, \omega) \end{bmatrix} \begin{bmatrix} S_0^{(i)}(\mathbf{r}, \omega) \\ S_1^{(i)}(\mathbf{r}, \omega) \\ S_2^{(i)}(\mathbf{r}, \omega) \\ S_3^{(i)}(\mathbf{r}, \omega) \end{bmatrix}. \tag{6.16}$$

Just like with a cascaded system of Jones matrices, Mueller matrix of the N aligned devices characterized by matrices $\overleftrightarrow{M}^{(n)}$, $n = 1, \ldots, N$

is given as the product of the individual matrices,

$$\overleftrightarrow{M}^{(t)}(\mathbf{r},\omega) = \prod_{n=1}^{N} \overleftrightarrow{M}^{(n)}(\mathbf{r},\omega), \tag{6.17}$$

where matrix multiplication is again performed from right to left.

6.2.2 Stokes parameters' determination

Practically, the measurement of the Stokes vector can be made by a system consisting of a QWP whose fast axis makes angle φ_R with the x axis and a polarizer aligned at angle φ_P with respect to the x axis. The well-known Stokes formula can then be directly used (Stokes, 1852):

$$S^{(m)}_{(\varphi_R,\varphi_P)}(\mathbf{r},\omega) = \frac{1}{2}\Big[S_0(\mathbf{r},\omega) + S_1(\mathbf{r},\omega)\cos 2\varphi_P$$
$$+ S_2(\mathbf{r},\omega)\cos\varphi_R\sin 2\varphi_P + S_3(\mathbf{r},\omega)\sin\varphi_R\sin 2\varphi_P\Big], \tag{6.18}$$

where $S^{(m)}$ is the measured spectral density of the beam for fixed values of φ_R and φ_P. Four measurements,

$$\begin{aligned} S^{(m)}_1(\mathbf{r},\omega) &= S_{(--,0°)}(\mathbf{r},\omega), \\ S^{(m)}_2(\mathbf{r},\omega) &= S_{(--,45°)}(\mathbf{r},\omega), \\ S^{(m)}_3(\mathbf{r},\omega) &= S_{(--,90°)}(\mathbf{r},\omega), \\ S^{(m)}_4(\mathbf{r},\omega) &= S_{(90°,45°)}(\mathbf{r},\omega), \end{aligned} \tag{6.19}$$

provide access to the linear combinations of the Stokes parameters,

$$\begin{aligned} S^{(m)}_1(\mathbf{r},\omega) &= \frac{1}{2}[S_0(\mathbf{r},\omega) + S_1(\mathbf{r},\omega)], \\ S^{(m)}_2(\mathbf{r},\omega) &= \frac{1}{2}[S_0(\mathbf{r},\omega) + S_2(\mathbf{r},\omega)], \\ S^{(m)}_3(\mathbf{r},\omega) &= \frac{1}{2}[S_0(\mathbf{r},\omega) - S_1(\mathbf{r},\omega)], \\ S^{(m)}_4(\mathbf{r},\omega) &= \frac{1}{2}[S_0(\mathbf{r},\omega) + S_3(\mathbf{r},\omega)]. \end{aligned} \tag{6.20}$$

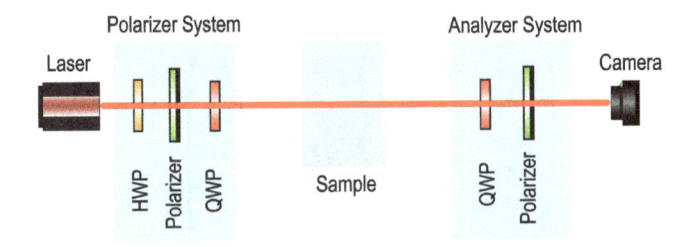

Light Interaction with Devices of Polarization Optics 159

Fig. 6.2 An optical system for measurment of the Mueller matrix elements.

On inverting, one arrives at the relations

$$\begin{aligned}
S_0(\mathbf{r},\omega) &= S_1^{(m)}(\mathbf{r},\omega) + S_3^{(m)}(\mathbf{r},\omega), \\
S_1(\mathbf{r},\omega) &= S_1^{(m)}(\mathbf{r},\omega) - S_3^{(m)}(\mathbf{r},\omega), \\
S_2(\mathbf{r},\omega) &= 2S_2^{(m)}(\mathbf{r},\omega) - S_1^{(m)}(\mathbf{r},\omega) - S_3^{(m)}(\mathbf{r},\omega), \\
S_3(\mathbf{r},\omega) &= 2S_4^{(m)}(\mathbf{r},\omega) - S_1^{(m)}(\mathbf{r},\omega) - S_3^{(m)}(\mathbf{r},\omega).
\end{aligned} \qquad (6.21)$$

6.2.3 *Mueller matrix determination*

In order to determine the Mueller matrix of a polarization device, an optical system or a slice of a random medium, one must synthesize four independent polarization states by the *polarizer system* placed before the studied sample and filter the output by the *analyzer system* placed after it (see Fig. 6.2). Let us consider a version of such a system that uses horizontally polarized illumination. Then the polarizer system consists of three aligned devices: HWP which controls the linear polarization direction of the laser beam (not needed for unpolarized laser source), polarizer P and QWP. The analyzer system consists of another QWP and a polarizer. The order of the devices should not be altered.

Mueller matrix $\overleftrightarrow{M}$ can then be obtained as a solution of a linear system with 16 unknowns, in a number of ways. One of the possibilities is to generate six independent states corresponding to the following Stokes vectors: (1) linear horizontal, (2) linear vertical,

(3) linear 45°, (4) linear −45°, (5) right circular, and (6) left circular:

$$\vec{S}_{in}^{(1)} = \begin{bmatrix} 1 \\ 1 \\ 0 \\ 0 \end{bmatrix}, \quad \vec{S}_{in}^{(2)} = \begin{bmatrix} 1 \\ -1 \\ 0 \\ 0 \end{bmatrix},$$

$$\vec{S}_{in}^{(3)} = \begin{bmatrix} 1 \\ 0 \\ 1 \\ 0 \end{bmatrix}, \quad \vec{S}_{in}^{(4)} = \begin{bmatrix} 1 \\ 0 \\ -1 \\ 0 \end{bmatrix}, \quad (6.22)$$

$$\vec{S}_{in}^{(5)} = \begin{bmatrix} 1 \\ 0 \\ 0 \\ 1 \end{bmatrix}, \quad \vec{S}_{in}^{(6)} = \begin{bmatrix} 1 \\ 0 \\ 0 \\ -1 \end{bmatrix},$$

and finding the solutions for the Mueller matrix elements using any 16 out of 24 relations from the system

$$\vec{S}_{out}^{(n)}(\mathbf{r},\omega) = \overleftrightarrow{M}(\mathbf{r},\omega)\vec{S}_{in}^{(n)}(\mathbf{r},\omega), \quad (n = 1, ..., 6). \quad (6.23)$$

Practically, the generation of the six input states can be made by aligning the polarization of the source with horizontal direction (set as the reference for the whole procedure) and controlling the orientation angles: HWP(φ_R^H), P(φ_P) and QWP(φ_R^Q) in the polarizer system,

$$\vec{S}_{in}^{(1)} = (0°, 0°, --), \quad \vec{S}_{in}^{(2)} = (45°, 90°, --),$$
$$\vec{S}_{in}^{(3)} = (22.5°, 45°, --), \quad \vec{S}_{in}^{(4)} = (-22.5°, -45°, --), \quad (6.24)$$
$$\vec{S}_{in}^{(5)} = (0°, 0°, 45°), \quad \vec{S}_{in}^{(6)} = (0°, 0°, -45°).$$

Here, "−" indicates that the device is removed for the particular step.

Filtering of the four sets of Stokes vectors $\vec{S}_{out}^{(n)}$ can then be made on the analyzer end by adjusting angles of QWP (φ_R) and P (φ_P). The Stokes formula (Stokes, 1852) can then be directly employed [see Eq. (6.18)]:

$$S_{\text{out}}^{(n)}(\mathbf{r},\omega;\varphi_R,\varphi_P) = \frac{1}{2}[S_0^{(n)}(\mathbf{r},\omega) + S_1^{(n)}(\mathbf{r},\omega)\cos 2\varphi_P$$
$$+ S_2^{(n)}(\mathbf{r},\omega)\cos\varphi_R \sin 2\varphi_P$$
$$+ S_3^{(n)}(\mathbf{r},\omega)\sin\varphi_R \sin 2\varphi_P], \quad (6.25)$$

where $S_{\text{out}}^{(n)}$ is the measured average intensity of the beam ($n = 1,\ldots,6$). From four sequential measurements $S_{\text{out},1}^{(n)} = S_{\text{out}}^{(n)}(--,0°)$, $S_{\text{out},2}^{(n)} = S_{\text{out}}^{(n)}(--,45°)$, $S_{\text{out},3}^{(n)} = S_{\text{out}}^{(n)}(--,90°)$, $S_{\text{out},4}^{(n)} = S_{\text{out}}^{(n)}(90°,45°)$, one can determine four linear combinations,

$$S_{\text{out},1}^{(n)}(\mathbf{r},\omega) = \frac{1}{2}[S_0^{(n)}(\mathbf{r},\omega) + S_1^{(n)}(\mathbf{r},\omega)],$$
$$S_{\text{out},2}^{(n)}(\mathbf{r},\omega) = \frac{1}{2}[S_0^{(n)}(\mathbf{r},\omega) + S_2^{(n)}(\mathbf{r},\omega)],$$
$$S_{\text{out},3}^{(n)}(\mathbf{r},\omega) = \frac{1}{2}[S_0^{(n)}(\mathbf{r},\omega) - S_1^{(n)}(\mathbf{r},\omega)],$$
$$S_{\text{out},4}^{(n)}(\mathbf{r},\omega) = \frac{1}{2}[S_0^{(n)}(\mathbf{r},\omega) + S_3^{(n)}(\mathbf{r},\omega)], \quad (n = 1,\ldots,6), \quad (6.26)$$

which, on solving for the Stokes parameters, return

$$S_0^{(n)}(\mathbf{r},\omega) = S_{\text{out},1}^{(n)}(\mathbf{r},\omega) + S_{\text{out},3}^{(n)}(\mathbf{r},\omega),$$
$$S_1^{(n)}(\mathbf{r},\omega) = S_{\text{out},1}^{(n)}(\mathbf{r},\omega) - S_{\text{out},3}^{(n)}(\mathbf{r},\omega),$$
$$S_2^{(n)}(\mathbf{r},\omega) = 2S_{\text{out},2}^{(n)}(\mathbf{r},\omega) - S_{\text{out},1}^{(n)}(\mathbf{r},\omega) - S_{\text{out},3}^{(n)}(\mathbf{r},\omega), \quad (6.27)$$
$$S_3^{(n)}(\mathbf{r},\omega) = 2S_{\text{out},4}^{(n)}(\mathbf{r},\omega) - S_{\text{out},1}^{(n)}(\mathbf{r},\omega) - S_{\text{out},3}^{(n)}(\mathbf{r},\omega),$$
$$(n = 1,\ldots,6).$$

Next, using Eqs. (6.22) for the Stokes vectors prepared by the polarizer system and Eqs. (6.27) for those filtered by the analyzer system in Eq. (6.23), and solving the linear system, delivers the Mueller matrix elements. In particular, the first two columns of the matrix can be obtained from initial states $n = 1$ and $n = 2$ (horizontal and vertical polarizations):

$$\begin{aligned}
m_{00} &= \frac{1}{2}[S_0^{(1)} + S_0^{(2)}], & m_{01} &= \frac{1}{2}[S_0^{(1)} - S_0^{(2)}], \\
m_{10} &= \frac{1}{2}[S_1^{(1)} + S_1^{(2)}], & m_{11} &= \frac{1}{2}[S_1^{(1)} - S_1^{(2)}], \\
m_{20} &= \frac{1}{2}[S_2^{(1)} + S_2^{(2)}], & m_{21} &= \frac{1}{2}[S_2^{(1)} - S_2^{(2)}], \\
m_{30} &= \frac{1}{2}[S_3^{(1)} + S_3^{(2)}], & m_{31} &= \frac{1}{2}[S_3^{(1)} - S_3^{(2)}].
\end{aligned} \quad (6.28)$$

The third column can be obtained from states $n = 3$ and $n = 4$ (the $\pm 45°$ polarizations); and the fourth column can be found from states $n = 5$ and $n = 6$ (the circular polarizations):

$$\begin{aligned}
m_{02} &= \frac{1}{2}[S_0^{(3)} - S_0^{(4)}], & m_{03} &= \frac{1}{2}[S_0^{(5)} - S_0^{(6)}], \\
m_{12} &= \frac{1}{2}[S_1^{(3)} - S_1^{(4)}], & m_{13} &= \frac{1}{2}[S_1^{(5)} - S_1^{(6)}], \\
m_{22} &= \frac{1}{2}[S_2^{(3)} - S_2^{(4)}], & m_{23} &= \frac{1}{2}[S_2^{(5)} - S_2^{(6)}], \\
m_{32} &= \frac{1}{2}[S_3^{(3)} - S_3^{(4)}], & m_{33} &= \frac{1}{2}[S_3^{(5)} - S_3^{(6)}].
\end{aligned} \quad (6.29)$$

Since this system of equations is overdetermined, there are other possible (but consistent) ways of expressing the Mueller matrix elements. We reiterate that unlike the Jones matrices, the Mueller matrices are capable of characterizing polarization changes of deterministic and random devices, systems and media, such as particulate dense collections suspended in a liquid or biological tissues, producing both deterministic and random effects on an incident light beam. In Fig. 6.3, the Mueller matrix of the Young interference experiment performed with laser light by interferometric method with a calcite crystal illustrates the deterministic system case (Arteaga et al., 2017) and in Fig. 6.4 the Mueller matrix of a cancerous biological cell suspension demonstrates the random medium case (Hielscher et al., 1997).

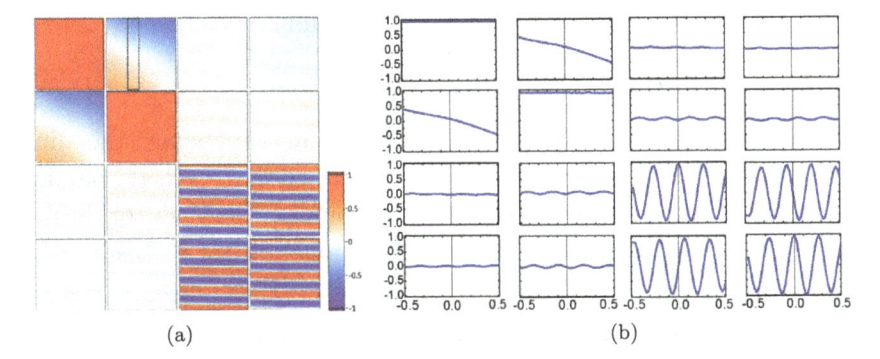

Light Interaction with Devices of Polarization Optics 163

Fig. 6.3 Mueller matrix of the Young interference experiment: images (a) and cross-sections taken from the rectangular area marked on element m_{01} (b). Reprinted with permission from Arteaga *et al.* (2017). © The Optical Society.

Fig. 6.4 Diffuse-backscattering Mueller matrix of cancerous biological cells. Reprinted with permission from Hielscher *et al.* (1997). © The Optical Society.

6.3 Two-point Stokes–Mueller Calculus

6.3.1 *Two-point Mueller matrix*

The conventional (single-point) Mueller matrix being the linear transformation of classic (single-point) Stokes vectors can be extended to that for generalized (two-point) Stokes vector introduced in Eq. (2.138). Let the incident generalized Stokes vector $\vec{S}_l^{(i)}(\mathbf{r}_1, \mathbf{r}_2, \omega)$ be transformed into the generalized Stokes vector $\vec{S}_l^{(t)}(\mathbf{r}_1, \mathbf{r}_2, \omega)$ as (Korotkova and Wolf, 2005c,d).

$$\vec{S}_l^{(t)}(\mathbf{r}_1, \mathbf{r}_2, \omega) = \overleftrightarrow{M}(\mathbf{r}_1, \mathbf{r}_2, \omega) \vec{S}_l^{(i)}(\mathbf{r}_1, \mathbf{r}_2, \omega), \quad (l = 0, 1, 2, 3), \tag{6.30}$$

or, more explicitly (omitting the frequency dependence),

$$\begin{bmatrix} S_0^{(t)}(\mathbf{r}_1, \mathbf{r}_2) \\ S_1^{(t)}(\mathbf{r}_1, \mathbf{r}_2) \\ S_2^{(t)}(\mathbf{r}_1, \mathbf{r}_2) \\ S_3^{(t)}(\mathbf{r}_1, \mathbf{r}_2) \end{bmatrix} = \begin{bmatrix} m_{00}(\mathbf{r}_1, \mathbf{r}_2) & m_{01}(\mathbf{r}_1, \mathbf{r}_2) & m_{02}(\mathbf{r}_1, \mathbf{r}_2) & m_{03}(\mathbf{r}_1, \mathbf{r}_2) \\ m_{10}(\mathbf{r}_1, \mathbf{r}_2) & m_{11}(\mathbf{r}_1, \mathbf{r}_2) & m_{12}(\mathbf{r}_1, \mathbf{r}_2) & m_{13}(\mathbf{r}_1, \mathbf{r}_2) \\ m_{20}(\mathbf{r}_1, \mathbf{r}_2) & m_{21}(\mathbf{r}_1, \mathbf{r}_2) & m_{31}(\mathbf{r}_1, \mathbf{r}_2) & m_{41}(\mathbf{r}_1, \mathbf{r}_2) \\ m_{40}(\mathbf{r}_1, \mathbf{r}_2) & m_{41}(\mathbf{r}_1, \mathbf{r}_2) & m_{42}(\mathbf{r}_1, \mathbf{r}_2) & m_{43}(\mathbf{r}_1, \mathbf{r}_2) \end{bmatrix}$$

$$\times \begin{bmatrix} S_0^{(i)}(\mathbf{r}_1, \mathbf{r}_2) \\ S_1^{(i)}(\mathbf{r}_1, \mathbf{r}_2) \\ S_2^{(i)}(\mathbf{r}_1, \mathbf{r}_2) \\ S_3^{(i)}(\mathbf{r}_1, \mathbf{r}_2) \end{bmatrix}. \tag{6.31}$$

For the cascaded systems of optical elements, the same product rule is applied as for classic Mueller matrices

$$\overleftrightarrow{M}^{(t)}(\mathbf{r}_1, \mathbf{r}_2, \omega) = \prod_{n=1}^{N} \overleftrightarrow{M}^{(n)}(\mathbf{r}_1, \mathbf{r}_2, \omega). \tag{6.32}$$

The generalized Mueller matrices carry information about the ability of optical systems to correlate the two electric field components at a pair of positions in space. This infomation becomes of importance if the optical field has to be further analyzed on propagation or interaction with other optical devices or media. One-point Mueller matrices (and one-point Stokes parameters) do not contain this information. In fact, it was shown in Salem *et al.* (2006) that two optical beams with the same set of classic Stokes vectors but different generalized Stokes vectors would evolve into two beams with different Stokes vectors. The two-point Stokes–Mueller calculus has been recently employed for characterization of spectral properties of spatially depolarized superluminescent diodes (Partanen *et al.*, 2019), nematic liquid crystals (Yakovlev and Yakovlev, 2019) and birefringent biological tissues (Peyvasteh *et al.*, 2020).

6.3.2 Analytic example: Spatial light modulator

The generalized Stokes parameters became particularly useful in characterization of the effects of the nematic liquid crystal Spatial Light Modulators (SLMs) on optical beams. An SLM is a large two-dimensional array of pixels each introducing a digitally controllable phase delay on the incoming beam. Hence, by assigning a two-dimensional phase distribution with a prescribed two-point correlation function, and then cycling through the ensemble of realizations, one may generate a variety of random beams with finely controllable statistics. At each pixel, the SLM affects only one component of the electric field (or it is made to do so by placing a polarizer in front of it). For example, a realization of a Jones matrix of an SLM affecting the y field component has the form

$$\overleftrightarrow{J}^{(\varphi)}(\mathbf{r},\omega) = \begin{bmatrix} 1 & 0 \\ 0 & \exp[i\varphi_S(\mathbf{r},\omega)] \end{bmatrix}, \tag{6.33}$$

where φ_S is a phase realization of a random process with zero average value,

$$\langle \varphi_S(\mathbf{r},\omega) \rangle_\varphi = 0, \tag{6.34}$$

and a given two-point correlation function $C^{(\varphi)}$,

$$\langle \exp[i(\varphi_S(\mathbf{r}_2,\omega) - \varphi_S(\mathbf{r}_1,\omega))] \rangle_\varphi = C^{(\varphi)}(\mathbf{r}_1,\mathbf{r}_2). \tag{6.35}$$

Since in this case the Jones matrix is not deterministic, but rather represents a member of a statistical ensemble, we first find, on taking correlations, that the transformation law for the cross-spectral density has the form

$$\overleftrightarrow{W}^{(t)}(\mathbf{r}_1,\mathbf{r}_2,\omega) = \langle \overleftrightarrow{J}^{(\varphi)*}(\mathbf{r}_1,\omega) \overleftrightarrow{W}^{(i)}(\mathbf{r}_1,\mathbf{r}_2,\omega) \overleftrightarrow{J}^{(\varphi)T}(\mathbf{r}_2,\omega) \rangle_\varphi, \tag{6.36}$$

where superscript T stands for matrix transpose. Further, on directly using definitions (2.138) and (6.31), we find, after straightforward algebra, that the generalized Mueller matrix of the SLM takes the form (Korotkova and Wolf, 2005c)

$$\overleftrightarrow{M}^{(\varphi)}(\mathbf{r}_1,\mathbf{r}_2,\omega) = \frac{1}{2}\begin{bmatrix} 1+C^{(\varphi)}(\mathbf{r}_1,\mathbf{r}_2,\omega) & 1-C^{(\varphi)}(\mathbf{r}_1,\mathbf{r}_2,\omega) & 0 & 0 \\ 1-C^{(\varphi)}(\mathbf{r}_1,\mathbf{r}_2,\omega) & 1+C^{(\varphi)}(\mathbf{r}_1,\mathbf{r}_2,\omega) & 0 & 0 \\ 0 & 0 & 2 & 0 \\ 0 & 0 & 0 & -2 \end{bmatrix}. \tag{6.37}$$

In order to modulate both components of the electric field, independently, the system of two consecutive SLMs may be set. For instance, the beam can be first modulated by the SLM described above (randomizing the y component) followed by another SLM randomizing the x-component with Jones matrix realization

$$\overleftrightarrow{J}^{(\psi)}(\mathbf{r},\omega) = \begin{bmatrix} \exp[i\psi_S(\mathbf{r},\omega)] & 0 \\ 0 & 1 \end{bmatrix}. \tag{6.38}$$

On assuming as well that phase ψ_S is a random process with zero mean and phase correlation $C^{(\psi)}(\mathbf{r}_1,\mathbf{r}_2,\omega)$, we find at once that its generalized Mueller matrix is

$$\overleftrightarrow{M}^{(\psi)}(\mathbf{r}_1,\mathbf{r}_2,\omega) = \frac{1}{2}\begin{bmatrix} C^{(\psi)}(\mathbf{r}_1,\mathbf{r}_2,\omega)+1 & C^{(\psi)}(\mathbf{r}_1,\mathbf{r}_2,\omega)-1 & 0 & 0 \\ C^{(\psi)}(\mathbf{r}_1,\mathbf{r}_2,\omega)+1 & C^{(\psi)}(\mathbf{r}_1,\mathbf{r}_2,\omega)-1 & 0 & 0 \\ 0 & 0 & 2 & 0 \\ 0 & 0 & 0 & -2 \end{bmatrix}. \tag{6.39}$$

The generalized Mueller matrix of the whole system then becomes

$$\overleftrightarrow{M}^{(\varphi\psi)}(\mathbf{r}_1,\mathbf{r}_2,\omega) = \overleftrightarrow{M}^{(\psi)}(\mathbf{r}_1,\mathbf{r}_2,\omega)\overleftrightarrow{M}^{(\varphi)}(\mathbf{r}_1,\mathbf{r}_2,\omega)$$

$$= \frac{1}{2}\begin{bmatrix} C^{(\psi)}+C^{(\varphi)} & C^{(\psi)}-C^{(\varphi)} & 0 & 0 \\ C^{(\psi)}+C^{(\varphi)} & C^{(\psi)}-C^{(\varphi)} & 0 & 0 \\ 0 & 0 & 2 & 0 \\ 0 & 0 & 0 & -2 \end{bmatrix}, \quad (6.40)$$

where the arguments in the last line were omitted.

Chapter 7

Image Formation with Random Light

The classic (intensity based), polarimetric and ghost (intensity–intensity correlation based) imaging systems use various statistical properties of illumination for capturing and carrying information about objects to distant locations. In the most general case, not only the illumination but also the object and the imaging system may be random, while in the variety of particular cases some of these entities may be deterministic. The linear system approach applied to the second- or the fourth-order correlation functions of the stationary optical field is capable of delivering the rigorous predictions about the properties of the formed images under a rather general set of assumptions. The aim of this chapter is to build the basic understanding regarding the image formation mechanisms by these systems and, in particular, to highlight the role of the correlation properties of illumination in improving the image resolution.

7.1 Classic Imaging Systems

7.1.1 *Linear system approach*

We begin the discussion on imaging with the case of a scalar, beam-like, stationary illumination and a linear optical system that consists of a sequence of optical devices located between input (object) plane (ξ, η) and output (image) plane (x, y), and operating in the paraxial regime, i.e., that preserves the incident light's high directionality (see Fig. 7.1). The result of a single realization of the illumination field

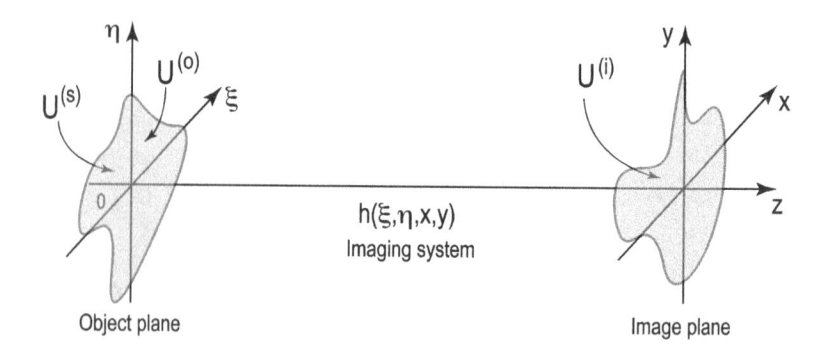

Fig. 7.1 Notations relating to classic image formation.

transformed by such a system can be expressed by the integral

$$U^{(i)}(x,y) = \int_{-\infty}^{\infty}\int_{-\infty}^{\infty} h(\xi,\eta,x,y)U^{(o)}(\xi,\eta)d\xi d\eta, \qquad (7.1)$$

representing the linear response to disturbance $U^{(o)}(\xi,\eta)$. Here, $U^{(o)}$ and $U^{(i)}$ are the optical fields oscillating at frequency ω (the dependence omitted throughout the chapter), across the object plane, after interaction with the object, and the image plane, respectively. Integration in Eq. (7.1) extends over the whole object plane. The Green function of the system $h(\xi,\eta,x,y)$ is also known as the *amplitude spread function*. Let us assume here that it is generally a complex-valued, scalar, deterministic function of its spatial arguments and is given at frequency ω (omitted). The expression (7.1) can be obtained as the solution of the Helmholtz equation discussed in Chapter 4, in which the refractive index distribution in the volume between the object and the image planes can be simply related to $h(\xi,\eta,x,y)$, the specific relation is to be discussed in what follows.

The field in the object plane can be regarded as a product (local transformation) of the incident illumination field $U^{(s)}(\xi,\eta)$ and the scalar, complex-valued object transparency $o(\xi,\eta)$

$$U^{(o)}(\xi,\eta) = o(\xi,\eta)U^{(s)}(\xi,\eta). \qquad (7.2)$$

We assume at this point that the transparency is a deterministic function of space and is given at frequency ω (omitted).

On substituting from Eq. (7.2) into Eq. (7.1), we obtain the basic relation between the illumination field in the object plane, just before interaction with the object, and in the image plane:

$$U^{(i)}(x,y) = \int_{-\infty}^{\infty}\int_{-\infty}^{\infty} h(\xi,\eta,x,y)o(\xi,\eta)U^{(s)}(\xi,\eta)d\xi d\eta. \quad (7.3)$$

On forming the second-order correlations for the object and the image fields in Eq. (7.1) and using the definition of the cross-spectral density (2.77), we find that

$$W^{(i)}(x_1,y_1,x_2,y_2)$$
$$= \int_{-\infty}^{\infty}\int_{-\infty}^{\infty}\int_{-\infty}^{\infty}\int_{-\infty}^{\infty} H(\xi_1,\eta_1,x_1,y_1,\xi_2,\eta_2,x_2,y_2)$$
$$\times W^{(o)}(\xi_1,\eta_1,\xi_2,\eta_2)d\xi_1 d\eta_1 d\xi_2 d\eta_2, \quad (7.4)$$

where integration is performed twice over the object plane. This result was first revealed in Hopkins (1951, 1952) and was reviewed in Thompson (1969). The kernel of the integral in Eq. (7.4),

$$H(\xi_1,\eta_1,x_1,y_1,\xi_2,\eta_2,x_2,y_2) = h^*(\xi_1,\eta_1,x_1,y_1)h(\xi_2,\eta_2,x_2,y_2), \quad (7.5)$$

is the generalization of the amplitude spread function of the deterministic system that may be termed the *correlation spread function*.

In the case of the partially coherent illumination, the relation between the correlations in the incident field, the object transparency field and the correlations in the outcoming field becomes (see Eq. (7.2)):

$$W^{(o)}(\xi_1,\eta_1,\xi_2,\eta_2) = O(\xi_1,\eta_1,\xi_2,\eta_2)W^{(s)}(\xi_1,\eta_1,\xi_2,\eta_2), \quad (7.6)$$

where

$$O(\xi_1,\eta_1,\xi_2,\eta_2) = o^*(\xi_1,\eta_1)o(\xi_2,\eta_2) \quad (7.7)$$

is the *object correlation* function. Substitution from Eq. (7.6) into Eq. (7.4) yields a rather general relation

$$W^{(i)}(x_1,y_1,x_2,y_2)$$
$$= \iiiint_{-\infty}^{\infty} H(\xi_1,\eta_1,x_1,y_1,\xi_2,\eta_2,x_2,y_2)$$
$$\times O(\xi_1,\eta_1,\xi_2,\eta_2)W^{(s)}(\xi_1,\eta_1,\xi_2,\eta_2)d\xi_1 d\eta_1 d\xi_2 d\eta_2, \quad (7.8)$$

occuring among the correlations in the source, the object, the system and the image. Recall that in establishing this relation we have assumed that only the illumination is random but the system and the object transparency are deterministic. In order to generalize the analysis to random systems and/or objects, one simply needs to consider, instead of deterministic H and O, their ensemble averages:

$$H(\xi_1, \eta_1, x_1, y_1, \xi_2, \eta_2, x_2, y_2) = \langle h^*(\xi_1, \eta_1, x_1, y_1) h(\xi_2, \eta_2, x_2, y_2) \rangle, \tag{7.9}$$

and

$$O(\xi_1, \eta_1, \xi_2, \eta_2) = \langle o^*(\xi_1, \eta_1) o(\xi_2, \eta_2) \rangle, \tag{7.10}$$

where the angular brackets stand for the ensembles of linear systems and object realizations in Eqs. (7.9) and (7.10), respectively. Thus, for random objects and systems, Eq. (7.8) still holds in form while H and O given by Eqs. (7.9) and (7.10) are implied. In order to achieve separation of three averages, one must assume that the light, the object and the system are governed by statistically independent random processes.

7.1.2 *Isoplanatic systems, homogeneous illumination*

Of particular interest are the frequently occuring in practice *isoplanatic* systems whose amplitude spread functions' realizations depend on the difference of coordinates in the object and the image planes:

$$h(\xi, \eta, x, y) = h(x - \xi, y - \eta). \tag{7.11}$$

In this case, a realization of the image field can be expressed as convolution

$$U^{(i)}(x, y) = \int_{-\infty}^{\infty} \int_{-\infty}^{\infty} h(x - \xi, y - \eta) U^{(o)}(\xi, \eta) d\xi d\eta$$

$$= h(\xi, \eta) \circledast U^{(o)}(\xi, \eta) \bigg|_{(x,y)}. \tag{7.12}$$

The amplitude spread function is then obtained as the impulse response of the linear system, i.e., by setting the disturbance produced by the object to the two-dimensional Dirac delta-function, $U^{(o)}(\xi,\eta) = \delta^{(2)}(\xi,\eta)$ and using its sifting property for finding $U^{(i)}(x,y)$.

We will now take the two-dimensional Fourier transform of $U^{(i)}(x,y)$ with respect to spatial coordinates. In this chapter, it will be convenient to use the definition of the Fourier transform having exponential kernel with coefficient $2\pi i$ and the positive sign in the kernel of the direct transform. In order to distinguish functions in the direct space and the Fourier space (spatial-frequency domain), we will use the old German script for the latter ones. It then follows from Eq. (7.12) and the convolution theorem expressed by Eq. (1.16) that

$$\mathfrak{U}^{(i)}(\kappa_x, \kappa_y) = \mathfrak{h}(\kappa_x, \kappa_y)\mathfrak{U}^{(o)}(\kappa_x, \kappa_y), \tag{7.13}$$

where $\mathfrak{h}$ is known as the *amplitude transfer function* of the imaging system depending on the two-dimensional spatial-frequency vector $\boldsymbol{\kappa} = (\kappa_x, \kappa_y)$. Further, for isoplanatic systems

$$H(x_1 - \xi_1, y_1 - \eta_1, x_2 - \xi_2, y_2 - \eta_2)$$
$$= \langle h^*(x_1 - \xi_1, y_1 - \eta_1)h(x_2 - \xi_2, y_2 - \eta_2)\rangle, \tag{7.14}$$

where the angular brackets take place for random systems and can be omitted for deterministic systems. Therefore, the relation between the two cross-spectral densities in Eq. (7.4) reduces to their convolution

$$W^{(i)}(x_1, y_1, x_2, y_2)$$
$$= \int_{-\infty}^{\infty}\int_{-\infty}^{\infty}\int_{-\infty}^{\infty}\int_{-\infty}^{\infty} H(x_1 - \xi_1, y_1 - \eta_1, x_2 - \xi_2, y_2 - \eta_2)$$
$$\times W^{(o)}(\xi_1, \eta_1, \xi_2, \eta_2) d\xi_1 d\eta_1 d\xi_2 d\eta_2$$
$$= H(\xi_1, \eta_1, \xi_2, \eta_2) \circledast W^{(o)}(\xi_1, \eta_1, \xi_2, \eta_2)\Big|_{(x_1, y_1, x_2, y_2)}. \tag{7.15}$$

Let us now take the four-dimensional Fourier transform of the correlation spread function, known as the *Correlation Transfer Function*

(CTF) of the isoplanatic system,

$$\mathfrak{H}(\kappa_{x1}, \kappa_{y1}, \kappa_{x2}, \kappa_{y2})$$
$$= FT[H(\xi_1, \eta_1, \xi_2, \eta_2)]$$
$$= \int_{-\infty}^{\infty}\int_{-\infty}^{\infty}\int_{-\infty}^{\infty}\int_{-\infty}^{\infty} H(\xi_1, \eta_1, \xi_2, \eta_2)$$
$$\times \exp[2\pi i(\xi_1\kappa_{x1} + \eta_1\kappa_{y1}$$
$$+ \xi_2\kappa_{x2} + \eta_2\kappa_{y2})]d\xi_1 d\eta_1 d\xi_2 d\eta_2. \qquad (7.16)$$

It factorizes at once in the Cartesian coordinate system as

$$\mathfrak{H}(\kappa_{x1}, \kappa_{y1}, \kappa_{x2}, \kappa_{y2}) = \langle \mathfrak{h}^*(-\kappa_{x1}, -\kappa_{y1})\mathfrak{h}(\kappa_{x2}, \kappa_{y2})\rangle. \qquad (7.17)$$

This implies that relation (7.15) between the object and the image cross-spectral densities in the Fourier domain can be written as

$$\mathfrak{W}^{(i)}(\kappa_{x1}, \kappa_{y1}, \kappa_{x2}, \kappa_{y2}) = \mathfrak{H}(-\kappa_{x1}, -\kappa_{y1}, \kappa_{x2}, \kappa_{y2})$$
$$\times \mathfrak{W}^{(o)}(\kappa_{x1}, \kappa_{y1}, \kappa_{x2}, \kappa_{y2}). \qquad (7.18)$$

We can refer to functions $\mathfrak{W}^{(o)}$ and $\mathfrak{W}^{(i)}$ as the *spatial cross-spectra* of light in the object and image planes.

In particular, if the optical source is homogeneous, i.e.,

$$W^{(s)}(\xi_1, \eta_1, \xi_2, \eta_2) = W^{(s)}(\xi_1 - \xi_2, \eta_1 - \eta_2), \qquad (7.19)$$

and the object is deterministic, viz., object correlation O factorizes as in Eq. (7.7), then, on taking the four-dimensional Fourier transform of Eq. (7.6), we find that

$$\mathfrak{W}^{(o)}(\kappa_{x1}, \kappa_{y1}, \kappa_{x2}, \kappa_{y2})$$
$$= \int_{-\infty}^{\infty}\int_{-\infty}^{\infty}\int_{-\infty}^{\infty}\int_{-\infty}^{\infty} O(\xi_1, \eta_1, \xi_2, \eta_2)W^{(s)}(\xi_1 - \xi_2, \eta_1 - \eta_2)$$
$$\times \exp[2\pi i(\xi_1\kappa_{x1} + \eta_1\kappa_{y1} + \xi_2\kappa_{x2} + \eta_2\kappa_{y2})]d\xi_1 d\eta_1 d\xi_2 d\eta_2. \qquad (7.20)$$

Introducing difference variables

$$\xi_d = \xi_1 - \xi_2, \quad \eta_d = \eta_1 - \eta_2, \qquad (7.21)$$

Image Formation with Random Light 175

and using them in Eq. (7.20) yields

$$\mathfrak{W}^{(o)}(\kappa_{x1}, \kappa_{y1}, \kappa_{x2}, \kappa_{y2})$$
$$= \int_{-\infty}^{\infty}\int_{-\infty}^{\infty}\int_{-\infty}^{\infty}\int_{-\infty}^{\infty} O(\xi_2 + \xi_d, \eta_2 + \eta_d, \xi_2, \eta_2)$$
$$\times W^{(s)}(\xi_d, \eta_d) \exp\Big[2\pi i[(\xi_2 + \xi_d)\kappa_{x1} + (\eta_2 + \eta_d)\kappa_{y1}$$
$$+ \xi_2\kappa_{x2} + \eta_2\kappa_{y2}]\Big] d\xi_d d\eta_d d\xi_2 d\eta_2. \tag{7.22}$$

Regrouping the individual integrals yields the product

$$\mathfrak{W}^{(o)}(\kappa_{x1}, \kappa_{y1}, \kappa_{x2}, \kappa_{y2})$$
$$= I \int_{-\infty}^{\infty}\int_{-\infty}^{\infty} o(\xi_2, \eta_2) \exp[2\pi i(\xi_2\kappa_{x1} + \eta_2\kappa_{y1}$$
$$+ \xi_2\kappa_{x2} + \eta_2\kappa_{y2})] d\xi_d d\eta_d. \tag{7.23}$$

Here, integral

$$I = \int_{-\infty}^{\infty}\int_{-\infty}^{\infty} W^{(s)}(\xi_d, \eta_d) o^*(\xi_2 + \xi_d, \eta_2 + \eta_d)$$
$$\times \exp[2\pi i(\xi_d\kappa_{x1} + \eta_d\kappa_{y1})] d\xi_d d\eta_d \tag{7.24}$$

may be written as the convolution

$$I = \mathfrak{W}^{(s)}(\xi_d, \eta_d) \circledast \{o^*(\xi_2, \eta_2)\exp[2\pi i(\xi_d\kappa_{x1} + \eta_d\kappa_{y1})]\}$$
$$= \int_{-\infty}^{\infty}\int_{-\infty}^{\infty} \mathfrak{W}^{(s)}(K_x, K_y) o^*(K_x - \kappa_x, K_y - \kappa_y)$$
$$\times \exp\Big[2\pi i[\xi_2(K_x - \kappa_{x1}) + \eta_2(K_y - \kappa_{y1})]\Big] dK_x dK_y, \tag{7.25}$$

where K_x and K_y are the integration variables. Hence, Eq. (7.23) becomes

$$\mathfrak{W}^{(o)}(\kappa_{x1}, \kappa_{y1}, \kappa_{x2}, \kappa_{y2})$$
$$= \int_{-\infty}^{\infty}\int_{-\infty}^{\infty} o(\xi_2, \eta_2) \exp[2\pi i(\xi_2\kappa_{x2} + \eta_2\kappa_{y2})] d\xi_2 d\eta_2$$
$$\times \int_{-\infty}^{\infty}\int_{-\infty}^{\infty} \mathfrak{W}^{(s)}(K_x, K_y) o^*(K_x - \kappa_{x1}, K_y - \kappa_{y1})$$
$$\times \exp[2\pi i(\xi_2 K_x + \eta_2 K_y)] dK_x dK_y. \tag{7.26}$$

Changing the order of integrations implies that

$$\mathfrak{W}^{(o)}(\kappa_{x1}, \kappa_{y1}, \kappa_{x2}, \kappa_{y2})$$
$$= \int_{-\infty}^{\infty} \int_{-\infty}^{\infty} o(\xi_2, \eta_2) \exp[2\pi i(\xi_2(\kappa_{x2} + K_x)$$
$$+ \eta_2(\kappa_{y2} + K_y)]d\xi_2 d\eta_2$$
$$\times \int_{-\infty}^{\infty} \int_{-\infty}^{\infty} \mathfrak{W}^{(s)}(K_x, K_y)$$
$$\times o^*(K_x - \kappa_{x1}, K_y - \kappa_{y1})dK_x dK_y. \quad (7.27)$$

On recognizing that the first integral is the two-dimensional Fourier transform of $o(\xi_2, \eta_2)$, we finally arrive at the expression

$$\mathfrak{W}^{(o)}(\kappa_{x1}, \kappa_{y1}, \kappa_{x2}, \kappa_{y2}) = \int_{-\infty}^{\infty} \int_{-\infty}^{\infty} \mathfrak{W}^{(s)}(K_x, K_y)$$
$$\times \mathfrak{O}(\kappa_{x1} - K_x, \kappa_{y1} - K_y, \kappa_{x2}$$
$$+ K_x, \kappa_{y2} + K_y)dK_x dK_y, \quad (7.28)$$

where we denoted

$$\mathfrak{O}(\kappa_{x1}, \kappa_{y1}, \kappa_{x2}, \kappa_{y2}) = o^*(-\kappa_{x1}, -\kappa_{y1})o(\kappa_{x1}, \kappa_{y1}), \quad (7.29)$$

$\mathfrak{O}$ being the *object's spatial cross-spectrum*. We may generalize this result at once to random (stationary) object transparencies by applying statistical averaging over the ensemble of object's realizations:

$$\mathfrak{O}(\kappa_{x1}, \kappa_{y1}, \kappa_{x2}, \kappa_{y2}) = \langle o^*(-\kappa_{x1}, -\kappa_{y1})o(\kappa_{x1}, \kappa_{y1})\rangle. \quad (7.30)$$

Thus, Eq. (7.28) expresses the spatial cross-spectrum of the image as the double convolution of those of the source and of the object. Together with Eq. (7.18), the two equations (7.29) and (7.30) characterize general relations among the source's, object's and image's spatial cross-spectra, and the correlation transfer function of an isoplanatic system.

We will now proceed to calculation of the spectral density (average intensity) of the image in the space and in the spatial-frequency domains in terms of the properties of the source, object and the

Image Formation with Random Light   177

system in these domains. In the real space, the image spectral density can be calculated as the following inverse Fourier transform:

$$S^{(i)}(x,y) = \int_{-\infty}^{\infty}\int_{-\infty}^{\infty}\int_{-\infty}^{\infty}\int_{-\infty}^{\infty} \mathfrak{W}^{(i)}(\kappa_{x1},\kappa_{y1},\kappa_{x2},\kappa_{y2})$$
$$\times \exp[-2\pi i[x(\kappa_{x1}+\kappa_{x2})+y(\kappa_{y1}+\kappa_{y2})]]$$
$$\times d\kappa_{x1}d\kappa_{y1}d\kappa_{x2}d\kappa_{y2}. \quad (7.31)$$

On the other hand, it can be related to its own spatial spectrum $\mathfrak{S}^{(i)}$ as the two-dimensional Fourier transform pair

$$S^{(i)}(x,y) = \int_{-\infty}^{\infty}\int_{-\infty}^{\infty} \mathfrak{S}^{(i)}(\kappa_x,\kappa_y)\exp[-2\pi i(\kappa_x x+\kappa_y y)]d\kappa_x d\kappa_y, \quad (7.32)$$

and

$$\mathfrak{S}^{(i)}(\kappa_x,\kappa_y) = \int_{-\infty}^{\infty}\int_{-\infty}^{\infty} S^{(i)}(x,y)\exp[2\pi i(\kappa_x x+\kappa_y y)]dxdy. \quad (7.33)$$

On introducing sum variables

$$\kappa_{xs} = \kappa_{x1}+\kappa_{x2}, \quad \kappa_{ys} = \kappa_{y1}+\kappa_{y2}, \quad (7.34)$$

we may write Eq. (7.31) as

$$S^{(i)}(x,y) = \int_{-\infty}^{\infty}\int_{-\infty}^{\infty}\int_{-\infty}^{\infty}\int_{-\infty}^{\infty} \mathfrak{W}^{(i)}(\kappa_{x1},\kappa_{y1},\kappa_{xs}-\kappa_{x1},\kappa_{ys}-\kappa_{y1})$$
$$\times \exp[-2\pi i(x\kappa_{xs}+y\kappa_{ys})]d\kappa_{x1}d\kappa_{y1}d\kappa_{xs}d\kappa_{ys}$$
$$= \int_{-\infty}^{\infty}\int_{-\infty}^{\infty}\left[\int_{-\infty}^{\infty}\int_{-\infty}^{\infty} \mathfrak{W}^{(i)}(\kappa_{x1},\kappa_{y1},\kappa_{xs}\right.$$
$$\left.-\kappa_{x1},\kappa_{ys}-\kappa_{y1})d\kappa_{x1}d\kappa_{y1}\right]$$
$$\times \exp[-2\pi i(x\kappa_{xs}+y\kappa_{ys})]d\kappa_{xs}d\kappa_{ys}. \quad (7.35)$$

The comparison of Eqs. (7.35) and (7.32) makes it possible to write:

$$\mathfrak{S}^{(i)}(\kappa_x,\kappa_y) = \int_{-\infty}^{\infty}\int_{-\infty}^{\infty} \mathfrak{W}^{(i)}(\kappa_{x1},\kappa_{y1},\kappa_x-\kappa_{x1},\kappa_y-\kappa_{y1})d\kappa_{x1}d\kappa_{y1}. \quad (7.36)$$

Substitution from Eqs. (7.18) and (7.28) into Eq. (7.36) gives

$$\mathfrak{S}^{(i)}(\kappa_x, \kappa_y) = \int_{-\infty}^{\infty}\int_{-\infty}^{\infty} \mathfrak{H}(\kappa_{x1}, \kappa_{y1}, \kappa_x - \kappa_{x1}, \kappa_y - \kappa_{y1}) d\kappa_{x1} d\kappa_{y1}$$

$$\times \left[\int_{-\infty}^{\infty}\int_{-\infty}^{\infty} \mathfrak{W}^{(s)}(K_x, K_y) \mathfrak{O}(\kappa_{x1} - K_x, \kappa_{y1} \right.$$

$$\left. - K_y, K_x + \kappa_x - \kappa_{x1}, K_y + \kappa_y - \kappa_{y1}) dK_x dK_y \right].$$
(7.37)

Next, with the change of variables

$$X = K_x + \kappa_x - \kappa_{x1}, \quad Y = K_y + \kappa_y - \kappa_{y1}, \tag{7.38}$$

expression (7.37) becomes

$$\mathfrak{S}^{(i)}(\kappa_x, \kappa_y) = \int_{-\infty}^{\infty}\int_{-\infty}^{\infty} \mathfrak{H}(-X + K_x + \kappa_x, -Y$$

$$+ K_y + \kappa_y, X - K_x, Y - K_y)$$

$$\times \left[\int_{-\infty}^{\infty}\int_{-\infty}^{\infty} \mathfrak{W}^{(s)}(K_x, K_y) \mathfrak{O}(\kappa_x - X, \kappa_y \right.$$

$$\left. - Y, X, Y) dX dY \right] d[K_x + \kappa_x - X]$$

$$\times d[K_y + \kappa_y - Y].$$
(7.39)

It can also be written as

$$\mathfrak{S}^{(i)}(\kappa_x, \kappa_y) = \int_{-\infty}^{\infty}\int_{-\infty}^{\infty} \mathfrak{O}(\kappa_x - X, \kappa_y - Y, X, Y)$$

$$\times \mathfrak{I}(X, \kappa_x, Y, \kappa_y) dX dY, \tag{7.40}$$

where we set

$$\mathfrak{I}(X, \kappa_x, Y, \kappa_y)$$

$$= \int_{-\infty}^{\infty}\int_{-\infty}^{\infty} \mathfrak{H}(K_x + \kappa_x - X, K_y + \kappa_y - Y, X - K_x, Y - K_y)$$

$$\times \mathfrak{W}^{(s)}(K_x, K_y) dK_x dK_y. \tag{7.41}$$

The two expressions above constitute the main relation of this section. They characterize the dependence of the spatial spectrum $\mathfrak{S}^{(i)}$ of the image on the spatial cross-spectra of the source $\mathfrak{W}^{(s)}$, of the object transparency, $\mathfrak{O}$, and of the imaging system $\mathfrak{H}$. Function $\mathfrak{J}$ can be viewed as the *transmission cross-coefficient* of the imaging system. It only depends on the coherence properties of the source $\mathfrak{W}^{(s)}$ and on the correlation transfer function $\mathfrak{H}$ of the optical system while being independent of the properties of the object.

7.1.3 *Coherent and incoherent imaging systems*

We will now deduce from the general relations (7.40)–(7.41) derived in the previous section the results pertaining to two limiting cases — of *coherent* and *incoherent imaging systems*. This terminology only refers to the coherence properties of illumination. At the same time, we will only consider these results in the limiting case when both the object transparency and the imaging system are deterministic. As we will now see, for coherent imaging systems the relations (7.40) and (7.41) will reduce to those established in the beginning of the chapter. Indeed, in this case the cross-spectral density of the source is constant:

$$W^{(s)}(\xi_d, \eta_d) = S_0. \tag{7.42}$$

Hence, the image spectral density becomes

$$S^{(i)}(x,y) = S_0 \int_{-\infty}^{\infty}\int_{-\infty}^{\infty}\int_{-\infty}^{\infty}\int_{-\infty}^{\infty} H(x-\xi_1, y-\eta_1, x-\xi_2, y-\eta_2)$$
$$\times O(\xi_1, \eta_1, \xi_2, \eta_2) d\xi_1 d\eta_1 d\xi_2 d\eta_2. \tag{7.43}$$

Such an integral factorizes as a product of two complex conjugate functions and, hence, can be expressed as

$$S^{(i)}(x,y) = S_0 \left| \int_{-\infty}^{\infty}\int_{-\infty}^{\infty} h(x-\xi, y-\eta) o(\xi, \eta) d\xi d\eta \right|^2. \tag{7.44}$$

Therefore, in the coherent limit, the imaging process becomes linear on the order of the field. In the spatial frequency domain, Eq. (7.42)

becomes
$$\mathfrak{W}^{(s)}(\kappa_x, \kappa_y) = S_0 \delta^{(2)}(\kappa_x, \kappa_y). \tag{7.45}$$

Substitution of this result into Eqs. (7.40) and (7.41) implies that
$$\mathfrak{S}^{(i)}(\kappa_x, \kappa_y) = S_0 \int_{-\infty}^{\infty} \int_{-\infty}^{\infty} \mathfrak{O}(\kappa_x - X, \kappa_y - Y, X, Y)$$
$$\times \mathfrak{H}(\kappa_x - X, \kappa_y - Y, X - K_x, Y - K_y) dX dY. \tag{7.46}$$

Since both functions in the integrand of the last expression factorize, it is possible to represent the spatial spectrum of the image as autocorrelation
$$\mathfrak{S}^{(i)}(\kappa_x, \kappa_y) = \mathfrak{A}^{(i)}(\kappa_x, \kappa_y) \otimes \mathfrak{A}^{(i)}(\kappa_x, \kappa_y), \tag{7.47}$$
where
$$\mathfrak{A}^{(i)}(\kappa_x, \kappa_y) = \sqrt{S_0} \mathfrak{o}(\kappa_x, \kappa_y) \mathfrak{h}(\kappa_x, \kappa_y), \tag{7.48}$$
i.e., it can be regarded as the spatial amplitude spectrum of the image. Factorization (7.48) implies that the transfer function of the coherent imaging system can be just defined as the *Amplitude Transfer Function* (ATF)
$$\text{ATF}(\kappa_x, \kappa_y) = \mathfrak{h}(\kappa_x, \kappa_y). \tag{7.49}$$

It is also of convenience (for making comparisons, for example) to employ its normalized version
$$\text{ATF}_n(\kappa_x, \kappa_y) = \frac{\mathfrak{h}(\kappa_x, \kappa_y)}{\mathfrak{h}(0,0)}. \tag{7.50}$$

Let us now proceed to the incoherent imaging system. In this limit,
$$W^{(s)}(\xi_d, \eta_d) = S_0 \delta^{(2)}(\xi_d, \eta_d). \tag{7.51}$$

First, on substituting this expression into Eq. (7.4), we find that
$$S^{(i)}(x, y) = \int_{-\infty}^{\infty} \int_{-\infty}^{\infty} \int_{-\infty}^{\infty} \int_{-\infty}^{\infty} H(x - \xi_1, y - \eta_1, x - \xi_2, y - \eta_2)$$
$$\times O(\xi_1, \eta_1, \xi_2, \eta_2) S_0 \delta^{(0)}(\xi_1 - \xi_2, \eta_1 - \eta_2) d\xi_1 d\xi_2 d\eta_1 d\eta_2. \tag{7.52}$$

Using the sifting property of the delta-function leads to the following result:
$$S^{(i)}(x,y) = \int_{-\infty}^{\infty}\int_{-\infty}^{\infty} |h(x-\xi, y-\eta)|^2 |o(\xi,\eta)|^2 d\xi d\eta. \qquad (7.53)$$

We therefore conclude that in the limit of incoherent illumination, the image formation process occurs on the order of intensity.

In the space-frequency domain, Eq. (7.51) becomes
$$\mathfrak{W}^{(s)}(\kappa_x, \kappa_y) = S_0. \qquad (7.54)$$

Then on using this result in Eq. (7.40), we arrive at the expression
$$S^{(i)}(x,y) = S_0 \int_{-\infty}^{\infty}\int_{-\infty}^{\infty} \mathfrak{O}(\kappa_x - X, \kappa_y - Y, X, Y) dX dY$$
$$\times \int_{-\infty}^{\infty}\int_{-\infty}^{\infty} \mathfrak{H}(K_x + \kappa_x - X, K_y + \kappa_y - Y, X$$
$$- K_x, Y - K_y) dK_x dK_y. \qquad (7.55)$$

The second integral only depends on the imaging system and is independent from the illumination and object properties. It can be expressed in a more concise form, with the help of variables
$$q_x = X - K_x, \quad q_y = Y - K_y, \qquad (7.56)$$
as
$$\mathrm{OTF}(\kappa_x, \kappa_y) = \int_{-\infty}^{\infty}\int_{-\infty}^{\infty} \mathfrak{H}(\kappa_x - q_x, \kappa_y - q_y, q_x, q_y) dq_x dq_y, \qquad (7.57)$$

where correlation transfer function $\mathfrak{H}$ was defined in Eq. (7.16). This function is known as the *Optical Transfer Function* (OTF) (Duffieux, 1946). It is generally a complex-valued function, with its magnitude and phase termed as the *Modulation Transfer Function* (MTF) and the *Phase Transfer Function* (PTF), respectively. A normalized version of the OTF,
$$\mathrm{OTF}_n(\kappa_x, \kappa_y) = \frac{\mathrm{OTF}(\kappa_x, \kappa_y)}{\mathrm{OTF}(0,0)}$$
$$= \frac{\int_{-\infty}^{\infty}\int_{-\infty}^{\infty} \mathfrak{H}(\kappa_x - q_x, \kappa_y - q_y, q_x, q_y) dq_x dq_y}{\int_{-\infty}^{\infty}\int_{-\infty}^{\infty} \mathfrak{H}(-q_x, -q_y, q_x, q_y) dq_x dq_y}, \qquad (7.58)$$

is also frequently used for imaging systems' comparisons.

In summary, imaging systems that use illumination with any state of coherence can be characterized, in the spatial-frequency domain, with the help of the transmission cross-coefficient (see Eq. (7.41)). In this case, the correlation properties of illumination and the transfer properties of the optical system cannot be separated. However, if the illumination becomes completely coherent or completely incoherent, then transmission cross-coefficient reduces to functions depending only on the optical system. Then in the coherent/incoherent limit, the optical system forms images on the order of the field/intensity. Further, in the coherent and incoherent limits the spatial-frequency content of the system is given by the amplitude transfer function (see Eq. (7.49)) and the optical transfer function (see Eq. (7.57)), respectively. Although the detailed discusssion about the spatial-frequency content of imaging systems is beyond the scope of our monograph and can be found elsewhere (Goodman, 2000), we must still share the essential observations. The transmission cross-coefficient (7.41) generally carries information about the ability of illumination and the optical system to form the image of the object at all spatial frequencies $0 < \kappa_x, \kappa_y < \infty$. In practice, both the illumination and the optical system act as spatial-frequency filters by adjusting the weights of various frequencies carried by the object transparency. In particular, since the OTF of an incoherent system is the convolution of an ATF of the corresponding coherent system (formed with the same amplitude spread function), the spatial-frequency range of the incoherent system doubles as compared to that of the coherent system. Thus, using incoherent systems leads to substantial improvement of the image quality (Considine, 1966).

7.1.4 *ABCD matrices*

Suppose that an optical image-forming device is situated between two planes, say, α and β, and has its axis of symmetry coinciding with direction z of light propagation. Let a light ray be incident onto the device in plane α at position $\boldsymbol{\rho}' = x'\hat{x} + y'\hat{y}$ while making angle $\boldsymbol{\theta}' = \theta'_x\hat{x} + \theta'_y\hat{y}$ with respect to the z-axis. Then according to the refractive index distribution of the device, the ray must exit in plane β at position $\boldsymbol{\rho} = x\hat{x} + y\hat{y}$ and angle $\boldsymbol{\theta} = \theta_x\hat{x} + \theta_y\hat{y}$. Within the validity of paraxial approximation, the relation between vectors $\vec{R}'$ and $\vec{R}$, which we form as

$$\vec{R}' = \begin{bmatrix} \rho' \\ \theta' \end{bmatrix}, \quad \vec{R} = \begin{bmatrix} \rho \\ \theta \end{bmatrix}, \tag{7.59}$$

can be represented as a 2×2 linear operator

$$\vec{R} = \overleftrightarrow{\Upsilon} \vec{R}', \quad \overleftrightarrow{\Upsilon} = \begin{bmatrix} A & B \\ C & D \end{bmatrix}, \tag{7.60}$$

widely known as the ray matrix or the $ABCD$ matrix with elements evaluated from formulas (Gerrard, 1994)

$$A = \left.\frac{\rho}{\rho'}\right|_{\theta'=0}, \quad B = \left.\frac{\rho}{\theta'}\right|_{\rho'=0}, \quad C = \left.\frac{\theta}{\rho'}\right|_{\theta'=0}, \quad D = \left.\frac{\theta}{\theta'}\right|_{\rho'=0}. \tag{7.61}$$

Here, ρ, θ, ρ' and θ' are the magnitudes of the vectors defined above. Elements A and D represent spatial and angular magnification, while B and C are the coefficients of mapping angle to position and position to angle, respectively. The systems characterized by the $ABCD$ matrix are classified into *imaging* if $(A, B, C, D) = (A, 0, C, 1/A)$, *collimating* if $(A, B, C, D) = (A, B, -1/B, 0)$, *focusing* if $(A, B, C, D) = (0, B, -1/B, D)$ and *afocal* if $(A, B, C, D) = (A, B, 0, 1/A)$.

For example, the specific $ABCD$ matrices of the free-space propagation at distance z, of the plane mirror, of the retro-reflector and of the interface between media with refractive indices n_1 and n_2 have the forms

$$\overleftrightarrow{\Upsilon}_f = \begin{bmatrix} 1 & z \\ 0 & 1 \end{bmatrix}, \quad \overleftrightarrow{\Upsilon}_m = \begin{bmatrix} 1 & 0 \\ 0 & 1 \end{bmatrix},$$

$$\overleftrightarrow{\Upsilon}_r = \begin{bmatrix} -1 & 0 \\ 0 & -1 \end{bmatrix}, \quad \overleftrightarrow{\Upsilon}_b = \begin{bmatrix} 1 & 0 \\ 0 & n_1/n_2 \end{bmatrix}, \tag{7.62}$$

respectively. Further, a thin, Gaussian lens with soft radius W_G ($1/\sqrt{2}$ off the maximum value) and focal distance F_G can be represented by the $ABCD$ matrix

$$\overleftrightarrow{\Upsilon}_l = \begin{bmatrix} 1 & 0 \\ i\alpha_G & 1 \end{bmatrix}, \quad \alpha_G = \frac{2}{kW_G^2} + \frac{i}{F_G}. \tag{7.63}$$

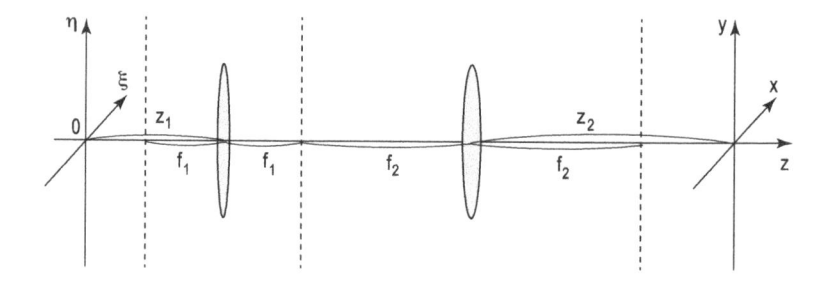

Fig. 7.2 Notations relating to the telescopic imaging system.

The optical system consisting of N elements aligned along the same axis can be characterized by the product of the $ABCD$ matrices of the individual elements:

$$\overleftrightarrow{\Upsilon} = \overleftrightarrow{\Upsilon}_N \cdot \overleftrightarrow{\Upsilon}_{N-1} \cdots \cdots \overleftrightarrow{\Upsilon}_2 \cdot \overleftrightarrow{\Upsilon}_1, \qquad (7.64)$$

where matrix multiplication from the right side is applied.

A *telescopic* system constitutes an important example of the imaging system. It consists of two Gaussian lenses with infinite radii and focal lengths f_1 and f_2 separated exactly by distance $f_1 + f_2$ (see Fig. 7.2). If the distance from the object to the first lens is z_1 and from the second lens to the image is z_2, then the ABCD matrix of the total system can be determined as follows:

$$\begin{bmatrix} A & B \\ C & D \end{bmatrix} = \begin{bmatrix} 1 & z_2 \\ 0 & 1 \end{bmatrix} \cdot \begin{bmatrix} 1 & 0 \\ -f_2^{-1} & 1 \end{bmatrix} \cdot \begin{bmatrix} 1 & f_1 + f_2 \\ 0 & 1 \end{bmatrix} \cdot \begin{bmatrix} 1 & 0 \\ -f_1^{-1} & 1 \end{bmatrix} \cdot \begin{bmatrix} 1 & z_1 \\ 0 & 1 \end{bmatrix}$$

$$= \begin{bmatrix} -\dfrac{f_2}{f_1} & f_1 + f_2 - \dfrac{1}{f_1 f_2}(z_1 f_2^2 + z_2 f_1^2) \\ 0 & -\dfrac{f_1}{f_2} \end{bmatrix}. \qquad (7.65)$$

The imaging condition $D = 1/A$ is satisfied automatically and imaging condition $B = 0$ holds if $z_1 = f_1$ and $z_2 = f_2$. The magnification of the system is given by $A = -f_2/f_1$. Hence, in the special case, when $f_1 = f_2$, it delivers the perfect image and is known as the *4f-system*.

7.1.5 Generalized Huygens–Fresnel integral

The evolution of the optical field from the object to the image plane in the presense of an $ABCD$ optical system can be found with the help of the generalized Huygens–Fresnel integral (Collins, 1970). In this case, the amplitude response function was shown to have the form

$$h(\xi, \eta, x, y) = \frac{e^{ikz}}{i\lambda B} \exp\left[\frac{i\pi D}{\lambda B}(x^2 + y^2)\right]$$
$$\times \exp\left[\frac{i\pi A}{\lambda B}(\xi^2 + \eta^2)\right] \exp\left[-\frac{2\pi i}{\lambda B}(\xi x + \eta y)\right], \quad (7.66)$$

where z is the total path through the system along the optical axis. Hence, according to Eq. (7.1) the relation between the object and the image fields takes the form

$$U^{(i)}(x, y) = \frac{e^{ikz}}{i\lambda B} \exp\left[\frac{i\pi D}{\lambda B}(x^2 + y^2)\right]$$
$$\times \int_{-\infty}^{\infty} \int_{-\infty}^{\infty} U^{(o)}(\xi, \eta) \exp\left[\frac{i\pi A}{\lambda B}(\xi^2 + \eta^2)\right]$$
$$\times \exp\left[-\frac{2\pi i}{\lambda B}(\xi x + \eta y)\right] d\xi d\eta. \quad (7.67)$$

We immediately recognize that the integral in Eq. (7.67) represents convolution of the object field with the $ABCD$ system's impulse response:

$$U^{(i)}(x, y) = \frac{e^{ikz}}{A} \exp\left[\frac{i\pi(D - A^{-1})}{\lambda B}(x^2 + y^2)\right]$$
$$\times U^{(o)}(\xi, \eta) \circledast \left\{\frac{A}{i\lambda B} \exp\left[\frac{i\pi A(\xi^2 + \eta^2)}{\lambda B}\right]\right\}\Bigg|_{(x/A, y/A)}. \quad (7.68)$$

In particular, for imaging systems $B = 0$ and, hence, the expression in the curly brackets reduces to delta-function $\delta^{(2)}(\xi, \eta)$ making the image perfect (apart from a phase factor and magnification A). Since the determinant of the $ABCD$ matrix is unity, $(D - A^{-1})/B = C/A$, and because for $4f$-systems $C/A = 0$, only the constant phase factor e^{ikL_0}/A is acquired in this case.

On taking correlations of the optical field in Eq. (7.67) in the object and image planes, we obtain the relation between the corresponding cross-spectral density functions:

$$W^{(i)}(x_1, y_1, x_2, y_2) = \frac{1}{\lambda^2 B^2} \exp\left[\frac{i\pi D}{\lambda B}(x_2^2 + y_2^2 - x_1^2 - y_1^2)\right]$$

$$\times \int_{-\infty}^{\infty}\int_{-\infty}^{\infty}\int_{-\infty}^{\infty}\int_{-\infty}^{\infty} W^{(o)}(\xi_1, \eta_1, \xi_2, \eta_2)$$

$$\times \exp\left[\frac{i\pi A}{\lambda B}(\xi_2^2 + \eta_2^2 - \xi_1^2 - \eta_1^2)\right]$$

$$\times \exp\left[-\frac{2\pi i}{\lambda B}(\xi_2 x_2 + \eta_2 y_2 - \xi_1 x_1 - \eta_1 y_1)\right]$$

$$\times d\xi_1 d\eta_1 d\xi_2 d\eta_2. \tag{7.69}$$

At coinciding arguments $x_1 = x_2 = x$ and $y_1 = y_2 = y$, we at once obtain the image spectral density

$$S^{(i)}(x, y) = \frac{1}{\lambda^2 B^2} \int_{-\infty}^{\infty}\int_{-\infty}^{\infty}\int_{-\infty}^{\infty}\int_{-\infty}^{\infty} W^{(o)}(\xi_1, \eta_1, \xi_2, \eta_2)$$

$$\times \exp\left[\frac{i\pi A}{\lambda B}(\xi_2^2 + \eta_2^2 - \xi_1^2 - \eta_1^2)\right]$$

$$\times \exp\left[-\frac{2\pi i}{\lambda B}[(\xi_2 - \xi_1)x + (\eta_2 - \eta_1)y]\right]$$

$$\times d\xi_1 d\eta_1 d\xi_2 d\eta_2. \tag{7.70}$$

The $ABCD$ matrix does not necessarily need to involve the imaging elements confined to planes and separated by free-space propagation paths. The effect of some deterministic media with continuously changing refractive index on propagating light can also be taken into account by introducing the $ABCD$ matrices with the elements depending on transverse variables x and y and on propagation distance z. The practically significant examples of such media are Gradient-Index (GRIN) fibers and thick lenses, such as the crystalline lens of the human eye (Korotkova, 2013).

7.2 Polarization Imaging

7.2.1 *Linear system approach*

The analysis presented in the previous sections can be readily generalized to the case when characterization of the polarization properties of the illumination, the object transparency and/or the imaging system is required. Indeed, if on interaction with the incident illumintaion the object transforms its polarization state, this information can be carried to the image plane and used for image formation. This type of imaging is known as *polarization imaging* and is widely used in nature by a variety of animals and in man-made imaging systems, such as telescopes, microscopes, etc. We will use the same notations relating to the system geometry as in the scalar case (see Fig. 7.1).

Consider first the case when all three entities: the light, the object and the system, are deterministic. Let the electric beam-like vector-fields in the object plane, before and after interaction with the object, be

$$\vec{E}^{(\alpha)}(\xi,\eta) = \begin{bmatrix} E_x^{(\alpha)}(\xi,\eta) \\ E_y^{(\alpha)}(\xi,\eta) \end{bmatrix}, \quad \alpha = (s,o), \quad (7.71)$$

respectively, and that in the image plane be

$$\vec{E}^{(i)}(x,y) = \begin{bmatrix} E_x^{(i)}(x,y) \\ E_y^{(i)}(x,y) \end{bmatrix}. \quad (7.72)$$

Assume also that the deterministic *polarimetric object transparency* is characterized by the 2×2 matrix:

$$\overleftrightarrow{o}(\xi,\eta) = \begin{bmatrix} o_{xx}(\xi,\eta) & o_{xy}(\xi,\eta) \\ o_{yx}(\xi,\eta) & o_{yy}(\xi,\eta) \end{bmatrix}, \quad (7.73)$$

and can be simply regarded as its Jones matrix (see Chapter 6). Then the local transformation of the electric field vector on interaction with the transparency is given by the product

$$\vec{E}^{(o)}(\xi,\eta) = \overleftrightarrow{o}(\xi,\eta) \cdot \vec{E}^{(s)}(\xi,\eta). \quad (7.74)$$

Suppose further that the optical system is deterministic and is characterized by 2×2 matrix

$$\overleftrightarrow{h}(\xi, \eta, x, y) = \begin{bmatrix} h_{xx}(\xi, \eta, x, y) & h_{xy}(\xi, \eta, x, y) \\ h_{yx}(\xi, \eta, x, y) & h_{yy}(\xi, \eta, x, y) \end{bmatrix}, \quad (7.75)$$

that can be viewed as the *amplitude spread matrix*. Then the passage of the electric vector-field through this system obeys the law:

$$\vec{E}^{(i)}(x, y) = \int_{-\infty}^{\infty} \int_{-\infty}^{\infty} \overleftrightarrow{h}(\xi, \eta, x, y) \cdot \vec{E}^{(o)}(\xi, \eta) d\xi d\eta. \quad (7.76)$$

Substitution from Eq. (7.74) into Eq. (7.76) implies that

$$\vec{E}^{(i)}(x, y) = \int_{-\infty}^{\infty} \int_{-\infty}^{\infty} \overleftrightarrow{h}(\xi, \eta, x, y) \cdot \overleftrightarrow{o}(\xi, \eta) \cdot \vec{E}^{(s)}(\xi, \eta) d\xi d\eta, \quad (7.77)$$

expression relating the illumination, the object and the optical system properties in the most general deterministic case.

Let the illumination, the object and the system no longer be deterministic but be rather represented by an ensemble of realizations of the corresponding stationary processes. Then on taking correlations in the object (after transparency) and the image electric fields, at a pair of points, we find the relation between the two cross-spectral density matrices:

$$\overleftrightarrow{W}^{(i)}(x_1, y_1, x_2, y_2)$$
$$= \langle \vec{E}^{(i)*}(x_1, y_1) \cdot \vec{E}^{(i)T}(x_2, y_2) \rangle$$
$$= \left\langle \int_{-\infty}^{\infty} \int_{-\infty}^{\infty} \int_{-\infty}^{\infty} \int_{-\infty}^{\infty} \overleftrightarrow{h}^*(\xi_1, \eta_1, x_1, y_1) \cdot \vec{E}^{(o)*}(\xi_1, \eta_1) \right.$$
$$\left. \cdot \vec{E}^{(o)T}(\xi_2, \eta_2) \cdot \overleftrightarrow{h}^T(\xi_2, \eta_2, x_2, y_2) d\xi_1 d\eta_1 d\xi_2 d\eta_2 \right\rangle, \quad (7.78)$$

where superscript T stands for matrix transpose. Unlike in the scalar case, the angular brackets cannot be simply passed to the cross-spectral density matrix that appearing in the integrand. This is only true if the optical system remains deterministic, in which case we deduce that

$$\overleftrightarrow{W}^{(i)}(x_1, y_1, x_2, y_2) = \int_{-\infty}^{\infty}\int_{-\infty}^{\infty}\int_{-\infty}^{\infty}\int_{-\infty}^{\infty} \overleftrightarrow{h}^*(\xi_1, \eta_1, x_1, y_1)$$
$$\cdot \overleftrightarrow{W}^{(o)}(\xi_1, \eta_1, \xi_2, \eta_2)$$
$$\cdot \overleftrightarrow{h}^T(\xi_2, \eta_2, x_2, y_2) d\xi_1 d\eta_1 d\xi_2 d\eta_2. \quad (7.79)$$

For individual cross-spectral density matrix elements, we can also write

$$W^{(i)}_{xx} = \iiiint [H^{xx}_{xx} W^{(o)}_{xx} + H^{xx}_{xy} W^{(o)}_{xy} + H^{xx}_{yx} W^{(o)}_{yx} + H^{xx}_{yy} W^{(o)}_{yy}]$$
$$\times d\xi_1 d\eta_1 d\xi_2 d\eta_2,$$

$$W^{(i)}_{xy} = \iiiint [H^{xy}_{xx} W^{(o)}_{xx} + H^{xy}_{xy} W^{(o)}_{xy} + H^{xy}_{yx} W^{(o)}_{yx} + H^{xy}_{yy} W^{(o)}_{yy}]$$
$$\times d\xi_1 d\eta_1 d\xi_2 d\eta_2,$$

$$W^{(i)}_{yx} = \iiiint [H^{yx}_{xx} W^{(o)}_{xx} + H^{yx}_{xy} W^{(o)}_{xy} + H^{yx}_{yx} W^{(o)}_{yx} + H^{yx}_{yy} W^{(o)}_{yy}]$$
$$\times d\xi_1 d\eta_1 d\xi_2 d\eta_2,$$

$$W^{(i)}_{yy} = \iiiint [H^{yy}_{xx} W^{(o)}_{xx} + H^{yy}_{xy} W^{(o)}_{xy} + H^{yy}_{yx} W^{(o)}_{yx} + H^{yy}_{yy} W^{(o)}_{yy}]$$
$$\times d\xi_1 d\eta_1 d\xi_2 d\eta_2, \quad (7.80)$$

where the limits of integration and the arguments of the W-matrix on the left-hand and on the right-hand sides are the same as in Eq. (7.79) but were dropped for brevity. Relation (7.80) can also be expressed as

$$\vec{W}^{(i)}_V(x_1, y_1, x_2, y_2)$$
$$= \int_{-\infty}^{\infty}\int_{-\infty}^{\infty}\int_{-\infty}^{\infty}\int_{-\infty}^{\infty} \overleftrightarrow{H}_V(\xi_1, \eta_1, x_1, y_1, \xi_2, \eta_2, x_2, y_2)$$
$$\cdot \vec{W}^{(o)}_V(\xi_1, \eta_1, \xi_2, \eta_2) d\xi_1 d\eta_1 d\xi_2 d\eta_2, \quad (7.81)$$

where $\vec{W}^{(i)}_V$ and $\vec{W}^{(o)}_V$ are the four-dimensional vectors formed from the cross-spectral density matrix components and matrix $\overleftrightarrow{H}_V$ is a 4×4 matrix.

The coefficient $H^{pq}_{kl}(x_1,y_1,\xi_1,\eta_1,x_2,y_2,\xi_2,\eta_2)$ establishes the linear scalar map from element $W^{(0)}_{kl}$ of the object matrix to element $W^{(i)}_{pq}$ of the image matrix. Explicitly, they are given by the following products:

$$\begin{aligned}
H^{xx}_{xx} &= h^*_{xx}h_{xx}, & H^{xx}_{xy} &= h^*_{xx}h_{xy}, & H^{xx}_{yx} &= h^*_{xy}h_{xx} & H^{xx}_{yy} &= h^*_{xy}h_{xy}, \\
H^{xy}_{xx} &= h^*_{xx}h_{yx}, & H^{xy}_{xy} &= h^*_{xx}h_{yy}, & H^{xy}_{yx} &= h^*_{xy}h_{yx}, & H^{xy}_{yy} &= h^*_{xy}h_{yy}, \\
H^{yx}_{xx} &= h^*_{yx}h_{xx}, & H^{yx}_{xy} &= h^*_{yx}h_{xy}, & H^{yx}_{yx} &= h^*_{yy}h_{xx}, & H^{yx}_{yy} &= h^*_{yy}h_{xy}, \\
H^{yy}_{xx} &= h^*_{yx}h_{yx}, & H^{yy}_{xy} &= h^*_{yx}h_{yy}, & H^{yy}_{yx} &= h^*_{yy}h_{yx}, & H^{yy}_{yy} &= h^*_{yy}h_{yy},
\end{aligned}$$
(7.82)

the arguments being omitted. If the imaging system is random, then the coefficients become $\langle H^{pq}_{kl}(x_1,y_1,\xi_1,\eta_1,x_2,y_2,\xi_2,\eta_2)\rangle$ where averaging is performed over the ensemble of the system realizations. We then define matrix

$$\overleftrightarrow{H}_V(x_1,y_1,\xi_1,\eta_1,x_2,y_2,\xi_2,\eta_2) \\ = [\langle H^{pq}_{kl}(x_1,y_1,\xi_1,\eta_1,x_2,y_2,\xi_2,\eta_2)\rangle], \qquad (7.83)$$

which is applicable in both deterministic and random cases.

We will now elucidate the relation among the cross-spectral density matrices of the field before and after interaction with the object polarization transparency, on correlating the electric fields in Eq. (7.74):

$$\begin{aligned}
\overleftrightarrow{W}^{(o)}(\xi_1,\eta_1,\xi_2,\eta_2) &= \langle \vec{E}^{(o)*}(\xi_1,\eta_1)\cdot\vec{E}^{(o)T}(\xi_2,\eta_2)\rangle \\
&= \langle \overleftrightarrow{o}^*(\xi_1,\eta_1)\cdot\vec{E}^{(s)*}(\xi_1,\eta_1)\cdot\vec{E}^{(s)T}(\xi_2,\eta_2)\cdot\overleftrightarrow{o}^T(\xi_2,\eta_2)\rangle.
\end{aligned}$$
(7.84)

In the case of a deterministic object, we then deduce:

$$\overleftrightarrow{W}^{(o)}(\xi_1,\eta_1,\xi_2,\eta_2) = \overleftrightarrow{o}^*(\xi_1,\eta_1)\cdot\overleftrightarrow{W}^{(s)}(\xi_1,\eta_1,\xi_2,\eta_2)\cdot\overleftrightarrow{o}^T(\xi_2,\eta_2). \qquad (7.85)$$

This expression can be written element by element as

$$W_{xx}^{(o)} = O_{xx}^{xx}W_{xx}^{(s)} + O_{xy}^{xx}W_{xy}^{(s)} + O_{yx}^{xx}W_{yx}^{(s)} + O_{yy}^{xx}W_{yy}^{(s)},$$
$$W_{xy}^{(o)} = O_{xx}^{xy}W_{xx}^{(s)} + O_{xy}^{xy}W_{xy}^{(s)} + O_{yx}^{xy}W_{yx}^{(s)} + O_{yy}^{xy}W_{yy}^{(s)},$$
$$W_{yx}^{(o)} = O_{xx}^{yx}W_{xx}^{(s)} + O_{xy}^{yx}W_{xy}^{(s)} + O_{yx}^{yx}W_{yx}^{(s)} + O_{yy}^{yx}W_{yy}^{(s)},$$
$$W_{yy}^{(o)} = O_{xx}^{yy}W_{xx}^{(s)} + O_{xy}^{yy}W_{xy}^{(s)} + O_{yx}^{yy}W_{yx}^{(s)} + O_{yy}^{yy}W_{yy}^{(s)}.$$

(7.86)

Alternatively, using four-dimensional vectors $\vec{W}_V^{(o)}$ and $\vec{W}_V^{(s)}$ and 4×4 matrix $\overleftrightarrow{O}_V$, we can express relation (7.86) in the compact form as

$$\vec{W}_V^{(o)}(\xi_1, \eta_1, \xi_2, \eta_2) = \overleftrightarrow{O}_V(\xi_1, \eta_1, \xi_2, \eta_2) \cdot \vec{W}_V^{(s)}(\xi_1, \eta_1, \xi_2, \eta_2). \quad (7.87)$$

Here, each of the 16 coefficients of the form $O_{kl}^{pq}(\xi_1, \eta_1, \xi_2, \eta_2)$ represents the linear map from the source cross-spectral density matrix element $W_{kl}^{(s)}$ incident onto transparency to element $W_{pq}^{(o)}$ emerging from it:

$$\begin{array}{llll}
O_{xx}^{xx} = o_{xx}^* o_{xx}, & O_{xy}^{xx} = o_{xx}^* o_{xy}, & O_{yx}^{xx} = o_{xy}^* o_{xx}, & O_{yy}^{xx} = o_{xy}^* o_{xy}, \\
O_{xx}^{xy} = o_{xx}^* o_{yx}, & O_{xy}^{xy} = o_{xx}^* o_{yy}, & O_{yx}^{xy} = o_{xy}^* o_{yx}, & O_{yy}^{xy} = o_{xy}^* o_{yy}, \\
O_{xx}^{yx} = o_{yx}^* o_{xx}, & O_{xy}^{yx} = o_{yx}^* o_{xy}, & O_{yx}^{yx} = o_{yy}^* o_{xx}, & O_{yy}^{yx} = o_{yy}^* o_{xy}, \\
O_{xx}^{yy} = o_{yx}^* o_{yx}, & O_{xy}^{yy} = o_{yx}^* o_{yy}, & O_{yx}^{yy} = o_{yy}^* o_{yx}, & O_{yy}^{yy} = o_{yy}^* o_{yy}.
\end{array}$$

(7.88)

We will now directly relate the cross-spectral density matrices of the illumination field, of the image field and the correlation in the object transparency. This can be achieved by correlating the left-hand and the right-hand sides of relation (7.77) at a pair of positions:

$$\overleftrightarrow{W}^{(i)}(x_1, y_1, x_2, y_2) = \left\langle \int_{-\infty}^{\infty}\int_{-\infty}^{\infty}\int_{-\infty}^{\infty}\int_{-\infty}^{\infty} \overleftrightarrow{h}^*(\xi_1, \eta_1, x_1, y_1) \right.$$
$$\cdot \overleftrightarrow{o}^*(\xi_1, \eta_1) \cdot \vec{E}^{(s)*}(\xi_1, \eta_1) \cdot \vec{E}^{(s)T}(\xi_2, \eta_2)$$
$$\left. \cdot \overleftrightarrow{o}^T(\xi_2, \eta_2) \cdot \overleftrightarrow{h}^T(\xi_2, \eta_2, x_2, y_2) d\xi_1 d\eta_1 d\xi_2 d\eta_2 \right\rangle.$$

(7.89)

When both the object and the optical system are deterministic, Eq. (7.89) reduces to that between the two cross-spectral density matrices:

$$\overleftrightarrow{W}^{(i)}(x_1, y_1, x_2, y_2)$$
$$= \int_{-\infty}^{\infty}\int_{-\infty}^{\infty}\int_{-\infty}^{\infty}\int_{-\infty}^{\infty} \overleftrightarrow{h}^*(\xi_1, \eta_1, x_1, y_1) \cdot \overleftrightarrow{o}^*(\xi_1, \eta_1)$$
$$\cdot \overleftrightarrow{W}^{(s)}(\xi_1, \eta_1, \xi_2, \eta_2) \cdot \overleftrightarrow{o}^T(\xi_2, \eta_2)$$
$$\cdot \overleftrightarrow{h}^T(\xi_2, \eta_2, x_2, y_2) d\xi_1 d\eta_1 d\xi_2 d\eta_2. \tag{7.90}$$

Formula (7.89) is the most general result of this section. In this compact form it cannot be further factorized into the product of correlations in the illumination, the object and the optical system. We can again use the 4-dimensional vector and the 4 × 4 matrix approach. Indeed, the combination of expressions (7.81) and (7.87) results in

$$\overrightarrow{W}_V^{(i)}(x_1, y_1, x_2, y_2)$$
$$= \int_{-\infty}^{\infty}\int_{-\infty}^{\infty}\int_{-\infty}^{\infty}\int_{-\infty}^{\infty} \overleftrightarrow{\Xi}_V(\xi_1, \eta_1, x_1, y_1, \xi_2, \eta_2, x_2, y_2)$$
$$\cdot \overrightarrow{W}_V^{(s)}(\xi_1, \eta_1, \xi_2, \eta_2) d\xi_1 d\eta_1 d\xi_2 d\eta_2, \tag{7.91}$$

where

$$\overleftrightarrow{\Xi}_V(\xi_1, \eta_1, x_1, y_1, \xi_2, \eta_2, x_2, y_2)$$
$$= \overleftrightarrow{H}_V(\xi_1, \eta_1, x_1, y_1, \xi_2, \eta_2, x_2, y_2) \cdot \overleftrightarrow{O}_V(\xi_1, \eta_1, \xi_2, \eta_2) \tag{7.92}$$

is the 4 × 4 matrix describing the properties of the object and the imaging system as a whole. Depending on whether the object and/or the system are deterministic or random, coefficients H_{kl}^{pq} and O_{kl}^{pq} may factorize or represent statistical averages of their realizations. Here as before the subscript kl and superscript pq indicate the elements of the cross-spectral density matrix of the source and of the image, respectively. Its 16 coefficients have the forms

$$
\begin{aligned}
\Xi^{xx}_{xx} &= H^{xx}_{xx}O^{xx}_{xx} + H^{xx}_{xy}O^{xy}_{xx} + H^{xx}_{yx}O^{yx}_{xx} + H^{xx}_{yy}O^{yy}_{xx}, \\
\Xi^{xx}_{xy} &= H^{xx}_{xx}O^{xx}_{xy} + H^{xx}_{xy}O^{xy}_{xy} + H^{xx}_{yx}O^{yx}_{xy} + H^{xx}_{yy}O^{yy}_{xy}, \\
\Xi^{xx}_{yx} &= H^{xx}_{xx}O^{xx}_{yx} + H^{xx}_{xy}O^{xy}_{yx} + H^{xx}_{yx}O^{yx}_{yx} + H^{xx}_{yy}O^{yy}_{yx}, \\
\Xi^{xx}_{yy} &= H^{xx}_{xx}O^{xx}_{yy} + H^{xx}_{xy}O^{xy}_{yy} + H^{xx}_{yx}O^{yx}_{yy} + H^{xx}_{yy}O^{yy}_{yy}, \\
\Xi^{xy}_{xx} &= H^{xy}_{xx}O^{xx}_{xx} + H^{xy}_{xy}O^{xy}_{xx} + H^{xy}_{yx}O^{yx}_{xx} + H^{xy}_{yy}O^{yy}_{xx}, \\
\Xi^{xy}_{xy} &= H^{xy}_{xx}O^{xx}_{xy} + H^{xy}_{xy}O^{xy}_{xy} + H^{xy}_{yx}O^{yx}_{xy} + H^{xy}_{yy}O^{yy}_{xy}, \\
\Xi^{xy}_{yx} &= H^{xy}_{xx}O^{xx}_{yx} + H^{xy}_{xy}O^{xy}_{yx} + H^{xy}_{yx}O^{yx}_{yx} + H^{xy}_{yy}O^{yy}_{yx}, \\
\Xi^{xy}_{yy} &= H^{xy}_{xx}O^{xx}_{yy} + H^{xy}_{xy}O^{xy}_{yy} + H^{xy}_{yx}O^{yx}_{yy} + H^{xy}_{yy}O^{yy}_{yy}, \\
\Xi^{yx}_{xx} &= H^{yx}_{xx}O^{xx}_{xx} + H^{yx}_{xy}O^{xy}_{xx} + H^{yx}_{yx}O^{yx}_{xx} + H^{yx}_{yy}O^{yy}_{xx}, \\
\Xi^{yx}_{xy} &= H^{yx}_{xx}O^{xx}_{xy} + H^{yx}_{xy}O^{xy}_{xy} + H^{yx}_{yx}O^{yx}_{xy} + H^{yx}_{yy}O^{yy}_{xy}, \\
\Xi^{yx}_{yx} &= H^{yx}_{xx}O^{xx}_{yx} + H^{yx}_{xy}O^{xy}_{yx} + H^{yx}_{yx}O^{yx}_{yx} + H^{yx}_{yy}O^{yy}_{yx}, \\
\Xi^{yx}_{yy} &= H^{yx}_{xx}O^{xx}_{yy} + H^{yx}_{xy}O^{xy}_{yy} + H^{yx}_{yx}O^{yx}_{yy} + H^{yx}_{yy}O^{yy}_{yy}, \\
\Xi^{yy}_{xx} &= H^{yy}_{xx}O^{xx}_{xx} + H^{yy}_{xy}O^{xy}_{xx} + H^{yy}_{yx}O^{yx}_{xx} + H^{yy}_{yy}O^{yy}_{xx}, \\
\Xi^{yy}_{xy} &= H^{yy}_{xx}O^{xx}_{xy} + H^{yy}_{xy}O^{xy}_{xy} + H^{yy}_{yx}O^{yx}_{xy} + H^{yy}_{yy}O^{yy}_{xy}, \\
\Xi^{yy}_{yx} &= H^{yy}_{xx}O^{xx}_{yx} + H^{yy}_{xy}O^{xy}_{yx} + H^{yy}_{yx}O^{yx}_{yx} + H^{yy}_{yy}O^{yy}_{yx}, \\
\Xi^{yy}_{yy} &= H^{yy}_{xx}O^{xx}_{yy} + H^{yy}_{xy}O^{xy}_{yy} + H^{yy}_{yx}O^{yx}_{yy} + H^{yy}_{yy}O^{yy}_{yy}.
\end{aligned} \quad (7.93)
$$

The relation (7.91) could also be rewritten for the generalized Stokes parameters, since they are linear combinations of the cross-spectral density matrix elements (see Chapter 2). Such modification, being rather technical and cumbersome, however would make it possible to directly prescribe images for the four Stokes parameters modulated by the object trasparency.

7.2.2 Isoplanatic polarimetric systems

Some special cases of Eq. (7.91) are easily deduced. First of all, the expressions similar to Eqs. (7.40) and (7.41) can be obtained for each of the four elements of the vector on the left-hand side of Eq. (7.91), at the coinciding spatial arguments, under the assumptions that the elements of the illumination's cross-spectral density matrix are homogeneous and the polarization-varying imaging system is isoplanatic. Further, the cases of coherent and incoherent imaging can also be deduced for the electromagnetic case, but they depend on the adopted definition of the degree of electromagnetic coherence (see Chapter 2). Let us outline the basic approach to the treatment of the polarimetrically isoplanatic systems.

If the optical system is isoplanatic in all four components of matrix $\overleftrightarrow{h}$:

$$\overleftrightarrow{h}(\xi, \eta, x, y) = \overleftrightarrow{h}(x - \xi, y - \eta), \tag{7.94}$$

then the electric field in the image plane is given by the expression

$$\begin{aligned}\vec{E}^{(i)}(x,y) &= \int_{-\infty}^{\infty}\int_{-\infty}^{\infty} \overleftrightarrow{h}(x-\xi, y-\eta) \cdot \overleftrightarrow{o}(\xi, \eta) \cdot \vec{E}^{(s)}(\xi, \eta) d\xi d\eta \\ &= \overleftrightarrow{h}(\xi, \eta) \circledast \{\overleftrightarrow{o}(\xi, \eta) \cdot \vec{E}^{(s)}(\xi, \eta)\}\Big|_{(x,y)} \\ &= \overleftrightarrow{h}(\xi, \eta) \circledast \vec{E}^{(o)}(\xi, \eta), \end{aligned} \tag{7.95}$$

where the convolution operator is applied "element by element".

On taking the two-dimensional Fourier transform of the expression above and applying the convolution theorem, we find that it becomes

$$\vec{\mathcal{E}}^{(i)}(\kappa_x, \kappa_y) = \overleftrightarrow{\mathfrak{h}}(\kappa_x, \kappa_y) \vec{\mathcal{E}}^{(o)}(\kappa_x, \kappa_y), \tag{7.96}$$

in the spatial-frequency domain, where $\overleftrightarrow{\mathfrak{h}}(\kappa_x, \kappa_y)$ is the generalization of the amplitude transfer function, and can then be termed the *amplitude transfer matrix*. It carries infomation about modulation of the spatial-frequency content of the electric field right after its interaction with the object and on passage through the optical system.

Further, on taking correlations in Eq. (7.95), we note that the H-matrix elements now have the shift-invariant dependence on the spatial arguments, i.e., $\langle H_{kl}^{pq}(x_1 - \xi_1, y_1 - \eta_1, x_2 - \xi_2, y_2 - \eta_2)\rangle$. In this case, the relation between the object and the image cross-spectral density matrices can be written via convolution operation as

$$\overleftrightarrow{W}_V^{(i)}(x_1, y_1, x_2, y_2) = \overleftrightarrow{H}_V(\xi_1, \eta_1, \xi_2, \eta_2) \circledast \overleftrightarrow{W}_V^{(o)}(\xi_1, \eta_1, \xi_2, \eta_2). \tag{7.97}$$

Now, let us define the 16-element matrix by using the four-dimensional Fourier transform

$$\overleftrightarrow{\mathfrak{H}}_V(\kappa_{x1}, \kappa_{y1}, \kappa_{x2}, \kappa_{y2})$$
$$= [\mathfrak{H}_{kl}^{pq}(\kappa_{x1}, \kappa_{y1}, \kappa_{x2}, \kappa_{y2})]$$
$$= FT[\langle H_{kl}^{pq}(x_1 - \xi_1, y_1 - \eta_1, x_2 - \xi_2, y_2 - \eta_2)\rangle], \tag{7.98}$$

that we may term the *Correlation Transfer Matrix* (CTM) since it can be viewed as the generalization of the correlation transfer function (see Eq. (7.16)) to the polarimetric domain. Each element of this matrix may be expressed in terms of the elements of the amplitude transfer matrix $\overleftrightarrow{\mathfrak{h}}(\kappa_x, \kappa_y)$. For example,

$$FT[\langle H_{xx}^{xx}(x_1 - \xi_1, y_1 - \eta_1, x_2 - \xi_2, y_2 - \eta_2)\rangle]$$
$$= \langle \mathfrak{H}_{xx}^{xx}(\kappa_{x1}, \kappa_{y1}, \kappa_{x2}, \kappa_{y2})\rangle$$
$$= \langle \mathfrak{h}_{xx}^*(-\kappa_{x1}, -\kappa_{y1}) \mathfrak{h}_{xx}(\kappa_{x2}, \kappa_{y2})\rangle. \tag{7.99}$$

The remaining fifteen elements can be similarly obtained using Eq. (7.82).

Next, on taking the four-dimensional Fourier transform of Eq. (7.97) and using the convolution theorem, we arrive at the relation

$$\overrightarrow{\mathfrak{W}}_V^{(i)}(\kappa_{x1}, \kappa_{y1}, \kappa_{x2}, \kappa_{y2})$$
$$= \overleftrightarrow{\mathfrak{H}}_V(-\kappa_{x1}, -\kappa_{y1}, \kappa_{x2}, \kappa_{y2}) \cdot \overrightarrow{\mathfrak{W}}_V^{(o)}(\kappa_{x1}, \kappa_{y1}, \kappa_{x2}, \kappa_{y2}). \tag{7.100}$$

If, in addition, the illumination is homogeneous, i.e., the elements of its cross-spectral density matrix obey the relation

$$W_{ij}^{(s)}(\xi_1, \eta_1, \xi_2, \eta_2) = W_{ij}^{(s)}(\xi_1 - \xi_2, \eta_1 - \eta_2), \quad (i, j = x, y), \tag{7.101}$$

then just like for the scalar fields (see Eqs. (7.39) and (7.40)) we can express the spatial spectrum vector of the image,

$$\vec{\mathfrak{S}}_V^{(i)}(\kappa_x, \kappa_y) = \vec{\mathfrak{W}}_V^{(i)}(\kappa_x, \kappa_y, \kappa_x, \kappa_y) \qquad (7.102)$$

as

$$\vec{\mathfrak{S}}_V^{(i)}(\kappa_x, \kappa_y) = \iint_{-\infty}^{\infty} \overleftrightarrow{\mathfrak{D}}_V(\kappa_x - X, \kappa_y - Y, X, Y)$$
$$\cdot \vec{\mathfrak{J}}_V(X, \kappa_x, Y, \kappa_y) dX dY, \qquad (7.103)$$

where

$$\vec{\mathfrak{J}}_V(X, \kappa_x, Y, \kappa_y)$$
$$= \iint_{-\infty}^{\infty} \overleftrightarrow{\mathfrak{H}}_V(K_x + \kappa_x - X, K_y + \kappa_y - Y, X - K_x, Y - K_y)$$
$$\cdot \vec{\mathfrak{W}}^{(s)}(K_x, K_y) dK_x dK_y. \qquad (7.104)$$

Vector $\vec{\mathfrak{J}}_V(X, \kappa_x, Y, \kappa_y)$ is the polarimetric generalization of the *transmission cross-coefficient* of the imaging system. It is independent of the properties of the object.

7.3 Two-point Resolution

7.3.1 *Imaging by structured illumination*

Before considering the classic problem of imaging of two identical point objects, let us first extend the results of the previous sections to the case of imaging systems which use inhomogeneous illumination, i.e., for which condition (7.19) is released. Adapting the Bochner representation (5.48) to classic imaging system notations yields:

$$W^{(s)}(\xi_1, \eta_1, \xi_2, \eta_2) = \int_{-\infty}^{\infty} \int_{-\infty}^{\infty} H_0^*(\xi_1, \eta_1, v_x, v_y) p(v_x, v_y)$$
$$\times H_0(\xi_2, \eta_2, v_x, v_y) dv_x dv_y, \qquad (7.105)$$

where $p(v_x, v_y)$ is a non-negative function and $H_0(\xi, \eta, v_x, v_y)$ is any, generally complex-valued, function. We assume here that the object is deterministic and, hence, expression (7.6) relating illumination before and after interaction with the object holds. Then on substituting from Eqs. (7.6) and (7.105) into Eq. (7.4) and calculating the result at the coinciding spatial arguments we obtain, after changing the order of integrals, the following expression for the image spectral density:

$$S^{(i)}(x,y) = \int_{-\infty}^{\infty}\int_{-\infty}^{\infty} p(v_x,v_y)dv_x dv_y$$
$$\times \left|\int_{-\infty}^{\infty}\int_{-\infty}^{\infty} H_0(v_x,v_y,\xi,\eta)h(\xi,\eta,x,y)o(\xi,\eta)d\xi d\eta\right|^2. \tag{7.106}$$

This expression implies that an image of a deterministic object formed with stationary light can be regarded as the incoherent superposition of images of object points carried by the imaging system specified by h with weights p via modes H_0 (Liang et al., 2021). In a special but broadly used symmetric case when $h(\xi,\eta,x,y) = h(-\xi-x,-\eta-y)$, the spectral density in the image plane reduces to the expression

$$S^{(i)}(x,y) = \int_{-\infty}^{\infty}\int_{-\infty}^{\infty} p(v_x,v_y)dv_x dv_y$$
$$\times \left|\int_{-\infty}^{\infty}\int_{-\infty}^{\infty} H_0(v_x,v_y,\xi,\eta)h(-\xi-x,-\eta-y)\right.$$
$$\left.\times o(\xi,\eta)d\xi d\eta\right|^2. \tag{7.107}$$

Alternatively,

$$S^{(i)}(x,y) = \int_{-\infty}^{\infty}\int_{-\infty}^{\infty} p(v_x,v_y)dv_x dv_y$$
$$\times \left|FT^{-1}\left[\mathfrak{G}(v_x,v_y,f_x,f_y)\mathfrak{h}(f_x,f_y)\right](-x,-y)\right|^2, \tag{7.108}$$

where

$$G(v_x,v_y,f_x,f_y) = H_0(v_x,v_y,\xi,\eta)o(\xi,\eta), \tag{7.109}$$

and $\mathfrak{G}$ is its two-dimensional Fourier transform. For Schell-like sources

$$H_0(v_x, v_y, \xi, \eta) = a(\xi, \eta)\exp[2\pi i(v_x\xi + v_y\eta)], \tag{7.110}$$

and Eq. (7.107) reduces further to the expression (Yamazoe, 2012)

$$S^{(i)}(x,y) = \int_{-\infty}^{\infty}\int_{-\infty}^{\infty} p(v_x, v_y)dv_x dv_y$$
$$\times \left|FT^{-1}\left[\mathfrak{G}_s(f_x - v_x, f_y - v_y)\mathfrak{h}(f_x, f_y)\right](-x, -y)\right|^2, \tag{7.111}$$

where $\mathfrak{G}_s$ is the two-dimensional Fourier transform of product

$$G_s(\xi, \eta) = a(\xi, \eta)o(\xi, \eta). \tag{7.112}$$

We have seen in Chapter 5 that with suitable structuring of the source coherence states, namely, by choosing specific profiles for p and H_0, it is possible to form any desirable spectral density of the propagating beam, at least in the far field. This opportunity can also be employed in imaging systems for image quality control or for highlighting specific object's details. Of course, the use of structured illumination can also be readily extended to the polarimetric domain and, therefore, the imaging of the single-point elements of the W-matrix (or the Stokes parameters) can be characterized and controlled.

7.3.2 Rayleigh resolution criterion

The general approach taken in the previous sections for characterization of image quality of classic linear imaging systems, which addresses their filtering capabilities for all involved spatial frequencies, while being exhaustive, is also very complex and is not required in a number of practical situations. For instance, the classic problem of resolving two spatially separated point objects by a given imaging system can be solved in a much simpler manner. Historically, such a resolution problem was first treated by Lord Rayleigh (1879) for the special case of an incoherent imaging system. This choice was made several decades prior to the invention of the laser and, hence, was dictated by the omnipresent contemporary usage of sunlight or candle light.

The classic Rayleigh criterion for resolving two identical hard apertures, can be set for free-space propagation (the simplest imaging system), whether in one or two dimensions, or for an optical system of any complexity, including telescopes, microspcopes, spectroscopes, etc. For a typical system, two small circular apertures produce the Airy-disk diffraction patterns, in the far field of the object (see Chapter 4). Both distributions have central maxima behind the two apertures and a number of smaller oscillations around them. Moreover, the widths of the two central peaks may be large enough to have a significant overlap. Hence, it is of a rather arbitrary choice when to qualify the two points as resolved or not resolved. According to the Rayleigh criterion, the *two (identical) point objects are just resolved if the first minimum of the image produced by one of them coincides with the central maximum produced by the other*. Other resolution criteria set for incoherent illumination have been proposed, including the one given in Sparrow (1916), mostly because of its adaptability to sensors being more refined than the human eye. The Sparrow criterion states that *for two (identical) point objects to be just resolved, the value of the intensity in the center of the image pattern must be equal to those at the individual intensity maxima*.

We will analyze the Rayleigh image resolution criterion for a telecentric imaging system (see Fig. 7.2) with a complex-valued transparency inserted in the pupil plane. It is a slightly modified system as compared with that discussed earlier in which no optical element was placed in the pupil plane. In the presence of such a transparency, the ABCD matrix method and the generalized Huygens–Fresnel integral cannot be directly applied. Instead, a two-stage calculation is carried out in which the cross-spectral density is evaluated on passing through the pupil plane and then used again as the secondary source cross-spectral density for the rest of the system (Goodman, 2000, Chapter 7). We omit this technical and somewhat cumbersome calculation here. It can be shown that the amplitude spread function $h(\xi, \eta, x, y)$ of such a system with $f_1 = f_2 = z_1 = z_2 = f$ has the form

$$h(\xi, \eta, x, y) = -\frac{1}{\lambda^2 f^2} \int_{-\infty}^{\infty} \int_{-\infty}^{\infty} P(x_p, y_p)$$
$$\times \exp\left[-\frac{ik}{f}[x_p(x+\xi) + y_p(y+\eta)]\right] dx_p dy_p, \quad (7.113)$$

where x_p and y_p are the Cartesian coordinates of a point in the pupil plane, and the complex-valued transparency is defined as

$$P(x_p, y_p) = |P(x_p, y_p)| \exp\{i \arg[P(x_p, y_p)]\}. \quad (7.114)$$

Suppose that the two small pinholes are placed on the ξ-axis symmetrically with respect to the origin and separated by distance d. We can then express the object transparency as

$$o(\xi, \eta) = \delta^{(2)}\left(\xi - \frac{d}{2}, \eta\right) + \delta^{(2)}\left(\xi + \frac{d}{2}, \eta\right), \quad (7.115)$$

where $\delta^{(2)}$ is the two-dimensional Dirac delta-function. On substituting from Eqs. (7.115) and (7.113) with

$$a(\xi, \eta) = \exp\left(-\frac{\xi^2 + \eta^2}{4\sigma^2}\right), \quad (7.116)$$

into Eqs. (7.111) and (7.112), we find that the spectral density in the image plane becomes

$$S^{(i)}(x, y) = \exp\left(-\frac{d^2}{8\sigma^2}\right) \int_{-\infty}^{\infty} \int_{-\infty}^{\infty} p(v_x, v_y)$$
$$\times \left(|S_+|^2 + |S_-|^2 + 2Re[S_+ S_-^* \mu(d, 0)]\right) dv_x dv_y, \quad (7.117)$$

where μ is the degree of coherence between the fields at the pinholes being the Fourier transform of P, normalized by its maximum value,

and
$$S_\pm = \mathfrak{P}\left(x \pm \frac{d}{2}, y\right), \qquad (7.118)$$

$\mathfrak{P}(x,y)$ being the Fourier transform of the pupil transparency $P(f_x, f_y)$, with $f_x = x_p/\lambda f$ and $f_y = y_p/\lambda f$. If P is a flat circular aperture of radius R centered on the optical axis, then

$$S_\pm = \frac{2\pi R^2}{\lambda^2 f^2} \frac{J_1\left[2\pi R \sqrt{(x \pm d/2)^2 + y^2}/\lambda f\right]}{2\pi R \sqrt{(x \pm d/2)^2 + y^2}/\lambda f}, \qquad (7.119)$$

where J_1 is the Bessel function of the first kind and order 1. In formula (7.117) terms, $|S_+|^2$ and $|S_-|^2$ are the images of the two points produced individually and the term containing $\mu(d,0)$ is the interference term, depending on the choice of $p(v_x, v_y)$. If illumination at the two pinholes is incoherent, i.e., if $\mu(d,0) = 0$ in Eq. (7.117), then, according to the Rayleigh criterion (in two dimensions), the minimum distance at which the two points can be resolved is

$$d_R = 0.61 \frac{\lambda f}{R}. \qquad (7.120)$$

If illumination is coherent or partially coherent: $0 < |\mu(d,0)| \leq 1$, then both the magnitude and the phase of $\mu(d,0)$ determine whether the image resolution is better or worse than that obtained in the incoherent limit. The seminal analysis of this dependence was performed in Grimes and Thompson (1967) for the case of a single lens (2f system). It is evident from Eq. (7.117) that the worst possible resolution occurs if $|\mu(d,0)| = 1$ and $\arg \mu(d,0) = 0$ and perfect resolution is expected if $|\mu(d,0)| = 1$ and $\arg \mu(d,0) = \pi$. In terms of the coherence curve (see Fig. 5.1), for perfect imaging the curve must pass through point $(-1, 0)$ of the complex plane, when separation distance parameter reaches value d. However, the non-negative definiteness condition that must always be obeyed by the illumination's cross-spectral density precludes from forming the degree of coherence corresponding to perfect resolution. However, the coherence states approaching the perfect state can be used: the coherence curve would then pass in the region of the unit circle very close to point $(-1, 0)$. In fact, several models for illumination with structured partial coherence have been applied and the results confirmed that the resolution limit can indeed exceed the Rayleigh limit. Among them are the Gaussian Schell-model illumination with a twist phase (Tong and Korotkova, 2012) and the Laguerre–Gaussian correlated illumination (Liang et al., 2017). In the former case, only the suitably situated pinholes were perfectly resolved, and in the latter case, only a slight improvement was illustrated but uniformly for all locations.

The idea of tayloring of illumination's coherence state to resolve the specific two-point and multi-point objects can also be employed. As was shown in Liang et al. (2021) using the Schell-like illumination with a suitable selection of function $p(v_x, v_y)$ it is possible to resolve two pinholes at a small fraction of the Rayleigh separation limit. For the same two pinholes placed on the ξ-axis, one may set

$$p(v_x, v_y) = \exp\left[-a^2\left[\left(v_x - \frac{b}{2}\right)^2 + v_y^2\right]\right]$$
$$+ \exp\left[-a^2\left[\left(v_x + \frac{b}{2}\right)^2 + v_y^2\right]\right], \quad (7.121)$$

where a and b are real parameters. Using the Fourier-transform relation between p and μ, we obtain the cos-Gaussian correlated degree of coherence

$$\mu(\xi_d, \eta_d) = \exp\left[-\frac{\pi^2(\xi_d^2 + \eta_d^2)}{a^2}\right] \cos\left(\frac{\pi \xi_d}{b}\right). \quad (7.122)$$

Figure 7.3 shows the degree of coherence (7.122) for several values of ratio a/b. For $a/b = 15$, the optimized (solid black) curve passes sufficiently close to value -1. Figure 7.4(a)–(c) shows the resolution of two pinholes with the degree of coherence in Fig. 7.3 for several values of ratio a/b. Figure 7.4(d) inset illustrates the corresponding image with incoherent illumination and Fig. 7.4(e) inset is the x-axis cross-lines for cases Fig. 7.4(a)–(d). As compared with the classic Rayleigh limit d_R, illumination in Eq. (7.122) leads to minimum separable distance of $0.2 d_R$ (for case $a/b = 15$).

Image Formation with Random Light 203

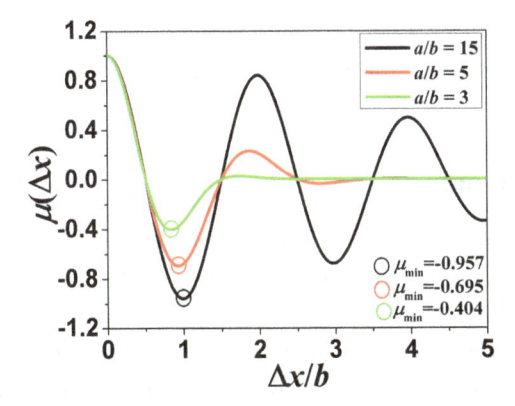

Fig. 7.3 Degree of coherence for optimized imaging. Variable Δx corresponds to ξ in our notation. From Liang *et al.* (2021).

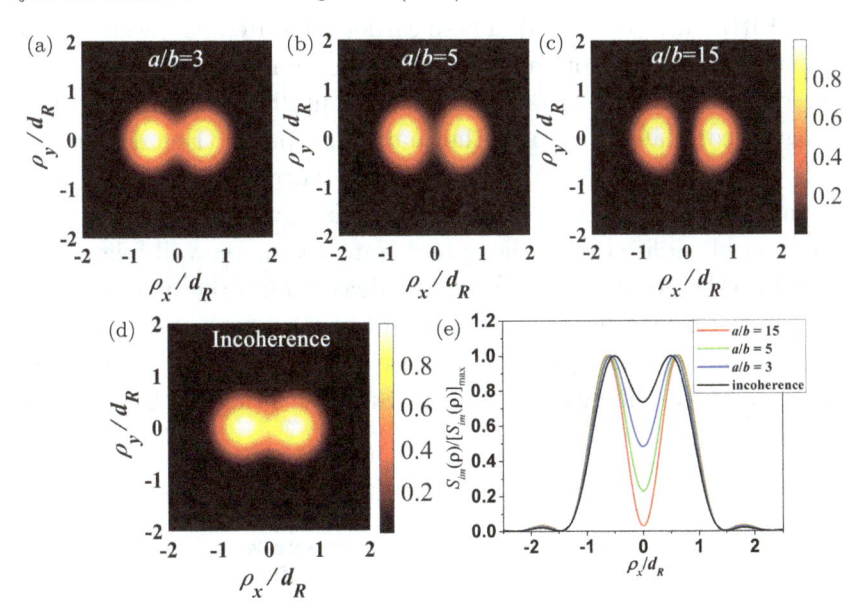

Fig. 7.4 Optimized imaging of two pinholes with structured degree of coherence, ρ_x and ρ_y correspond to x and y in our notation. From Liang *et al.* (2021).

For objects consisting of three or more pinholes, the enhancement in resolution obtained by structuring of the illumination's degree of coherence for a pair of pinholes is partially suppressed (Liang *et al.*, 2021). It was shown in particular that three pinholes placed in the vertices of an equilateral triangle can be resolved by illumination structured in a similar manner as in Eq. (7.122) with minimal separation $d = 0.4d_R$.

7.4 Ghost Imaging

The ghost imaging technique is based on the results of the Hanbury Brown and Twiss experiment described in Chapter 3. Unlike in the classic image formation process where the object's information is carried by the spectral density of light, the ghost image is obtained from the light field's intensity–intensity correlations. The theoretical background for ghost imaging was developed in Klyshko (1988a,b) while the first experimental verification of these ideas was carried out in Pittman et al. (1995). These earlier measurements were based on quantum formulation and used entangled photon pairs. However, later on ghost imaging was also experimentally realized by means of classic correlations with a thermal light source (Bennink et al., 2002). The theoretical explanation of how thermal light can be used for ghost image formation and its comparison with quantum theory were later elucidated (c.f. Cheng and Han, 2004). We will follow this paper in order to introduce the main ideas behind the method.

In Fig. 7.5, a schematic diagram of the lensless ghost imaging scheme is introduced. For simplicity, we will only discuss ghost image formation of one-dimensional objects. Suppose a thermal light source

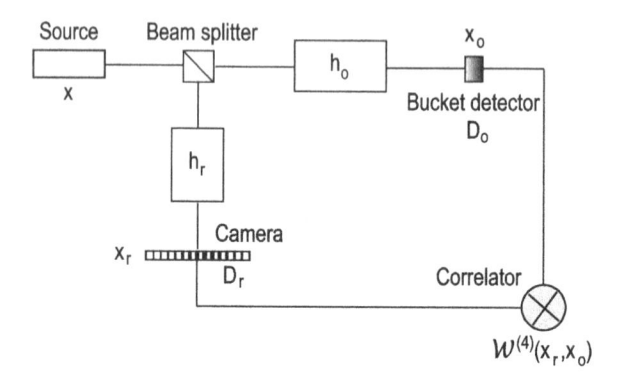

Fig. 7.5 Optical arrangement for lenseless ghost imaging.

radiates an optical field that is split into two branches. The light in the first (object) branch passes through the object located at distance z_o from the source and then is measured by the single-pixel bucket detector placed at distance z_b from the object. The light in the second (reference) branch is recorded by a spatially resolved camera placed at distance z_r from the source. Since in this scheme the source and the object are located in two different planes and the image information is obtained from two different planes, we will use variable x with various subscripts for transverse coordinates of points in these planes. The outputs of the two detectors are then sent to the coincidence circuit and the fourth-order field correlation function in the space–time domain is calculated as (see Chapter 2)

$$\mathcal{W}^{(4)}(x_r, x_o) = \langle \mathcal{U}^*(x_r)\mathcal{U}^*(x_o)\mathcal{U}(x_r)\mathcal{U}(x_o)\rangle$$
$$= \langle \mathcal{I}(x_r)\mathcal{I}(x_o)\rangle, \qquad (7.123)$$

Here, $\mathcal{U}(x_r)$ and $\mathcal{U}(x_o)$ are the optical fields reaching the correlator through the reference path and the object path, respectively. The optical fields propagated through the two branches are

$$\mathcal{U}(x_\alpha) = \int_{-\infty}^{\infty} \mathcal{U}(x) h_\alpha(x, x_\alpha) dx, \quad (\alpha = r, o), \qquad (7.124)$$

where h_α is the (deterministic) amplitude spread function of the branch. Since the reference branch only contains a free-space path, we get

$$h_r(x, x_r) = \frac{\exp[-ikz_r]}{i\lambda z_r} \exp\left[\frac{-i\pi(x - x_r)^2}{\lambda z_r}\right]. \qquad (7.125)$$

The amplitude spread function for the object branch containing free path of length z_o, interaction with object transparency $o(x')$ followed by free path of length z_b takes the form

$$h_o(x, x_o) = \int_{-\infty}^{\infty} \frac{\exp[-ikz_o]}{i\lambda z_o} \exp\left[\frac{-i\pi(x - x')^2}{\lambda z_o}\right]$$
$$\times o(x') \frac{\exp[-ikz_b]}{i\lambda z_b} \exp\left[\frac{-i\pi(x_o - x')^2}{\lambda z_b}\right] dx'. \qquad (7.126)$$

Substitution of Eq. (7.124) into Eq. (7.123) then yields the integral

$$\mathcal{W}^{(4)}(x_r, x_o) = \int_{-\infty}^{\infty}\int_{-\infty}^{\infty}\int_{-\infty}^{\infty}\int_{-\infty}^{\infty} h_r^*(x_1', x_r)$$
$$\times h_o^*(x_2', x_o) h_r(x_1, x_r) h_o(x_2, x_o)$$
$$\times \mathcal{W}^{(4)}(x_1, x_1', x_2, x_2') dx_1 dx_1' dx_2 dx_2', \quad (7.127)$$

where

$$\mathcal{W}^{(4)}(x_1, x_1', x_2, x_2') = \langle \mathcal{U}^*(x_1')\mathcal{U}^*(x_2')\mathcal{U}(x_1)\mathcal{U}(x_2)\rangle \quad (7.128)$$

is the fourth-order optical field correlation function at the source plane. Equation (7.127) indicates that the formed image directly depends on the fourth-order statistical properties of the source. In a number of cases, light source can be regarded as a complex-valued circular Gaussian random prosess with zero mean. Then its fourth-order field moment is uniquely defined by the second-order correlations (see Chapter 1):

$$\mathcal{W}^{(4)}(x_1, x_1', x_2, x_2') = \mathcal{W}(x_1, x_1')\mathcal{W}(x_2, x_2') + \mathcal{W}(x_1, x_2')\mathcal{W}(x_2, x_1'), \quad (7.129)$$

with

$$\mathcal{W}(x_i, x_j') = \langle \mathcal{U}^*(x_i)\mathcal{U}(x_j')\rangle, \quad (i, j = 1, 2), \quad (7.130)$$

being the mutual coherence function of the source. Combination of Eq. (7.129) with Eq. (7.127) implies that

$$\mathcal{W}^{(4)}(x_r, x_o) = \langle \Delta\mathcal{I}(x_r)\Delta\mathcal{I}(x_o)\rangle + \langle \mathcal{I}(x_r)\rangle\langle \mathcal{I}(x_o)\rangle, \quad (7.131)$$

where

$$\langle \mathcal{I}(x_r)\rangle = \int_{-\infty}^{\infty}\int_{-\infty}^{\infty} \mathcal{W}(x_1, x_1') h_r(x_1, x_r) h_r^*(x_1', x_r) dx_1 dx_1', \quad (7.132)$$

and

$$\langle \mathcal{I}(x_o)\rangle = \int_{-\infty}^{\infty}\int_{-\infty}^{\infty} \mathcal{W}(x_2, x_2') h_o(x_2, x_o) h_o^*(x_2', x_o) dx_2 dx_2', \quad (7.133)$$

are the average intensities at the two detectors and

$$\langle \Delta \mathcal{I}(x_r)\Delta \mathcal{I}(x_o)\rangle$$
$$= \left| \int_{-\infty}^{\infty}\int_{-\infty}^{\infty} \mathcal{W}(x_1, x_2')h_r(x_1, x_r)h_o^*(x_2', x_o)dx_1 dx_2' \right|^2 \quad (7.134)$$

is the intensity–intensity correlation function where

$$\Delta \mathcal{I}(x_\alpha) = \mathcal{I}(x_\alpha) - \langle \mathcal{I}(x_\alpha)\rangle, \quad (\alpha = r, o). \quad (7.135)$$

To derive this result, the Hermiticity of the mutial coherence function was used to confirm that

$$\int_{-\infty}^{\infty}\int_{-\infty}^{\infty} \mathcal{W}(x_1, x_2')h_r(x_1, x_r)h_o^*(x_2', x_o)dx_1 dx_2'$$
$$= \int_{-\infty}^{\infty}\int_{-\infty}^{\infty} \mathcal{W}(x_2, x_1')h_o(x_2, x_o)h_r^*(x_2', x_r)dx_2 dx_1'. \quad (7.136)$$

Let the source have uniform intensity of value $\mathcal{I}_0$ and be spatially incoherent. Then its mutual coherence function may be modeled as

$$\mathcal{W}(x_1, x_2) = \mathcal{I}_0 \delta(x_1 - x_2), \quad (7.137)$$

where $\delta(x)$ is the Dirac delta-function. Then on substituting from Eqs. (7.125), (7.126) and (7.137) into Eq. (7.134) and performing one integration results in the expression

$$\langle \Delta \mathcal{I}(x_r)\Delta \mathcal{I}(x_o)\rangle = \left| \int_{-\infty}^{\infty} \mathcal{I}_0 \frac{\exp[-ik(z_o - z_r)]}{i\lambda(z_o - z_r)} \exp\left[\frac{-i\pi(x_r - x')^2}{\lambda(z_o - z_r)}\right] \right.$$
$$\left. \times o(x') \frac{\exp[-ikz_b]}{i\lambda z_b} \exp\left[\frac{-i\pi(x_o - x')^2}{\lambda z_b}\right] dx' \right|^2.$$
$$(7.138)$$

For the special case when $z_r = z_o + z_b$, this expression simplifies to

$$\langle \Delta \mathcal{I}(x_r)\Delta \mathcal{I}(x_o)\rangle = \left| \int_{-\infty}^{\infty} \frac{\mathcal{I}_0}{\lambda^2 z_b^2} \exp\left[\frac{-i\pi(x_o^2 - x_r^2)}{\lambda z_b}\right] \right.$$
$$\left. \times o(x') \frac{\exp[2i\pi(x_o - x_r)x']}{\lambda z_b} dx' \right|^2$$
$$= \frac{\mathcal{I}_0^2}{\lambda^4 z_b^4} \left| FT\left[o\left(\frac{2\pi(x_o - x_r)}{\lambda z_b}\right)\right]\right|. \quad (7.139)$$

In a special case of a point-like detector placed at $x_o = 0$, Eq. (7.139) reduces further to

$$\langle \Delta \mathcal{I}(x_r)\Delta \mathcal{I}(x_o)\rangle = \frac{\mathcal{I}_0^2}{\lambda^4 z_b^4}\left|FT\left[o\left(\frac{-2\pi x_r}{\lambda z_b}\right)\right]\right|. \qquad (7.140)$$

This result indicates that for a sufficiently large, uniform, spatially δ-correlated source and in the absence of any optical elements, such as lenses, the ghost (coincidence) imaging of a point detector in the object branch and a spatially resolved detector in the reference branch constitute the Fourier-domain imaging system. The object information is carried in the intensity correlation function of the two detectors. More comprehensive schemes involving lenses can be shown to form ghost images in the direct space.

Ghost imaging was also extended to sources with any state of partial coherence (Cai and Wang, 2007), random light with a twist phase (Cai et al., 2009) and to the electromagnetic sources (Tong et al., 2010). With the increase of source coherence, the image's visibility and signal-to-noise ratio were shown to increase, but its edge blur was shown to decrease. The temporal analog of ghost imaging was also recently suggested for imaging the optical signals distributed in time (Ryczkowski et al., 2016). For obtaining more comprehensive information regarding the practical realizations and applications of ghost imaging, the reader is referred to two reviews (Moreau et al., 2018; Padgett and Boyd, 2017).

Chapter 8

Light Scattering from Three-Dimensional Media

In the previous chapters, we dealt with manipulation of the statistics of light fields on their free-space propagation and passage through various optical systems. Another broad class of problems studied in statistical optics is interactions between light fields and natural media occupying three-dimensional volumes which, in most of the cases, are random. A number of well-known optical phenomena fall into this cathegory: formation of rainbows, light scintillation on the bottom of a swimming pool, twinkling of the stars, hot air mirages, etc. In this and the following chapters, we will outline the current knowledge regarding the methods of description of three-dimensional media and the theoretical tools required for the analysis of interactions of light with them. This theory is also of utmost importance in solving the inverse problems of optical media characterization, such as remote sensing of atmosphere and oceans as well as medical diagnostics of soft tissues. In this chapter, we will be interested in formation of statistics of light outside of the medium's volume, referring to it as scattering.

8.1 The Scattering Phenomenon

Scattering may be generally understood as any observable change in the electromagnetic radiation from its original state resulting from its interaction with a material object (scatterer) having the distribution of the refractive index different from that of the surroundings. In this general sense, the concept of a scattering process encompasses diffraction from an aperture, reflection from a rough surface, diffusion through a continuous random medium, and other situations. In this chapter, we will be primarily interested in scattering of light by media occupying three-dimensional volumes to points located sufficiently far from them.

The physical reason for scattering is the interaction of the incident electric field with the bound electric charges present within the scattering volume. A perturbed charge starts to oscillate and produces a secondary electromagnetic wave known as the scattered field. In both regions, inside and outside of the scatterer, the incident and the scattered fields add up producing the total field. In the experiments, unless the incident field is a plane wave, the scattered field cannot be physically separated from the total field.

If the incident and the scattered fields are oscillating at the same frequency, then no energy is lost or produced on interaction with media. Such type of scattering is called *elastic*, in contrast with inelastic scattering, for which there might exist a positive or negative frequency shift. Only elastic scattering is considered in this text. If the incident field has a finite, while narrow, spectral composition, even in the elastic scattering regime the energy may be redistributed among different frequencies leading to the so-called correlation-induced spectral shifts, similar to those occurring in free-space propagation (see Chapter 4).

The main complexity associated with the theoretical aspects of scattering stems from the fact that the incident and the secondary fields produced in different parts of the scattering volume can interfere with each other a number of times both inside and outside of the scatterer. For a sufficiently large and optically strong scatterer, the cumulative effect of all these interference events leads to a scattered wave being entirely different from the incident one, and, hence, little tractable. Moreover, while in this case the scattered wave still contains a substantial amount of information about the scatterer's physical and optical properties, the inverse problem of finding the scatterer's refractive index distribution from the comparson of the incident and the scattered light no longer possesses the unique solution. Therefore, the scattering process is typically classified as single or multiple. In the former case, the scattered field can only be formed on interaction of the incident radiation with one element in the scattering volume. In the latter case, the fields produced on scattering from the first encountered element are assumed to be able to scatter again from other elements, making the total field a complex combination of the incident field, as well as the scattered ones of various orders. This chapter will be solely confined to single scattering phenomena. We will also restrict ourselves to stationary scatterers and stationary incident fields while setting aside the analysis of possible temporal effects arising on interaction of non-stationary light and/or media.

8.2 Potential Scattering for Scalar Fields

8.2.1 *The first Born approximation*

We begin by setting the theoretical basis for the potential light scattering by deriving the general expression for the scattered field and deducing its versions in cases of weak scattering, under the first Born approximation, and under the far-field approximation.

Suppose that a scattering medium is contained in volume V located between planes $z = 0$ and $z = L$ (see Fig. 8.1). Let a scalar field $U^{(i)}(\mathbf{r}, \omega)$ defined at a point with position vector $\mathbf{r} = x\hat{x} + y\hat{y} + z\hat{z}$,

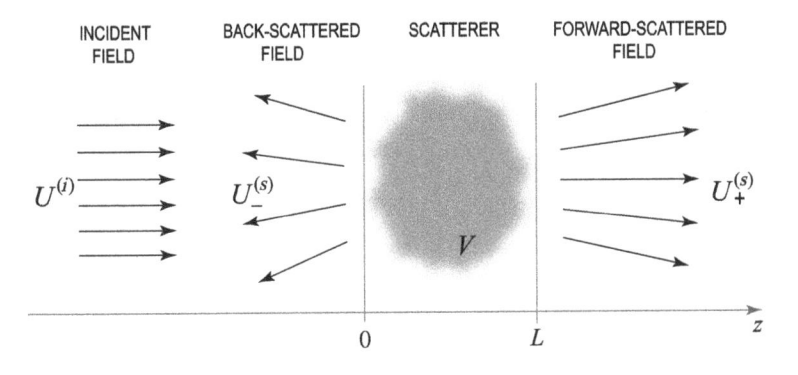

Fig. 8.1 Illustrating the notations relating to light scattering.

and oscillating at angular frequency ω, be incident onto the medium and scatters into field $U^{(s)}(\mathbf{r},\omega)$. The total field $U^{(t)}(\mathbf{r},\omega)$ produced on scattering is the sum of the incident and the scattered fields:

$$U^{(t)}(\mathbf{r},\omega) = U^{(i)}(\mathbf{r},\omega) + U^{(s)}_{\pm}(\mathbf{r},\omega). \tag{8.1}$$

Here, positive and negative signs correspond to scattering in the forward and backward directions. Under the assumption that the index of refraction $n(\mathbf{r},\omega)$ of the medium is a slowly-varying function of position over distances compared to the wavelength of light, the total field in Eq. (8.1) is shown to satisfy the Helmholtz equation with the space-dependent coefficient (Born and Wolf, 1999, p. 696)

$$\nabla^2 U^{(t)}(\mathbf{r},\omega) + k^2 n^2(\mathbf{r},\omega) U^{(t)}(\mathbf{r},\omega) = 0. \tag{8.2}$$

Equation (8.2) can also be expressed as

$$\nabla^2 U^{(t)}(\mathbf{r},\omega) + k^2 U^{(t)}(\mathbf{r},\omega) = -4\pi k^2 \eta_m(\mathbf{r},\omega) U^{(t)}(\mathbf{r},\omega), \tag{8.3}$$

where the *dielectric susceptibility* is defined as

$$\eta_m(\mathbf{r},\omega) = \frac{1}{4\pi}[n^2(\mathbf{r},\omega) - 1], \tag{8.4}$$

if $\mathbf{r} \in V$ and 0 otherwise (Cairns and Foley, 1993; Shirai and Asakura, 1996). Its scaled version,

$$F(\mathbf{r},\omega) = k^2 \eta_m(\mathbf{r},\omega), \tag{8.5}$$

is known as the *scattering potential*. It is usually assumed that the scattering medium is lossless and, hence, both the dielectric susceptibility and the potential are real-valued functions of space. Since the incident field satisfies the homogeneous Helmholtz equation

$$\nabla^2 U^{(i)}(\mathbf{r},\omega) + k^2 U^{(i)}(\mathbf{r},\omega) = 0, \tag{8.6}$$

Eq. (8.3) becomes

$$(\nabla^2 + k^2) U^{(s)}(\mathbf{r},\omega) = -4\pi F(\mathbf{r},\omega) U^{(t)}(\mathbf{r},\omega). \tag{8.7}$$

This is an inhomogeneous Helmholtz equation for the scattered field whose solution can be obtained by the method of Green's functions in the form of the Fredholm integral equation:

$$U^{(s)}(\mathbf{r},\omega) = \int_V G^{(S)}(\mathbf{r}-\mathbf{r}',\omega)F(\mathbf{r}',\omega)U^{(t)}(\mathbf{r}',\omega)d^3r', \qquad (8.8)$$

where $\mathbf{r}'$ and $\mathbf{r}$ are points within and outside of the scatterer, and $G^{(S)}(\mathbf{r}-\mathbf{r}',\omega)$ is the Green function of the three-dimensional Helmholtz equation in the form of the outgoing spherical wave (see Chapter 4).

For an arbitrary distribution of the refractive index within the scattering volume, it does not appear possible to solve Eq. (8.8) in the closed form. However, if the scatterer is weak, it is possible to approximate it either by the Born series (Born and Wolf, 1999, p. 708) or the Rytov series (Shirai and Asakura, 1995). In the former case, the following sum is to be constructed:

$$U^{(s)}_{[M]}(\mathbf{r},\omega) = \sum_{m=1}^{M} U^{(s)}_{(m)}(\mathbf{r},\omega), \qquad (8.9)$$

where terms $U^{(s)}_{(m)}(\mathbf{r},\omega)$ are the m-order contributions and $U^{(s)}_{[M]}(\mathbf{r},\omega)$ are the corresponding partial sums. If the scattered field is sufficiently weak as compared to the incident field,

$$|U^{(s)}| \ll |U^{(i)}|, \qquad (8.10)$$

the assumption known as the *first Born approximation*, then the approximate solution of Eq. (8.8) can be written as

$$U^{(s)}_{(1)}(\mathbf{r},\omega) = \int_V G^{(S)}(\mathbf{r}-\mathbf{r}',\omega)F(\mathbf{r}',\omega)U^{(i)}(\mathbf{r}',\omega)d^3r', \qquad (8.11)$$

which possesses the closed form solution for a number of model incident fields and scattering potentials. Further, substitution of $U^{(s)}_{(1)}$ in place of the field $U^{(i)}$ in the integrand of Eq. (8.11) yields

$$U^{(s)}_{(2)}(\mathbf{r},\omega) = \int_V G^{(S)}(\mathbf{r}-\mathbf{r}',\omega)F(\mathbf{r}',\omega)U^{(s)}_{(1)}(\mathbf{r}',\omega)d^3r', \qquad (8.12)$$

being the second Born approximation. On repeating the same procedure m times, the mth term in the Born series is found from the

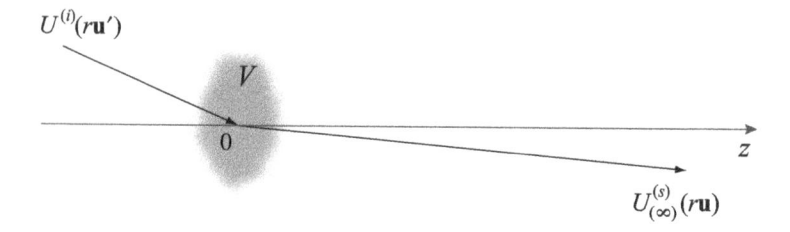

Fig. 8.2 Scattering to the far zone.

expression

$$U^{(s)}_{(m)}(\mathbf{r},\omega) = \int_V G^{(S)}(\mathbf{r}-\mathbf{r}',\omega)F(\mathbf{r}',\omega)U^{(s)}_{(m-1)}(\mathbf{r}',\omega)d^3r'. \quad (8.13)$$

8.2.2 Far-zone approximation

The formula describing the scattered field can be substantially simplified in the far zone of the scatterer, specified by condition $kr \to \infty$ where $r = |\mathbf{r}|$, with the help of approximation $|\mathbf{r}-\mathbf{r}'| \approx |\mathbf{r}| - \mathbf{u}\cdot\mathbf{r}'$, $\mathbf{u}$ being the unit vector along the direction of vector $\mathbf{r}$, i.e., $\mathbf{r} = r\mathbf{u}$ (see Fig. 8.2). In this case, the spherical wave Green's function can be approximated as (see Born and Wolf, 1999, p. 698)

$$G^{(S)}(\mathbf{r}-\mathbf{r}',\omega) = \frac{\exp(ikr)}{r}\exp(-ik\mathbf{u}\cdot\mathbf{r}'). \quad (8.14)$$

This implies that expression (8.8) for the scattered field evaluated at a point $\mathbf{r} = r\mathbf{u}$ reduces to

$$U^{(s)}_{(\infty)}(\mathbf{r},\omega) = \frac{\exp(ikr)}{r}\int_V \exp(-ik\mathbf{u}\cdot\mathbf{r}')F(\mathbf{r}',\omega)U^{(i)}(\mathbf{r}',\omega)d^3r'. \quad (8.15)$$

In a special case when the incident field is a plane wave with amplitude $a^{(i)}(\omega)$ and propagation direction $\mathbf{u}_0$

$$U^{(i)}(\mathbf{r},\omega) = a^{(i)}(\omega)\exp(ik\mathbf{u}_0\cdot\mathbf{r}), \quad (8.16)$$

Eq. (8.15) takes the form

$$U^{(s)}_{(\infty)}(\mathbf{r},\omega) = a^{(i)}(\omega)\frac{\exp(ikr)}{r}\mathfrak{F}(k(\mathbf{u}-\mathbf{u}_0),\omega), \quad (8.17)$$

where $\mathfrak{F}$ denotes the three-dimensional Fourier transform of the scattering potential, i.e.,

$$\mathfrak{F}(\mathbf{K},\omega) = \int_V F(\mathbf{r}',\omega) \exp[-i\mathbf{K}\cdot\mathbf{r}']d^3r'. \qquad (8.18)$$

Equation (8.17) implies that in the weak scattering regime, apart from the common geometrical factor $1/r$, the amplitude of the scattered field is equal to the Fourier component of the scattering potential at spatial frequency vector $\mathbf{K} = k(\mathbf{u} - \mathbf{u}_0)$ known as the *momentum transfer vector*.

8.2.3 Scattering matrix

The majority of studies on scattering use a plane wave as the incident radiation, but at times other non-trivial spatial profiles must be applied. In these cases, the decomposition of the incident field into the spectrum of plane waves having different amplitudes along different directions, known as the *angular spectrum*, makes it possible to obtain a simple expression for the scattered field. Indeed, we express the incident field $U^{(i)}(\mathbf{r},\omega)$ as (Korotkova and Wolf, 2007)

$$U^{(i)}(\mathbf{r},\omega) = \int a^{(i)}(\mathbf{u},\omega) \exp[ik(\mathbf{u}_\perp \mathbf{r} + u_z z)] d^2 u_\perp, \qquad (8.19)$$

where $\mathbf{u}_\perp = (u_x, u_y)$ is a two-dimensional projection of the unit vector $\mathbf{u}$. When $|\mathbf{u}_\perp| \leq 1$, $u_z = \sqrt{1 - u_\perp^2}$, the corresponding waves are known as *homogeneous* (propagating), otherwise, if $|\mathbf{u}_\perp| > 1$, then $u_z = i\sqrt{u_\perp^2 - 1}$, in which case they are called *evanescent* (decaying). The evanescent waves vanish at distances on the order of a wavelength of light and are pertinent only to near-field calculations (Novotny and Hetch, 2006). The plane waves with amplitudes $a^{(i)}(\mathbf{u},\omega)$ along directions $\mathbf{u}$ constitute the angular spectrum of the incident field $U^{(i)}(\mathbf{r},\omega)$.

The scattered and the total fields can be represented by their angular spectra as well

$$U^{(s)}_\pm(\mathbf{r},\omega) = \int a^{(s)}(\mathbf{u},\omega) \exp[ik(\mathbf{u}_\perp \mathbf{r} \pm u_z z)] d^2 u_\perp, \qquad (8.20)$$

and

$$U^{(t)}_\pm(\mathbf{r},\omega) = \int a^{(t)}(\mathbf{u},\omega)\exp[ik(\mathbf{u}_\perp \mathbf{r} \pm u_z z)]d^2 u_\perp, \qquad (8.21)$$

where the positive and the negative signs correspond to waves scattered to regions $z > L$ and $z < 0$, respectively (see Fig. 8.1).

When a single monochromatic plane wave is scattered by a deterministic elastic medium, the amplitudes of the incident and the total plane waves in the angular spectrum are proportional

$$a^{(t)}(\mathbf{u},\omega) = \mathfrak{Y}(\mathbf{u}',\mathbf{u},\omega)a^{(i)}(\mathbf{u}',\omega). \qquad (8.22)$$

Here, linear transformation $\mathfrak{Y}$, relating the waves along directions $\mathbf{u}'$ of incidence and $\mathbf{u}$ of scattering, is called the *spectral scattering matrix*. If the incident field consists of a continuum of homogeneous plane waves with amplitudes $a^{(i)}(\mathbf{u}',\omega)$ propagating along directions $\mathbf{u}'$, then Eq. (8.22) can be generalized to

$$a^{(t)}(\mathbf{u},\omega) = \int \mathfrak{Y}(\mathbf{u}',\mathbf{u},\omega)a^{(i)}(\mathbf{u}',\omega)d^2 u'_\perp. \qquad (8.23)$$

Using relation (8.1) together with Eqs. (8.21)–(8.23), we conclude that

$$a^{(t)}(\mathbf{u},\omega) = a^{(i)}(\mathbf{u},\omega) + a^{(s)}(\mathbf{u},\omega). \qquad (8.24)$$

Also, the relation between an incident and a scattered amplitude becomes

$$a^{(s)}(\mathbf{u},\omega) = [\mathfrak{Y}(\mathbf{u}',\mathbf{u},\omega) - 1]a^{(i)}(\mathbf{u},\omega), \qquad (8.25)$$

and, for more general incident fields, it is given by the integral

$$a^{(s)}(\mathbf{u},\omega) = \int [\mathfrak{Y}(\mathbf{u}',\mathbf{u},\omega) - 1]a^{(i)}(\mathbf{u}',\omega)d^2 u'_\perp. \qquad (8.26)$$

The total field and the scattered field produced on scattering can then be evaluated by taking integrals over all incident and all scattered directions:

$$U_\pm^{(t)}(\mathbf{r},\omega) = \iint \mathfrak{Y}(\mathbf{u}',\mathbf{u},\omega) a^{(i)}(\mathbf{u},\omega) \exp[ik(\mathbf{u}_\perp \mathbf{r} \pm u_z z)] d^2 u_\perp d^2 u'_\perp, \qquad (8.27)$$

and

$$U_\pm^{(s)}(\mathbf{r},\omega) = \iint [\mathfrak{Y}(\mathbf{u}',\mathbf{u},\omega) - 1] a^{(i)}(\mathbf{u},\omega)$$
$$\times \exp[ik(\mathbf{u}_\perp \mathbf{r} \pm u_z z)] d^2 u_\perp d^2 u'_\perp. \qquad (8.28)$$

These relations simplify substantially in the far field of the scatterer, with the help of the Weyl representation of a plane wave. As $kr \to \infty$, the incident, the scattered and the total far fields can be approximated as (see Mandel and Wolf, 1995), Eq. (3.2.22)

$$U_{(\infty)}^{(i)}(r\mathbf{u},\omega) \approx \frac{2\pi i u_z e^{ikr}}{kr} a^{(i)}(\mathbf{u},\omega), \qquad (8.29)$$

$$U_{(\infty)}^{(s)}(r\mathbf{u}) \approx \pm \frac{2\pi i u_z e^{ikr}}{kr} a^{(s)}(\mathbf{u},\omega), \qquad (8.30)$$

$$U_{(\infty)}^{(t)}(r\mathbf{u}) \approx \pm \frac{2\pi i u_z e^{ikr}}{kr} a^{(t)}(\mathbf{u},\omega), \qquad (8.31)$$

implying that the scattered and the total fields in the far zone can then be approximated by the expressions

$$U_{(\infty)}^{(s)}(r\mathbf{u},\omega) \approx \pm \frac{2\pi i u_z e^{ikr}}{kr} \int [\mathfrak{Y}(\mathbf{u}',\mathbf{u},\omega) - 1]$$
$$\times a^{(i)}(\mathbf{u}') d^2 u'_\perp, \qquad (8.32)$$

$$U_{(\infty)}^{(t)}(r\mathbf{u},\omega) \approx \pm \frac{2\pi i u_z e^{ikr}}{kr} \int \mathfrak{Y}(\mathbf{u}',\mathbf{u},\omega) a^{(i)}(\mathbf{u}') d^2 u'_\perp. \qquad (8.33)$$

It follows at once from Eqs. (8.1) and (8.22) that within the accuracy of the first Born approximation the scattering matrix relates to the three-dimensional Fourier transform (see Eq. (8.18)) of the scattering potential evaluated at the momentum transfer vector $\mathbf{K} = k(\mathbf{u} - \mathbf{u}')$, i.e.,

$$\mathfrak{Y}_{(1)}(\mathbf{u}',\mathbf{u},\omega) = \mathfrak{F}[\mathbf{K},\omega]. \qquad (8.34)$$

8.2.4 Random incident field and/or scatterer

Consider now a situation in which either the incident field or the scattering medium (or both) are random and stationary. We will also imply that the medium's fluctuations are weak. Let the cross-spectral density functions of the incident and the scattered fields be

$$W^{(i)}(\mathbf{r}_1, \mathbf{r}_2, \omega) = \langle U^{(i)*}(\mathbf{r}_1, \omega) U^{(i)}(\mathbf{r}_2, \omega) \rangle, \qquad (8.35)$$

and

$$W^{(s)}(\mathbf{r}_1, \mathbf{r}_2, \omega) = \langle U^{(s)*}(\mathbf{r}_1, \omega) U^{(s)}(\mathbf{r}_2, \omega) \rangle. \qquad (8.36)$$

On using correlation (8.36) in Eq. (8.11), we find that the cross-spectral densities of the incident and of the scattered fields are related as

$$W^{(s)}(\mathbf{r}_1, \mathbf{r}_2, \omega)$$
$$= \int_V \int_V G^{(S)*}(\mathbf{r}_1 - \mathbf{r}'_1, \omega) G^{(S)}(\mathbf{r}_2 - \mathbf{r}'_2, \omega) C_F(\mathbf{r}'_1, \mathbf{r}'_2, \omega)$$
$$\times W^{(i)}(\mathbf{r}'_1, \mathbf{r}'_2, \omega) d^3 r'_1 d^3 r'_2, \qquad (8.37)$$

where the integration is performed twice over the scattering volume. Here,

$$C_F(\mathbf{r}_1, \mathbf{r}_2, \omega) = \langle F^*(\mathbf{r}_1, \omega) F(\mathbf{r}_2, \omega) \rangle_m \qquad (8.38)$$

is the *correlation function of the scattering potential* with $\langle \cdot \rangle_m$ standing for the statistical average taken over the realizations of the medium. Function C_F can be used to define the *strength of the potential*

$$I_F(\mathbf{r}, \omega) = C_F(\mathbf{r}, \mathbf{r}, \omega) \qquad (8.39)$$

and the *degree of potential's correlation*

$$\mu_F(\mathbf{r}_2, \mathbf{r}_1, \omega) = \frac{C_F(\mathbf{r}_1, \mathbf{r}_2, \omega)}{\sqrt{I_F(\mathbf{r}_1, \omega)} \sqrt{I_F(\mathbf{r}_2, \omega)}}. \qquad (8.40)$$

In the limiting case of a deterministic medium, the correlation function factorizes as

$$C_F(\mathbf{r}_1, \mathbf{r}_2, \omega) = F^*(\mathbf{r}_1, \omega) F(\mathbf{r}_2, \omega). \qquad (8.41)$$

On the other hand, if the scattering medium is random, but the incident field is deterministic, then the scattered field is still random and formula (8.37) can be employed with

$$W^{(i)}(\mathbf{r}_1, \mathbf{r}_2, \omega) = U^{(i)*}(\mathbf{r}_1, \omega)U^{(i)}(\mathbf{r}_2, \omega) \tag{8.42}$$

under the integral sign. We note that since the total field produced on scattering cannot be physically separated into the incident and the scattered fields, its cross-spectral density

$$W^{(t)}(\mathbf{r}_1, \mathbf{r}_2, \omega) = \langle U^{(t)*}(\mathbf{r}_1, \omega)U^{(t)}(\mathbf{r}_2, \omega)\rangle, \tag{8.43}$$

in view of Eqs. (8.1), (8.35) and (8.36), takes the form

$$W^{(t)}(\mathbf{r}_1, \mathbf{r}_2, \omega) = W^{(i)}(\mathbf{r}_1, \mathbf{r}_2, \omega) + W^{(s)}(\mathbf{r}_1, \mathbf{r}_2, \omega) \\ + W^{(is)}(\mathbf{r}_1, \mathbf{r}_2, \omega) + W^{(si)}(\mathbf{r}_1, \mathbf{r}_2, \omega). \tag{8.44}$$

The two last terms are the mixed correlation functions

$$W^{(is)}(\mathbf{r}_1, \mathbf{r}_2, \omega) = \langle U^{(i)*}(\mathbf{r}_1, \omega)U^{(s)}(\mathbf{r}_2, \omega)\rangle, \\ W^{(si)}(\mathbf{r}_1, \mathbf{r}_2, \omega) = \langle U^{(s)*}(\mathbf{r}_1, \omega)U^{(i)}(\mathbf{r}_2, \omega)\rangle, \tag{8.45}$$

which are typically much weaker in comparison with self-correlations and can be neglected. Therefore, to a good approximation we get the expression

$$W^{(t)}(\mathbf{r}_1, \mathbf{r}_2, \omega) = W^{(i)}(\mathbf{r}_1, \mathbf{r}_2, \omega) + W^{(s)}(\mathbf{r}_1, \mathbf{r}_2, \omega). \tag{8.46}$$

It follows from Eq. (8.15) that in the far zone of the scatterer Eq. (8.37) reduces to the expression

$$W^{(s)}_{(\infty)}(r\mathbf{u}_1, r\mathbf{u}_2, \omega) = \frac{1}{r^2} \int_V \int_V C_F(\mathbf{r}'_1, \mathbf{r}'_2, \omega) W^{(i)}(\mathbf{r}'_1, \mathbf{r}'_2, \omega) \\ \times \exp[-ik(\mathbf{u}_2 \cdot \mathbf{r}'_2 - \mathbf{u}_1 \cdot \mathbf{r}'_1)] d^3 r'_1 d^3 r'_2. \tag{8.47}$$

Since the six-dimensional integral in Eq. (8.47) can be regarded as the Fourier transform of the product of two correlation functions, C_F

and $W^{(i)}$, calculated at points $-k\mathbf{u}'_1$ and $k\mathbf{u}'_2$, it can alternatively be written as convolution

$$W^{(s)}_{(\infty)}(r\mathbf{u}_1, r\mathbf{u}_2, \omega) = \frac{1}{r^2}\mathfrak{C}_F(\mathbf{r}'_1, \mathbf{r}'_2, \omega) \circledast \mathfrak{W}^{(i)}(\mathbf{r}'_1, \mathbf{r}'_2, \omega)\Big|_{(-k\mathbf{u}_1, k\mathbf{u}_2, \omega)}, \qquad (8.48)$$

where $\mathfrak{C}_F$ and $\mathfrak{W}^{(i)}$ are the six-dimensional Fourier transforms of C_F and $W^{(i)}$, respectively.

As evident from Eqs. (8.17) and (8.36), if the scatterer is illuminated by a plane wave with amplitude a and frequency ω, incident along direction $\mathbf{u}_0$, then Eq. (8.48) considerably simplifies as

$$W^{(s)}_{(\infty)}(r\mathbf{u}_1, r\mathbf{u}_2, \omega) = \frac{1}{r^2}|a^{(i)}(\omega)|^2 \mathfrak{C}_F(-\mathbf{K}_1, \mathbf{K}_2, \omega), \qquad (8.49)$$

where $\mathbf{K}_j = k(\mathbf{u}_j - \mathbf{u}_0)$, $(j = 1, 2)$. It will be useful for the subsequent analysis to show that

$$\mathfrak{C}_F(\mathbf{K}_1, \mathbf{K}_2, \omega) = \langle \mathfrak{F}^*(-\mathbf{K}_1, \omega)\mathfrak{F}(\mathbf{K}_2, \omega)\rangle_m. \qquad (8.50)$$

Indeed,

$$\mathfrak{C}_F(\mathbf{K}_1, \mathbf{K}_2, \omega)$$
$$= \int_V \int_V \langle F^*(\mathbf{r}'_1, \omega) F(\mathbf{r}'_2, \omega)\rangle_m \exp[-i(\mathbf{K}_1 \cdot \mathbf{r}'_1 + \mathbf{K}_2 \cdot \mathbf{r}'_2)] d^3 r'_1 d^3 r'_2$$
$$= \Big\langle \int_V F^*(\mathbf{r}'_1, \omega) \exp[-i(-\mathbf{K}_1 \cdot \mathbf{r}'_1)] d^3 r'_1$$
$$\times \int_V F(\mathbf{r}'_2, \omega) \exp[-i(\mathbf{K}_2 \cdot \mathbf{r}'_2)] d^3 r'_2 \Big\rangle_m$$
$$= \langle \mathfrak{F}^*(-\mathbf{K}_1, \omega)\mathfrak{F}(\mathbf{K}_2, \omega)\rangle_m. \qquad (8.51)$$

Hence,

$$\mathfrak{C}_F(-\mathbf{K}_1, \mathbf{K}_2, \omega) = \langle \mathfrak{F}^*(\mathbf{K}_1, \omega)\mathfrak{F}(\mathbf{K}_2, \omega)\rangle_m, \qquad (8.52)$$

and

$$\mathfrak{C}_F(-\mathbf{K}, \mathbf{K}, \omega) = \langle |\mathfrak{F}(\mathbf{K}, \omega)|^2 \rangle_m. \qquad (8.53)$$

From the knowledge of the cross-spectral density function, of the incident, the scattered or the total field, one can determine its spectral density and its spectral degree of coherence

$$S^{(\alpha)}(\mathbf{r},\omega) = W^{(\alpha)}(\mathbf{r},\mathbf{r},\omega), \quad (\alpha = i, s, t), \tag{8.54}$$

and

$$\mu^{(\alpha)}(\mathbf{r}_1,\mathbf{r}_2,\omega) = \frac{W^{(\alpha)}(\mathbf{r}_1,\mathbf{r}_2,\omega)}{\sqrt{W^{(\alpha)}(\mathbf{r}_1,\mathbf{r}_1,\omega)}\sqrt{W^{(\alpha)}(\mathbf{r}_2,\mathbf{r}_2,\omega)}}, \quad (\alpha = i, s, t). \tag{8.55}$$

In particular, for the plane wave scattered to the far zone of the random medium, we get

$$S^{(s)}_{(\infty)}(r\mathbf{u},\omega) = \frac{1}{r^2}|a^{(i)}(\omega)|^2 \langle |\mathfrak{F}(\mathbf{K},\omega)|^2 \rangle_m, \tag{8.56}$$

and

$$\mu^{(s)}_{(\infty)}(r\mathbf{u}_1, r\mathbf{u}_2,\omega) = \frac{\langle \mathfrak{F}^*(\mathbf{K}_1,\omega)\mathfrak{F}(\mathbf{K}_2,\omega)\rangle_m}{\sqrt{\langle |\mathfrak{F}(\mathbf{K}_1,\omega)|^2\rangle_m}\sqrt{\langle |\mathfrak{F}(\mathbf{K}_2,\omega)|^2\rangle_m}}. \tag{8.57}$$

8.2.5 *Deterministic mode representation of scatterers*

As any two-point correlation function, C_F can be represented via the Mercer series (Li and Korotkova, 2017):

$$C_F(\mathbf{r}_1,\mathbf{r}_2,\omega) = \sum_{n=0}^{\infty} \lambda_n(\omega)\phi_n^*(\mathbf{r}_1,\omega)\phi_n(\mathbf{r}_2,\omega). \tag{8.58}$$

Here, $\lambda_n(\omega)$ and $\phi_n(\mathbf{r},\omega)$ are the sets of real-valued, non-negative eigenvalues and real-valued, mutually orthonormal eigenfunctions, respectively, viz.,

$$\lambda_n(\omega) \geq 0, \quad \int_V \phi_m^*(\mathbf{r},\omega)\phi_n(\mathbf{r},\omega)d^3r = \delta_{nm}, \tag{8.59}$$

where integration is carried over the volume occupied by the scatterer and δ_{nm} stands for Kronecker symbol. The eigenvalues and the eigenfunctions must satisfy the homogeneous Fredholm integral equation of the second kind

$$\int_V C_F(\mathbf{r}_1,\mathbf{r}_2,\omega)\phi_n(\mathbf{r}_1,\omega)d^3r_1 = \lambda_n(\omega)\phi_n(\mathbf{r}_2,\omega). \tag{8.60}$$

Evidently, each mode in series (8.58), i.e., each contribution with a fixed index n is completely correlated. Further, one can express the strength of the potential I_F as series

$$I_F(\mathbf{r},\omega) = \sum_{n=0}^{\infty} \lambda_n(\omega)|\phi_n(\mathbf{r},\omega)|^2, \qquad (8.61)$$

providing the information about the contributions (weights) of the individual modes. Finally, on integrating both sides of (8.61) over the volume and using the orthonormality of the eigenvalues, we find that

$$\int_V I_F(\mathbf{r},\omega) d^3r = \sum_{n=0}^{\infty} \lambda_n(\omega), \qquad (8.62)$$

implying that the total scattering power of the medium is equal to the sum of the contributions of the individual deterministic modes. Thus scattering from random media can be analyzed by decomposing C_F into the deterministic modes, scattering them individually and then combining the results. This technique is especially useful in situations when the results of scattering are already developed for individual deterministic contributions.

8.2.6 Pair-scattering matrix

The concept of the scattering matrix that we introduced for the deterministic fields and scatterers can now be generalized to either random fields, scatterers, or both. On expressing the cross-spectral matrices of the incident and the total fields as

$$W^{(\alpha)}(\mathbf{r}_1,\mathbf{r}_2,\omega) = \iint \mathfrak{A}^{(\alpha)}(\mathbf{u}_1,\mathbf{u}_2,\omega)$$
$$\times \exp[ik(\mathbf{r}_2 \cdot \mathbf{u}_2 - \mathbf{r}_1 \cdot \mathbf{u}_1)] d^2u_{1\perp} d^2u_{2\perp}, \qquad (8.63)$$

where $(\alpha = i, t)$, and $\mathfrak{A}$ is known as the *angular correlation function*, and setting

$$\mathfrak{X}(\mathbf{u}_1,\mathbf{u}_2,\mathbf{u}_1',\mathbf{u}_2',\omega) = \langle \mathfrak{Y}^*(\mathbf{u}_1,\mathbf{u}_1',\omega)\mathfrak{Y}(\mathbf{u}_2,\mathbf{u}_2',\omega)\rangle_m, \qquad (8.64)$$

we can readily establish the relation

$$\mathfrak{A}^{(t)}(\mathbf{u}_1, \mathbf{u}_2, \omega) = \mathfrak{X}(\mathbf{u}_1, \mathbf{u}_2, \mathbf{u}_1', \mathbf{u}_2', \omega)\mathfrak{A}^{(i)}(\mathbf{u}_1', \mathbf{u}_2', \omega). \quad (8.65)$$

Note that functions $\mathfrak{A}$ are not exactly the same as function $\mathfrak{W}$ used in Chapter 7. Hence, for random fields the following general relation holds:

$$W^{(t)}(\mathbf{r}_1, \mathbf{r}_2, \omega)$$
$$= \iint \mathfrak{X}(\mathbf{u}_1, \mathbf{u}_2, \mathbf{u}_1', \mathbf{u}_2', \omega)\mathfrak{A}^{(i)}(\mathbf{u}_1, \mathbf{u}_2, \omega)$$
$$\times \exp[ik(\mathbf{r}_2 \cdot \mathbf{u}_2 - \mathbf{r}_1 \cdot \mathbf{u}_1)]d^2u_{1\perp}d^2u_{2\perp}d^2u_{1\perp}'d^2u_{2\perp}'. \quad (8.66)$$

Operator $\mathfrak{X}$ is the *pair-scattering matrix* of the scattering system defined at two incident and two scattered directions of the plane waves in their angular spectra. This matrix carries complete information about the transfer of correlations from the incident field and the scatterer to the total field produced on scattering. In the far zone, expression (8.66) reduces to

$$W^{(t)}_{(\infty)}(r\mathbf{u}_1, r\mathbf{u}_2, \omega) = \pm \frac{4\pi^2}{k^2 r^2} u_{1z} u_{2z} \iint \mathfrak{X}(\mathbf{u}_1, \mathbf{u}_2, \mathbf{u}_1', \mathbf{u}_2', \omega)$$
$$\times \mathfrak{A}^{(i)}(\mathbf{u}_1, \mathbf{u}_2, \omega) d^2u_{1\perp}' d^2u_{2\perp}'. \quad (8.67)$$

It also follows from Eqs. (8.34) and (8.65) that under the first Born approximation $\mathfrak{X}$ relates to C_F as

$$\mathfrak{X}_{(1)}(\mathbf{u}_1, \mathbf{u}_2, \mathbf{u}_1', \mathbf{u}_2', \omega) = \mathfrak{C}_F(-\mathbf{K}_1, \mathbf{K}_2, \omega). \quad (8.68)$$

8.3 Potential Scattering for Electromagnetic Fields

8.3.1 *Deterministic incident field and scatterer*

The scattering theory of the preceding sections can be readily extended to electromagnetic fields. Let the electric and the magnetic incident row-vector fields both have three components and assume that the medium is non-magnetic. Then the fields scattered to a point specified by position vector **r** may be expressed as (Born and Wolf, 1999, p. 730)

$$\vec{E}^{(s)}(\mathbf{r},\omega) = \nabla \times \nabla \times \vec{\Pi}_e(\mathbf{r},\omega), \tag{8.69}$$

and

$$\vec{B}^{(s)}(\mathbf{r},\omega) = -ik\nabla \times \vec{\Pi}_e(\mathbf{r},\omega), \tag{8.70}$$

where $\vec{\Pi}_e$ is the *electric Hertz potential* defined by the formula

$$\vec{\Pi}_e(\mathbf{r},\omega) = \int_V G^{(S)}(\mathbf{r}-\mathbf{r}',\omega)\vec{P}_e(\mathbf{r}',\omega)d^3r'. \tag{8.71}$$

Here, $\vec{P}_e(\mathbf{r},\omega)$ is the *polarization vector* of the scatterer and $\mathbf{r}'$ belongs to the scattering volume. Within the validity of the first-order Born approximation $\vec{P}_e(\mathbf{r},\omega)$ can be expressed as

$$\vec{P}_e(\mathbf{r}',\omega) = \frac{1}{k^2}F(\mathbf{r}',\omega)\vec{E}^{(i)}(\mathbf{r}',\omega), \tag{8.72}$$

where $F(\mathbf{r}',\omega)$ is the scattering potential defined in Eq. (8.5). On substituting from Eqs. (8.71) and (8.72) into Eqs. (8.69) and (8.70), respectively, we obtain the explicit formulas for the scattered electric and magnetic fields:

$$\vec{E}^{(s)}(\mathbf{r},\omega) = \frac{1}{k^2}\nabla \times \nabla \times \int_V G^{(S)}(\mathbf{r}-\mathbf{r}',\omega)F(\mathbf{r}')\vec{E}^{(i)}(\mathbf{r}',\omega)d^3r', \tag{8.73}$$

and

$$\vec{B}^{(s)}(\mathbf{r},\omega) = -\frac{i}{k}\nabla \times \int_V G^{(S)}(\mathbf{r}-\mathbf{r}',\omega)F(\mathbf{r}')\vec{E}^{(i)}(\mathbf{r}',\omega)d^3r'. \tag{8.74}$$

The total electric and magnetic fields produced on scattering are the linear superpositions of the corresponding incident and scattered fields:

$$\vec{E}^{(t)}(\mathbf{r},\omega) = \vec{E}^{(i)}(\mathbf{r},\omega) + \vec{E}^{(s)}(\mathbf{r},\omega), \tag{8.75}$$

and

$$\vec{B}^{(t)}(\mathbf{r},\omega) = \vec{B}^{(i)}(\mathbf{r},\omega) + \vec{B}^{(s)}(\mathbf{r},\omega). \tag{8.76}$$

Just like in the scalar case, unless the incident field is a plane wave, the incident and scattered fields cannot be experimentally discriminated.

8.3.2 Far-zone approximation

In the far field of the scatterer, the electric Hertz vector (8.71) can be approximated as

$$\vec{\Pi}_e(r\mathbf{u},\omega) = \frac{\exp(ikr)}{r}\vec{\mathfrak{P}}_e(r\mathbf{u},\omega), \qquad (8.77)$$

where $\vec{\mathfrak{P}}_e$ is the three-dimensional Fourier transform of $\vec{P}_e(\mathbf{u}',\omega)$

$$\vec{\mathfrak{P}}_e(r\mathbf{u},\omega) = \int_V \vec{P}_e(\mathbf{r}',\omega)\exp(-ik\mathbf{u}\cdot\mathbf{r}')d^3r'$$

$$= \frac{1}{k^2}\int_V F(\mathbf{r}',\omega)\vec{E}^{(i)}(\mathbf{r}')\exp(-ik\mathbf{u}\cdot\mathbf{r}')d^3r'. \qquad (8.78)$$

Further, on using vector identities and retaining only the terms in power of $1/r$, we can reduce (8.69) and (8.70) to the expressions

$$\vec{E}^{(s)}(r\mathbf{u},\omega) = -k^2\frac{e^{ikr}}{r}\mathbf{u}\times[\mathbf{u}\times\vec{\mathfrak{P}}_e(k\mathbf{u},\omega)], \qquad (8.79)$$

and

$$\vec{B}^{(s)}(r\mathbf{u},\omega) = ik\frac{e^{ikr}}{r}\mathbf{u}\times\vec{\mathfrak{P}}_e(k\mathbf{u},\omega). \qquad (8.80)$$

Finally, on substituting from Eq. (8.78) into Eqs. (8.79) and (8.80) and performing vector multiplications, we arrive at the following scattered far fields (Tong and Korotkova, 2010):

$$\vec{E}^{(s)}(r\mathbf{u},\omega) = \frac{\exp(ikr)}{r}\int_V F(\mathbf{r}',\omega)\vec{E}^{(i)}(\mathbf{r}',\omega)\cdot\overleftrightarrow{S}_1(\mathbf{u})$$

$$\times\exp(-ik\mathbf{u}\cdot\mathbf{r}')d^3r', \qquad (8.81)$$

and

$$\vec{B}^{(s)}(r\mathbf{u},\omega) = \frac{\exp(ikr)}{r}\int_V F(\mathbf{r}',\omega)\vec{E}^{(i)}(\mathbf{r}',\omega)\cdot\overleftrightarrow{S}_2(\mathbf{u})$$

$$\times\exp(-ik\mathbf{u}\cdot\mathbf{r}')d^3r', \qquad (8.82)$$

where

$$\overleftrightarrow{S}_1(\mathbf{u}) = \begin{bmatrix} u_x^2 - 1 & u_x u_y & u_x u_z \\ u_y u_x & u_y^2 - 1 & u_y u_z \\ u_z u_x & u_z u_y & u_z^2 - 1 \end{bmatrix},$$

$$\overleftrightarrow{S}_2(\mathbf{u}) = \begin{bmatrix} 0 & u_z & -u_y \\ -u_z & 0 & u_x \\ u_y & -u_x & 0 \end{bmatrix}. \qquad (8.83)$$

In the far field, $\mathbf{u} \cdot \vec{E}^{(s)}(r\mathbf{u}, \omega) = 0$, $\mathbf{u} \cdot \vec{B}^{(s)}(r\mathbf{u}, \omega) = 0$ and $\vec{B}^{(s)}(r\mathbf{u}, \omega) = \mathbf{u} \times \vec{E}^{(s)}(r\mathbf{u}, \omega)$, making the scattered fields propagating along direction $\mathbf{u}$ transverse. Then it is convenient to represent the scattered far fields in the spherical coordinate system, as functions of the polar and the azimuthal angles. The transformation between the Cartesian coordinate system relating to the scatterer and the spherical coordinate system of the scattered far field then takes the form (see Fig. 8.3)

$$\begin{bmatrix} e_r \\ e_\theta \\ e_\phi \end{bmatrix}^T = \begin{bmatrix} e_x \\ e_y \\ e_z \end{bmatrix}^T \cdot \begin{bmatrix} \sin\theta\cos\phi & \cos\theta\cos\phi & -\sin\phi \\ \sin\theta\sin\phi & \cos\theta\sin\phi & \cos\phi \\ \cos\theta & -\sin\theta & 0 \end{bmatrix}, \qquad (8.84)$$

or, in the inverse form, as

$$\begin{bmatrix} e_x \\ e_y \\ e_z \end{bmatrix}^T = \begin{bmatrix} e_r \\ e_\theta \\ e_\phi \end{bmatrix}^T \cdot \begin{bmatrix} \sin\theta\cos\phi & \sin\theta\sin\phi & \cos\theta \\ \cos\theta\cos\phi & \cos\theta\sin\phi & -\sin\theta \\ -\sin\phi & \cos\phi & 0 \end{bmatrix}, \qquad (8.85)$$

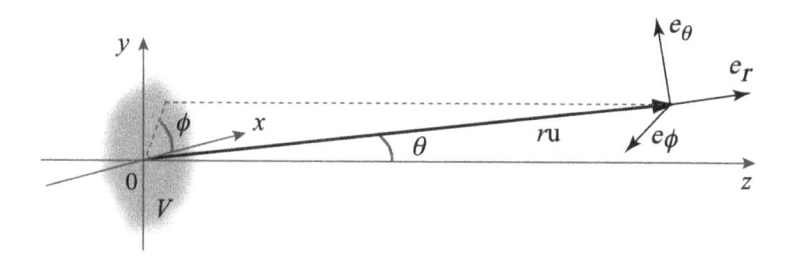

Fig. 8.3 Illustrating the spherical coordinate system and notation relating to the scatterer and the scattered field in the far zone.

Light Scattering from Three-Dimensional Media 227

where (e_x, e_y, e_z) and (e_r, e_θ, e_ϕ) are the unit vector sets of the Cartesian and the spherical coordinate systems. Then the electric field in Eq. (8.81) may be written as

$$\begin{bmatrix} E_r(\theta, \phi, \omega) \\ E_\theta(\theta, \phi, \omega) \\ E_\phi(\theta, \phi, \omega) \end{bmatrix}^T = \frac{\exp(ikr)}{r} \int_V F(\mathbf{r}', \omega) \begin{bmatrix} E_x(\mathbf{r}', \omega) \\ E_y(\mathbf{r}', \omega) \\ E_z(\mathbf{r}', \omega) \end{bmatrix}^T$$

$$\circ \begin{bmatrix} 0 & -\cos\theta\cos\phi & \sin\phi \\ 0 & -\cos\theta\sin\phi & -\cos\phi \\ 0 & \sin\theta & 0 \end{bmatrix}$$

$$\times \exp[-ik\mathbf{u}\cdot\mathbf{r}']d^3r', \qquad (8.86)$$

and a similar equation can be obtained for the magnetic field. Note that unlike in the previous chapters, it is a convention in scattering theory to use θ and ϕ as azimuthal and polar angles.

8.3.3 *Random incident field and scatterer*

In cases when the incident electromagnetic field involved in the scattering process is stationary, the number of correlation terms produced on scattering becomes substantial. Indeed, the electric, the magnetic and the mixed cross-spectral density matrices of the total scattered field take the forms

$$\overleftrightarrow{W}^{(t)}_{\alpha\beta}(\mathbf{r}_1, \mathbf{r}_2, \omega) = \overleftrightarrow{W}^{(ii)}_{\alpha\beta}(\mathbf{r}_1, \mathbf{r}_2, \omega) + \overleftrightarrow{W}^{(ss)}_{\alpha\beta}(\mathbf{r}_1, \mathbf{r}_2, \omega)$$
$$+ \overleftrightarrow{W}^{(is)}_{\alpha\beta}(\mathbf{r}_1, \mathbf{r}_2, \omega) + \overleftrightarrow{W}^{(si)}_{\alpha\beta}(\mathbf{r}_1, \mathbf{r}_2, \omega), \qquad (8.87)$$

where $(\alpha, \beta = E, B)$. Here, self-correlations with superscripts (ii) and (ss) and subscripts (EE) and (BB) define usual cross-spectral density matrices while the cross terms (EB) and (BE) are formed by correlating field components from different ensembles. Similarly to the scalar case, the various cross-terms are typically much weaker than self-correlations and are frequently neglected.

To illustrate the use of these expressions, let us consider a particular example of scattering of a polychromatic electromagnetic plane wave with amplitude $\vec{a}(\omega) = [a_x(\omega), a_y(\omega), 0]$ incident on a scatterer from direction $\mathbf{u}_0$:

$$\vec{E}^{(i)}(\mathbf{r},\omega) = \vec{a}(\omega)e^{ik\mathbf{u}_0\cdot\mathbf{r}}, \quad \vec{B}^{(i)}(\mathbf{r},\omega) = \vec{E}^{(i)}(\mathbf{r},\omega) \cdot \overleftrightarrow{S}_2(\mathbf{u}_0). \tag{8.88}$$

Here, the orthogonality relation $\vec{B}^{(i)}(\mathbf{r},\omega) = \mathbf{u}_0 \times \vec{E}^{(i)}(\mathbf{r},\omega)$ was used and

$$\overleftrightarrow{S}_2(\mathbf{u}_0) = \begin{bmatrix} 0 & 1 & 0 \\ -1 & 0 & 0 \\ 0 & 0 & 0 \end{bmatrix}. \tag{8.89}$$

On denoting the correlations in the incident plane wave as

$$\langle a_i^*(\omega) a_j(\omega) \rangle = A_i A_j B_{ij} S^{(i)}(\omega), \tag{8.90}$$

we determine that

$$\overleftrightarrow{W}_{EE}^{(ii)}(\mathbf{r}_1, \mathbf{r}_2, \omega) = S^{(i)}(\omega) \overleftrightarrow{S}_{22}(\mathbf{u}_0), \tag{8.91}$$

where

$$\overleftrightarrow{S}_{22} = \begin{bmatrix} A_x^2 & A_x A_y B_{xy} & 0 \\ A_x A_y B_{xy} & A_y^2 & 0 \\ 0 & 0 & 0 \end{bmatrix}. \tag{8.92}$$

Similarly, the correlations in the incident magnetic field take the form

$$\overleftrightarrow{W}_{BB}^{(ii)}(\mathbf{r}_1, \mathbf{r}_2, \omega) = -S^{(i)}(\omega) \overleftrightarrow{S}_2(\mathbf{u}_0) \cdot \overleftrightarrow{S}_{22} \cdot \overleftrightarrow{S}_2(\mathbf{u}_0). \tag{8.93}$$

If the medium is characterized by correlation C_F, then the self-correlation tensors of the scattered far fields can be shown to reduce to expressions (Tong and Korotkova, 2011)

$$\overleftrightarrow{W}_{EE}^{(ss)}(r\mathbf{u}_1, r\mathbf{u}_2, \omega)$$
$$= \frac{S^{(i)}(\omega)}{r^2} \mathfrak{C}_F(-K_1, K_2, \omega) \overleftrightarrow{S}_1(\mathbf{u}_1) \cdot \overleftrightarrow{S}_{22} \cdot \overleftrightarrow{S}_2(\mathbf{u}_2),$$
$$\overleftrightarrow{W}_{BB}^{(ss)}(r\mathbf{u}_1, r\mathbf{u}_2, \omega)$$
$$= \frac{-S^{(i)}(\omega)}{r^2} \mathfrak{C}_F(-K_1, K_2, \omega) \overleftrightarrow{S}_1(\mathbf{u}_2) \cdot \overleftrightarrow{S}_{22} \cdot \overleftrightarrow{S}_2(\mathbf{u}_2). \tag{8.94}$$

Other statistical quantities can be deduced from these expressions. For example, if we assume that the interference between the incident and the scattered fields can be neglected, the total Maxwell stress tensor becomes

$$\langle \overleftrightarrow{T}^{(t)}(\mathbf{r},\omega)\rangle = \langle \overleftrightarrow{T}^{(i)}(\mathbf{r},\omega)\rangle + \langle \overleftrightarrow{T}^{(s)}(\mathbf{r},\omega)\rangle, \qquad (8.95)$$

where

$$\langle \overleftrightarrow{T}^{(\alpha)}(\mathbf{r},\omega)\rangle$$
$$= \frac{1}{4\pi}\Bigg[\overleftrightarrow{W}^{(\alpha\alpha)}_{EE}(r\mathbf{u},r\mathbf{u},\omega) + \overleftrightarrow{W}^{(\alpha\alpha)}_{BB}(r\mathbf{u},r\mathbf{u},\omega)$$
$$- \frac{1}{2}\Big[Tr(\overleftrightarrow{W}^{(\alpha\alpha)}_{EE}(r\mathbf{u},r\mathbf{u},\omega) + \overleftrightarrow{W}^{(\alpha\alpha)}_{BB}(r\mathbf{u},r\mathbf{u},\omega))\overleftrightarrow{T}^{(3)}\Big]\Bigg]$$
(8.96)

with ($\alpha = i,s$). For the incident plane wave,

$$\langle \overleftrightarrow{T}^{(i)}(\mathbf{r},\omega)\rangle = -\frac{S^{(i)}(\omega)}{4\pi}\begin{bmatrix}0 & 0 & 0\\ 0 & 0 & 0\\ 0 & 0 & A_x^2 + A_y^2\end{bmatrix}. \qquad (8.97)$$

Using $\overleftrightarrow{T}$ one can evaluate the *momentum flow vector* $\vec{Q}_F^{(s)}$. For the incident field, it takes the form

$$\vec{Q}_F^{(i)}(r\mathbf{u},\omega) = \mathbf{u}\cdot\langle \overleftrightarrow{T}^{(i)}(\mathbf{r},\omega)\rangle$$
$$= -\frac{S^{(i)}(\omega)}{4\pi}[0\ 0\ (A_x^2 + A_y^2)u_z], \qquad (8.98)$$

and for the scattered field, it becomes

$$\vec{Q}_F^{(s)}(r\mathbf{u},\omega) = \mathbf{u}\cdot\langle \overleftrightarrow{T}^{(s)}(\mathbf{r},\omega)\rangle$$
$$= -\frac{S^{(i)}(\omega)}{4\pi r^2}\mathfrak{C}_F(-\mathbf{K},\mathbf{K},\omega)(Tr\overleftrightarrow{S}_{22}(\mathbf{u}_0) - \mathbf{u}\cdot\overleftrightarrow{S}_{22}\cdot\mathbf{u}^T)\mathbf{u}.$$
(8.99)

The last two expressions then yield for the total momentum flow vector

$$\vec{Q}_F^{(t)}(r\mathbf{u},\omega) = \vec{Q}_F^{(i)}(r\mathbf{u},\omega) + \vec{Q}_F^{(s)}(r\mathbf{u},\omega)$$
$$= -\frac{S^{(i)}(\omega)}{4\pi}\left\{[0\ 0\ (A_x^2 + A_y^2)u_z] + \frac{1}{r^2}\mathfrak{C}_F(-\mathbf{K},\mathbf{K},\omega)\right.$$
$$\left. \times (\text{Tr}\,\overleftrightarrow{S}_{22} - \mathbf{u}\cdot\overleftrightarrow{S}_{22}\cdot\mathbf{u}^T)\mathbf{u}\right\}. \tag{8.100}$$

8.4 Examples of Scattering from Deterministic Media

8.4.1 *Spherically symmetric media*

The simplest model for a scattering medium with the spherically symmetric potential is the solid sphere of radius r_0:

$$F(\mathbf{r},\omega) = F_0, \tag{8.101}$$

if $r \leq r_0$ and 0 otherwise. The classic theory that describes scattering of the electromagentic waves from solid spheres was developed by Mie (1908). While the hard-edge potential is a practically sound model, it lacks tractability in analytical calculations. Because of the Fourier transform relation of a pair of Gaussian functions, the soft-edged sphere

$$F(\mathbf{r},\omega) = F_c \exp\left[-\frac{x^2 + y^2 + z^2}{2\sigma_s^2}\right] \tag{8.102}$$

can often fill the gap in situations when a rough analytical prediction of the scattering process is sufficient. Here, F_c is the potential's maximum value and σ_s is its typical width.

Let a plane wave (8.16) oscillating at frequency ω and having a unit amplitude scatter from a Gaussian sphere with potential (8.102) of unit magnitude centered at point $\mathbf{r}_c = (0, 0, z_c)$ of the reference system. If the plane wave's incident direction is specified by unit vector $\mathbf{u}_0 = (0, 0, u_{0z})$, then Eqs. (8.17) and (8.18) imply that at position $r\mathbf{u}$ in the far field of the scatterer the scattered field becomes

$$U^{(s)}_{(\infty)}(r\mathbf{u},\omega) = \exp\left[-\frac{\sigma_s^2 k^2}{2}[u_x^2 + u_y^2 + (u_z - u_{0z})^2]\right]$$
$$\times \sigma_s^3 \frac{\exp[-ikz_c(u_z - u_{0z})]}{r}. \quad (8.103)$$

The scattered spectral density $|U^{(s)}(\mathbf{r},\omega)|^2$ is then also a Gaussian function. If the incident and the scattered directions are expressed in terms of polar angles ϕ_0, ϕ and azymuthal angles θ_0, θ in spherical coordinates, i.e.,

$$u_{0x} = \sin\phi_0 \cos\theta_0, \quad u_{0y} = \sin\phi_0 \sin\theta_0,$$
$$u_{0z} = \cos\theta_0, \quad (8.104)$$
$$u_x = \sin\phi \cos\theta, \quad u_y = \sin\phi \sin\theta, \quad u_z = \cos\theta, \quad (8.105)$$

then the spectral density of the field in Eq. (8.103) reduces to

$$S^{(s)}_{(\infty)}(\mathbf{r},\omega) = |U^{(s)}(\mathbf{r},\omega)|^2$$
$$= \exp\left[-2k^2\sigma_s^2(1-\cos\theta)\right]\frac{\sigma_s^6}{r^2}, \quad (8.106)$$

not depending either on polar angle ϕ or the value of z_c. Figure 8.4 shows the distribution of the scattered spectral density normalized by r^2/σ_s^6 as a function of θ for several values of $k\sigma$.

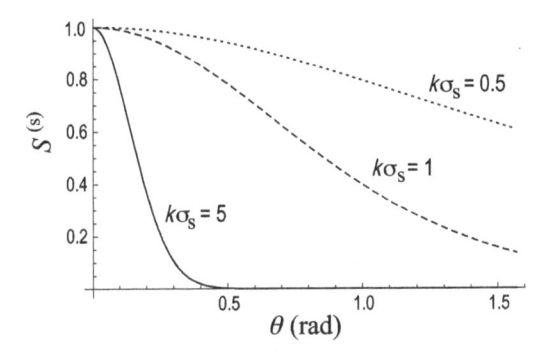

Fig. 8.4 Far-zone spectral density of a plane wave with $\lambda = 0.633\,\mu$m scattered from a Gaussian sphere with different radii.

A model that bridges the gap between the hard-edged and the Gaussian spheres and allows for edge sharpness adjustment has been introduced in Sahin et al. (2011) with the help of the three-dimensional multi-Gaussian functions whose analogs of lower dimensions have been also employed in modeling of aperture edges (Li et al., 2003), beam fields (Gori, 1994) and random beams' coherence states (see Chapter 5)

$$F(\mathbf{r},\omega) = \frac{1}{C_l} \sum_{l=1}^{L} (-1)^{l-1} \binom{L}{l} \exp\left[-l\frac{x^2+y^2+z^2}{2\sigma_s^2}\right], \quad (8.107)$$

where normalization factor C_l is the maximum value of the potential at $\mathbf{r} = 0$. Then the scattered plane wave becomes

$$U^{(s)}(\mathbf{r},\omega) = \frac{\sigma_s^3}{C_l r} \exp[-ikz_c(u_z - u_{0z})] \sum_{l=1}^{L} \frac{(-1)^{l-1}}{l^{3/2}} \binom{L}{l}$$
$$\times \exp\left[-\frac{\sigma_s^2 k^2}{2l}[u_x^2 + u_y^2 + (u_z - u_{0z})^2]\right]. \quad (8.108)$$

Figure 8.5 shows variation in the normalized spectral density of a plane wave scattered from a sphere with $k\sigma_s = 5$ as a function of θ for several values of L. With increasing L, the edges become harder and the scattering outcomes start resembling those of Mie theory ($L \to \infty$).

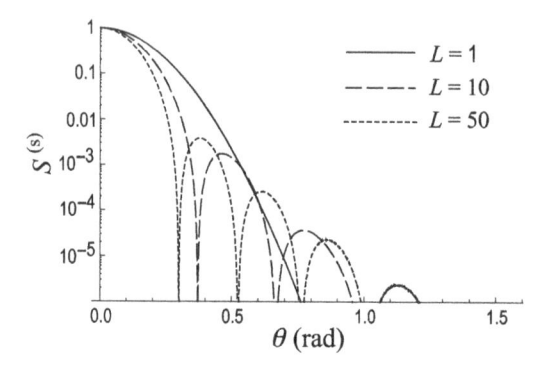

Fig. 8.5 Far-zone spectral density of a plane wave with wavelength $\lambda = 0.633\ \mu\mathrm{m}$ scattered from a semi-soft sphere with radius $k\sigma = 5$.

A hollow spherical shell can be modeled as (8.107)

$$F(\mathbf{r},\omega) = \frac{1}{C_l}\sum_{l=1}^{L}(-1)^{l-1}\binom{L}{l} \times \left(a_o \exp\left[-l\frac{x^2+y^2+z^2}{2\sigma_{so}^2}\right]\right.$$
$$\left. - a_i \exp\left[-l\frac{x^2+y^2+z^2}{2\sigma_{si}^2}\right]\right), \tag{8.109}$$

where $\sigma_{so} > \sigma_{si}$, $a_o \geq a_i$. By setting different number of terms in the two sums, one can adjust edge sharpness of the inner and the outer boundaries. When $a_o = a_i$, the potential vanishes at the scatterer's center. Then

$$U^{(s)}(\mathbf{r},\omega) = \frac{1}{C_l}\frac{\exp[-ikz_c(u_z-u_{0z})]}{r}\sum_{l=1}^{L}(-1)^{l-1}\binom{L}{l}\frac{1}{l^{3/2}}$$
$$\times \left\{\sigma_{so}^3 \exp\left[-\frac{l}{2}\sigma_{so}^2 k^2[u_x^2+u_y^2+(u_z-u_{0z})^2]\right]\right.$$
$$\left. - \sigma_{si}^3 \exp\left[-\frac{l}{2}\sigma_{si}^2 k^2[u_x^2+u_y^2+(u_z-u_{0z})^2]\right]\right\}. \tag{8.110}$$

The variation in the scattered spectral density corresponding to the field in Eq. (8.110) with angle θ is shown for several values of L in Fig. 8.6. The thickness of the boundary in this case affects the number and the strength of interference events: for thinner shells interference effects diminish.

8.4.2 Hard-edged ellipsoids, cylinders, parallelepipeds

The ellipse-, cylinder- and parallelepiped-like scatterers, with adjustable edge softness, can also be modeled using the multi-Gaussian functions (Korotkova et al., 2014) respectively, as

$$F_E(\mathbf{r},\omega) = \frac{1}{C_l}\sum_{l=1}^{L}\frac{(-1)^{l-1}}{l}\binom{L}{l}$$
$$\times \exp\left[-l\left(\frac{x^2}{2\sigma_x^2}+\frac{y^2}{2\sigma_y^2}+\frac{z^2}{2\sigma_z^2}\right)\right], \tag{8.111}$$

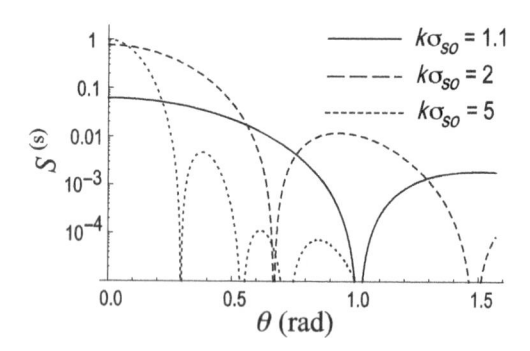

Fig. 8.6 Normalized far-zone spectral density of a plane wave at $\lambda = 0.633\,\mu m$ scattered from a semi-soft hollow sphere with $L = 50$ and radii $k\sigma_{si} = 1$; $k\sigma_{so} = 1.1$ (solid curve), $k\sigma_{so} = 2$ (dashed curve) and $k\sigma_{so} = 5$ (dotted curve).

$$F_C(\mathbf{r},\omega) = \frac{1}{C_l}\sum_{l=1}^{L}(-1)^{l-1}\binom{L}{l}\exp\left[-l\left(\frac{x^2}{2\sigma_x^2}+\frac{y^2}{2\sigma_y^2}\right)\right]$$

$$\times \frac{1}{C_m}\sum_{m=1}^{M}(-1)^{m-1}\binom{M}{m}\exp\left[-m\frac{z^2}{2\sigma_z^2}\right], \quad (8.112)$$

$$F_P(\mathbf{r},\omega) = \frac{1}{C_l}\sum_{l=1}^{L}(-1)^{l-1}\binom{L}{l}\exp\left[-l\frac{x^2}{2\sigma_x^2}\right]$$

$$\times \frac{1}{C_m}\sum_{m=1}^{M}(-1)^{m-1}\binom{M}{m}\exp\left[-m\frac{y^2}{2\sigma_y^2}\right]$$

$$\times \frac{1}{C_n}\sum_{n=1}^{N}(-1)^{n-1}\binom{N}{n}\exp\left[-n\frac{z^2}{2\sigma_z^2}\right]. \quad (8.113)$$

Figure 8.7 illustrates typical examples of these distributions.

If the plane wave scatters from a particle having shape different from spherical, both scattering angles; azymuthal and polar, become involved in representing the scattered field distributions. The fields scattered from ellipsoids, cylinders and parallelepipeds are

$$U_E^{(s)}(r\mathbf{u},\omega) = \frac{\sigma_x\sigma_y\sigma_z}{r^2 C_l}\sum_{l=1}^{L}\frac{(-1)^{l-1}}{l^{3/2}}\binom{L}{l}\exp\left[\frac{-k^2}{2l}(\sigma_x^2(u_x-u_{0x})^2\right.$$

$$\left. + \sigma_y^2(u_y-u_{0y})^2 + \sigma_z^2(u_z-u_{0z})^2)\right], \quad (8.114)$$

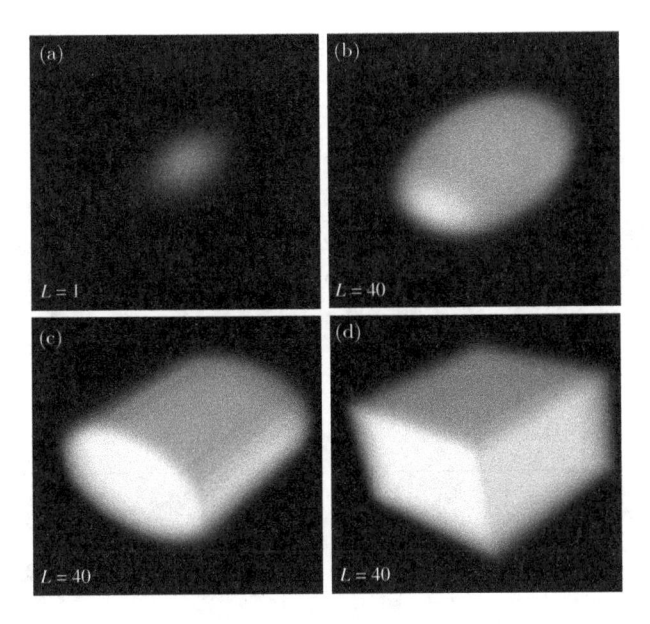

Light Scattering from Three-Dimensional Media   235

Fig. 8.7 Scattering potentials with $k\sigma_x = 1$, $k\sigma_y = 1.5$, $k\sigma_z = 2$ $L = M = N$; (a) $L = 1$; (b) $L = 40$ ellipsoid; (c) $L = 40$ cylinder; and (d) $L = 40$ parallelepiped.

$$U_C^{(s)}(r\mathbf{u},\omega) = \frac{\sigma_x\sigma_y\sigma_z}{r^2 C_l C_n} \sum_{l=1}^{L} \frac{(-1)^{l-1}}{l} \binom{L}{l} \exp\left[\frac{-k^2}{2l}(\sigma_x^2(u_x - u_{0x})^2\right.$$

$$\left. + \sigma_y^2(u_y - u_{0y})^2)\right] \sum_{m=1}^{M} \frac{(-1)^{m-1}}{\sqrt{m}} \binom{M}{m}$$

$$\times \exp\left[\frac{-k^2\sigma_z^2}{2m}(u_z - u_{0z})^2\right], \qquad (8.115)$$

$$U_P^{(s)}(r\mathbf{u},\omega) = \frac{\sigma_x\sigma_y\sigma_z}{r^2 C_l C_m C_n} \sum_{l=1}^{L} \frac{(-1)^{l-1}}{\sqrt{l}} \binom{L}{l} \exp\left[\frac{-k^2\sigma_x^2}{2l}(u_x - u_{0x})^2\right]$$

$$\times \sum_{m=1}^{M} \frac{(-1)^{m-1}}{\sqrt{m}} \binom{M}{m} \exp\left[\frac{-k^2\sigma_z^2}{2m}(u_z - u_{0z})^2\right]$$

$$\times \sum_{n=1}^{N} \frac{(-1)^{n-1}}{\sqrt{n}} \binom{N}{n} \exp\left[\frac{-k^2\sigma_y^2}{2n}(u_y - u_{0y})^2\right]. \qquad (8.116)$$

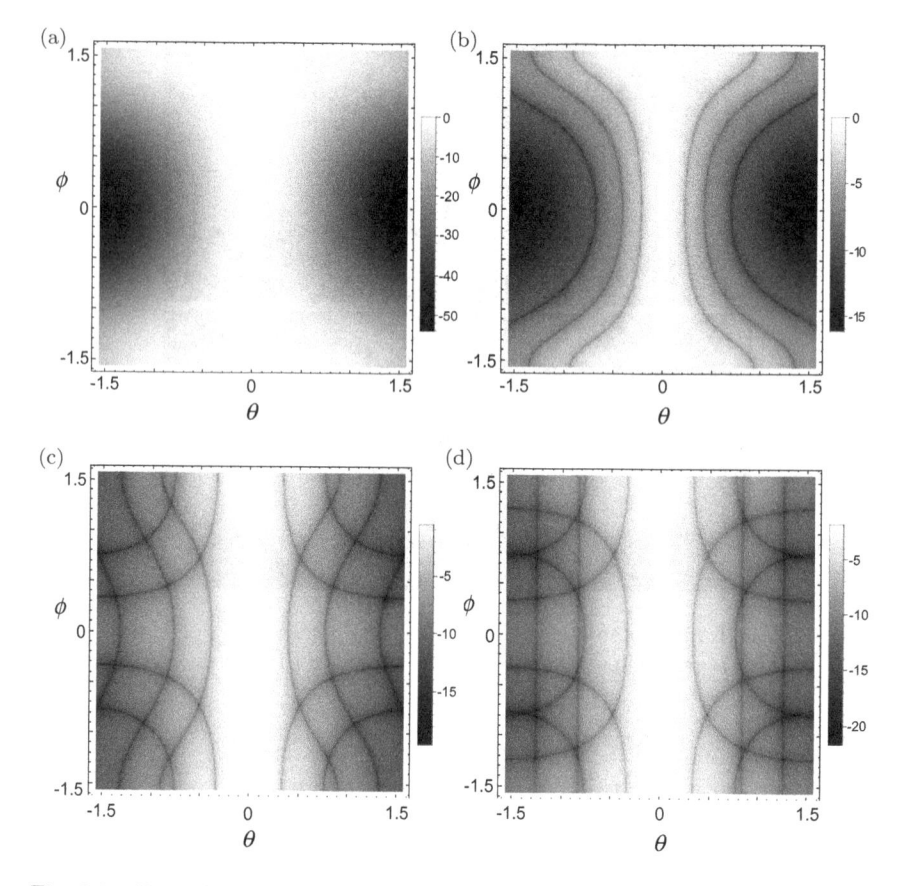

Fig. 8.8 Normalized far-zone spectral density of a plane wave at $\lambda = 0.633\,\mu$m scattered from a particle with $L = M = N$. (a) $L = 1$, $k\sigma_x = 10$, $k\sigma_x = 1$ and $k\sigma_z = 5$; (b) $L = 4$ $k\sigma_x = 10$, $k\sigma_x = 1$ and $k\sigma_z = 5$ ellipsoid; (c) $L = 4$ $k\sigma_x = 5$, $k\sigma_x = 5$ and $k\sigma_z = 5$, cylinder with symmetry about y-axis; (d) $L = 4$, $k\sigma_x = 5$, $k\sigma_x = 5$ and $k\sigma_z = 5$, parallelepiped.

Figure 8.8 presents typical far-field spectral density distributions of light scattered from the three types of particles described above for $L = M = L$. Unlike in the case when $L = 1$, in the three situations with $L = 4$ the sharp black lines appear which could be considered as generalizations of the Mie scattering minima to other three-dimensional shapes. Also, depending on whether the scatterer has elliptical, cylindrical or Cartesian symmetry, the dependence of such dark level curves on angles θ and ϕ differs qualitatively.

8.4.3 Deterministic collections of scatterers

For a scattering collection consisting of M identical particles with potentials $F(\mathbf{r})$, the net potential is given by the sum

$$F^{(\text{col})}(\mathbf{r},\omega) = \sum_{m=1}^{M} F(\mathbf{r} - \mathbf{r}_m, \omega), \qquad (8.117)$$

where $\mathbf{r}_m$ is the center of the particle with index m. If the collection consists of particles of L types, then its net potential becomes

$$F^{(\text{col})}(\mathbf{r},\omega) = \sum_{l=1}^{L} \sum_{m=1}^{M_l} F_l(\mathbf{r} - \mathbf{r}_{lm}, \omega). \qquad (8.118)$$

Here, F_l, $\mathbf{r}_{lm}$ and M_l are the potential of each particle, the center of a particle and the number of particles of type l, respectively. Hence, for a single particle-type collection and a collection with several particle types the Fourier transforms of the potentials become:

$$\mathfrak{F}^{(\text{col})}(\mathbf{K},\omega) = \sum_{m=1}^{M} \mathfrak{F}(\mathbf{K},\omega) \exp[-i\mathbf{K} \cdot \mathbf{r}_m], \qquad (8.119)$$

$$\mathfrak{F}^{(\text{col})}(\mathbf{K},\omega) = \sum_{l=1}^{L} \sum_{m=1}^{M_l} \mathfrak{F}_l(\mathbf{K},\omega) \exp[-i\mathbf{K} \cdot \mathbf{r}_{lm}]. \qquad (8.120)$$

For example, a field produced by a plane wave scattered from two identical Gaussian spheres with centers at $\mathbf{r}_{c1} = (x_c, 0, z_c)$ and $\mathbf{r}_{c2} = (-x_c, 0, z_c)$ and with soft size σ_s becomes

$$U^{(s)}(\mathbf{r},\omega) = \frac{\sigma_s^3}{r} \exp\left[-\frac{\sigma_s^2 k^2}{2}[u_x^2 + u_y^2 + (u_z - u_{0z})^2]\right]$$

$$\times \Big[\exp[-ik(x_c(u_x - u_{0x}) + z_c(u_z - u_{0z}))]$$

$$+ \exp[-ik(-x_c(u_x - u_{0x}) + z_c(u_z - u_{0z}))]\Big]. \qquad (8.121)$$

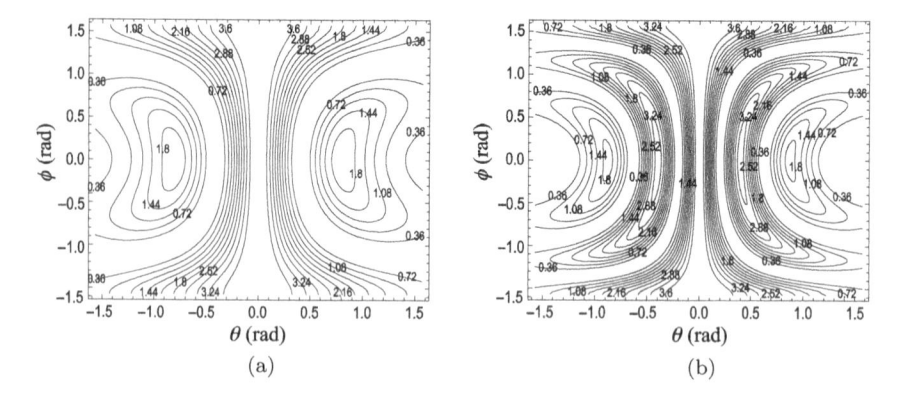

Fig. 8.9 Far-zone spectral density of a plane wave at wavelength $\lambda = 0.633\,\mu m$ scattered from two Gaussian spheres with radii $k\sigma = 1$: (a) $x_c = 4\sigma_s$; (b) $x_c = 8\sigma_s$.

The corresponding spectral density is shown in Fig. 8.9 in the spherical coordinate system for two values of the separation distance $2x_c$.

8.4.4 Effect of random incident field

On assuming that the scattering potential is given by the soft Gaussian sphere model (8.102) with $F_c = 1$ and centered at $\mathbf{r} = 0$, we will now consider a stationary incident field with a typical realization of spectral amplitude being a combination of two plane waves propagating along directions $\mathbf{u}'_1$ and $\mathbf{u}'_2$ (Korotkova and Wolf, 2007):

$$a^{(i)}(\mathbf{u}',\omega) = a^{(i)}(\mathbf{u}'_1,\omega)\delta^{(2)}(\mathbf{u}'-\mathbf{u}'_1) + a^{(i)}(\mathbf{u}'_2,\omega)\delta^{(2)}(\mathbf{u}'-\mathbf{u}'_2), \tag{8.122}$$

where $\delta^{(2)}$ is the delta function. Their angular correlation function becomes

$$\begin{aligned}\mathfrak{A}^{(i)}(\mathbf{u}'_1,\mathbf{u}'_2,\omega) =\;& \mathfrak{a}^{(i)}(\mathbf{u}'_1,\mathbf{u}'_1,\omega)\delta^{(2)}(\mathbf{u}'-\mathbf{u}'_1)\delta^{(2)}(\mathbf{u}'-\mathbf{u}'_1)\\ &+ \mathfrak{a}^{(i)}(\mathbf{u}'_2,\mathbf{u}'_2,\omega)\delta^{(2)}(\mathbf{u}'-\mathbf{u}'_2)\delta^{(2)}(\mathbf{u}'-\mathbf{u}'_2)\\ &+ \mathfrak{a}^{(i)}(\mathbf{u}'_1,\mathbf{u}'_2,\omega)\delta^{(2)}(\mathbf{u}'-\mathbf{u}'_1)\delta^{(2)}(\mathbf{u}'-\mathbf{u}'_2)\\ &+ \mathfrak{a}^{(i)}(\mathbf{u}'_2,\mathbf{u}'_1,\omega)\delta^{(2)}(\mathbf{u}'-\mathbf{u}'_2)\delta^{(2)}(\mathbf{u}'-\mathbf{u}'_1).\end{aligned} \tag{8.123}$$

Let the plane waves be Gaussian-correlated:

$$\mathfrak{a}^{(i)}(\mathbf{u}'_\alpha, \mathbf{u}'_\beta, \omega) = \langle a^{(i)*}(\mathbf{u}'_\alpha, \omega) a^{(i)}(\mathbf{u}'_\beta, \omega) \rangle$$

$$= \mathfrak{a}_{\alpha\beta} \exp[-(\mathbf{u}'_\alpha - \mathbf{u}'_\beta)^2 k^2 \delta_p^2 / 2], \quad (\alpha, \beta = 1, 2). \tag{8.124}$$

Equations (8.102) and (8.123) lead to the following cross-spectral density of the total field produced on scattering:

$$W^{(t)}(r\mathbf{u}_1, r\mathbf{u}_2, \omega) = \frac{4\pi^2 u_{1z} u_{2z}}{k^2 r^2} [\mathfrak{C}_F(\mathbf{u}_1, \mathbf{u}_2, \mathbf{u}'_1, \mathbf{u}'_1, \omega) \mathfrak{a}^{(i)}(\mathbf{u}'_1, \mathbf{u}'_1, \omega)$$

$$+ \mathfrak{C}_F(\mathbf{u}_1, \mathbf{u}_2, \mathbf{u}'_2, \mathbf{u}'_2, \omega) \mathfrak{a}^{(i)}(\mathbf{u}'_1, \mathbf{u}'_1, \omega)$$

$$+ \mathfrak{C}_F(\mathbf{u}_1, \mathbf{u}_2, \mathbf{u}'_1, \mathbf{u}'_2, \omega) \mathfrak{a}^{(i)}(\mathbf{u}'_1, \mathbf{u}'_2, \omega)$$

$$+ \mathfrak{C}_F(\mathbf{u}_1, \mathbf{u}_2, \mathbf{u}'_2, \mathbf{u}'_1, \omega) \mathfrak{a}^{(i)}(\mathbf{u}'_2, \mathbf{u}'_1, \omega)], \tag{8.125}$$

where

$$\mathfrak{C}_F(\mathbf{K}_1, \mathbf{K}_2, \omega) = (2\pi)^3 \sigma_s^6 \exp[-K_1^2 \sigma_s^2 / 2] \exp[-K_2^2 \sigma_s^2 / 2]. \tag{8.126}$$

The spectral density produced on scattering yields

$$S^{(t)}(r\mathbf{u}, \omega) = \frac{(2\pi)^5 \sigma_s^6 u_z^2}{k^2 r^2} [\mathfrak{a}_{11} \exp[-(\mathbf{u} - \mathbf{u}'_1)^2 k^2 \sigma_s^2]$$

$$+ \mathfrak{a}_{22} \exp[-(\mathbf{u} - \mathbf{u}'_2)^2 k^2 \sigma_s^2] + 2\mathfrak{a}_{12}$$

$$\times \exp[-(\mathbf{u} - \mathbf{u}'_1)^2 k^2 \sigma_s^2 / 2]$$

$$\times \exp[-(\mathbf{u} - \mathbf{u}'_2)^2 k^2 \sigma_s^2 / 2]$$

$$\times \exp[-(\mathbf{u}'_2 - \mathbf{u}'_1)^2 k^2 \delta_p^2 / 2]]. \tag{8.127}$$

Figure 8.10 shows the contours of a normalized version of $S^{(t)}$ for two incident plane waves with fixed directions, varying with polar and azimuthal angles of the spherical coordinate system. Note that (a) and (d) subfigures are identical.

8.5 Examples of Scattering from Random Media

8.5.1 *Gaussian-correlated particle*

In order to characterize scattering from a stationary random medium, correlation function $C_F(\mathbf{r}_1, \mathbf{r}_2, \omega)$ of the scattering potential at two

240 Theoretical Statistical Optics

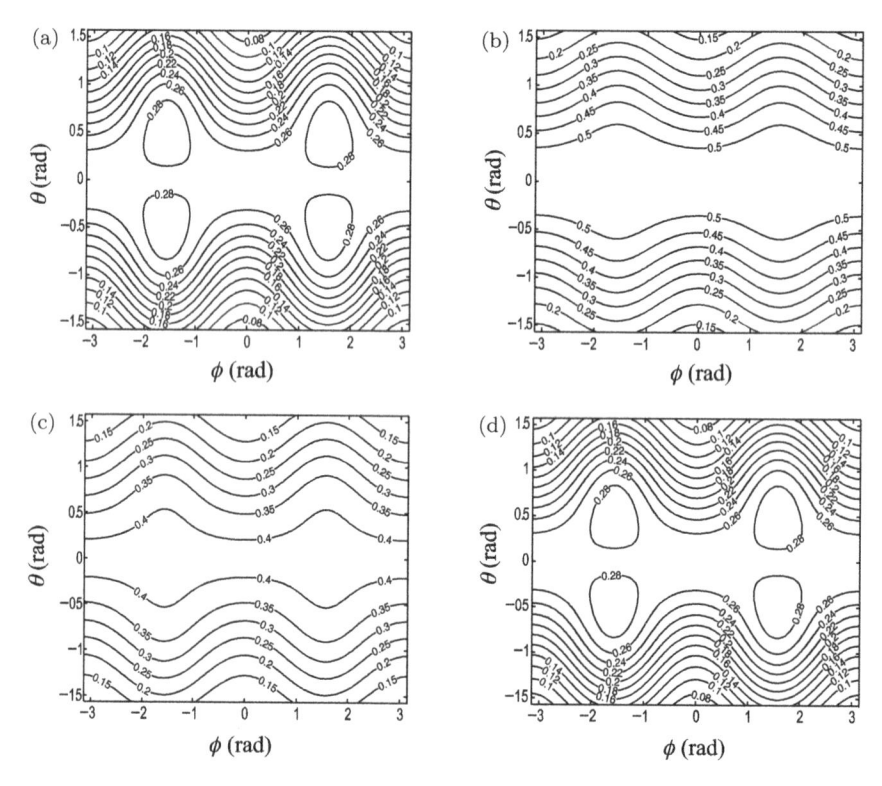

Fig. 8.10 Far-zone spectal density of two partially correlated plane waves with $k\sigma_s = 1$, $\phi'_1 = -\pi/2$, $\phi'_2 = \pi/2$, $\theta'_1 = \theta'_2 = \pi/4$. (a) $\mathfrak{a}_{12} = 1$, $k\delta_p = 0.1$; (b) $\mathfrak{a}_{12} = 1$, $k\delta_p = 10$; (c) $\mathfrak{a}_{12} = 0.5$, $k\delta_p = 10$; (d) $\mathfrak{a}_{12} = 0$, $k\delta_p = 10$.

positions, $\mathbf{r}_1$ and $\mathbf{r}_2$ and at angular frequency ω must be specified (see Eq. (8.38)). As any other two-point correlation function, C_F must satisfy a number of conditions, such as integrability with respect to space and frequency, quasi-Hermiticity, and non-negative-definiteness. In analytic calculations, one of the following models is then frequently used: the *Schell-model scatterer* (Wolf, 2007a)

$$C_F^{(SM)}(\mathbf{r}_1, \mathbf{r}_2, \omega) = \sqrt{I_F(\mathbf{r}_1, \omega)}\sqrt{I_F(\mathbf{r}_2, \omega)}\mu_F(\mathbf{r}_2 - \mathbf{r}_1, \omega), \quad (8.128)$$

and *the quasi-homogeneous scatterer* (Sylverman, 1958)

$$C_F^{(QH)}(\mathbf{r}_1, \mathbf{r}_2, \omega) = I_F\left(\frac{\mathbf{r}_1 + \mathbf{r}_2}{2}, \omega\right)\mu_F(\mathbf{r}_2 - \mathbf{r}_1, \omega). \quad (8.129)$$

The widely used models for the potential's strength I_F and the potential correlation function μ_F are the three-dimensional Gaussian functions (Jannson et al., 1988; Wang and Zhao, 2010):

$$I_F^{(G)}(\mathbf{r},\omega) = A_s \exp\left[-\frac{r^2}{2\sigma_s^2}\right], \qquad (8.130)$$

where $r = |\mathbf{r}|$ and

$$\mu_F^{(G)}(\mathbf{r}_1, \mathbf{r}_2, \omega) = \exp\left[-\frac{|\mathbf{r}_1 - \mathbf{r}_2|^2}{2\delta_s^2}\right], \qquad (8.131)$$

where A_s is its maximum strength, while σ_s and δ_s are the potential strength's and the potential correlation function's root-mean-square values, respectively. It is typically assumed that $\delta_s \ll \sigma_s$. On substituting from Eqs. (8.130) and (8.131) into the Schell-model scatterer (8.128), we obtain the Gaussian Schell-model medium

$$C_F^{(\text{GSM})}(\mathbf{r}_1, \mathbf{r}_2, \omega) = A_s \exp\left[-\frac{r_1^2 + r_2^2}{4\sigma_s^2}\right] \exp\left[-\frac{|\mathbf{r}_1 - \mathbf{r}_2|^2}{2\delta_s^2}\right]. \qquad (8.132)$$

Its Fourier transform becomes

$$\mathfrak{C}_F^{(\text{GSM})}(-\mathbf{K}_1, \mathbf{K}_2, \omega)$$

$$= \frac{64\pi^3 \sigma_s^4 \delta_s^2}{(4\sigma_s^2 + \delta_s^2)\sqrt{\left(\frac{1}{\sigma_s^2} + \frac{2}{\delta_s^2}\right)\left(\frac{1}{\sigma_s^2} + \frac{2}{2\sigma_s^2 + \delta_s^2}\right)}}$$

$$\times \exp\left[-\frac{2\sigma_s^4(K_1^2 + K_2^2)}{4\sigma_s^2 + \delta_s^2}\right] \exp\left[-\frac{\sigma_s^2(\mathbf{K}_2 - \mathbf{K}_1)^2}{4\sigma_s^2 + \delta_s^2}\right], \qquad (8.133)$$

and can be directly used for the scattered field statistical analysis.

The Gaussian Schell-model medium has deterministic mode representation separable in the Cartesian coordinates (Li and Korotkova, 2017):

$$C_F^{(\text{GSM})}(\mathbf{r}_1, \mathbf{r}_2, \omega) = \sum_{l,m,n=0}^{\infty} \lambda_{lmn}(\omega) \phi_{lmn}^*(x_1, y_1, z_1, \omega)$$
$$\times \phi_{lmn}(x_2, y_2, z_2, \omega), \quad (8.134)$$

where

$$\lambda_{lmn}^{(\text{GSM})}(\omega) = A_s \pi^{3/2} \frac{a_2^{l+m+n}}{(a_1 + a_2 + a_3)^{l+m+n+3/2}} \quad (8.135)$$

and

$$\phi_{lmn}^{(\text{GSM})}(\mathbf{r}, \omega) = \left(\frac{2a_3}{\pi}\right)^{3/2} \frac{1}{\sqrt{2^{l+m+n} l! m! n!}}$$
$$\times H_l(x\sqrt{2a_3}) H_m(y\sqrt{2a_3}) H_n(z\sqrt{2a_3})$$
$$\times \exp[-a_3(x^2 + y^2 + z^2)], \quad (8.136)$$

where H_α are the Hermite polynomials ($\alpha = l, m, n$) and

$$a_1 = \frac{1}{\sigma_s^2}, \quad a_2 = \frac{1}{2\delta_s^2}, \quad a_3 = \sqrt{a_1^2 + 2a_1 a_2}. \quad (8.137)$$

Figure 8.11 shows the strengths of the first several coherent modes

$$\Phi_{lmn}^{(\text{GSM})}(\mathbf{r}, \omega) = |\phi_{lmn}^{(\text{GSM})}(\mathbf{r}, \omega)|^2, \quad (8.138)$$

plotted from (8.136). The eigenvalues (8.135) can be shown to form monotonically decaying sequences with decay rate depending on ratio σ_s/δ_s.

8.5.2 Scatterers with structured correlations

Generally, the non-negative definiteness condition for C_F is somewhat hard to verify. We will now show how to establish an alternative simple condition for C_F to represent a physically realizable medium. Such a theoretical approach was successfully used for modeling of the cross-spectral densities of beam-like fields (see Chapter 5) and can be generalized to three dimensions in a straighforward manner.

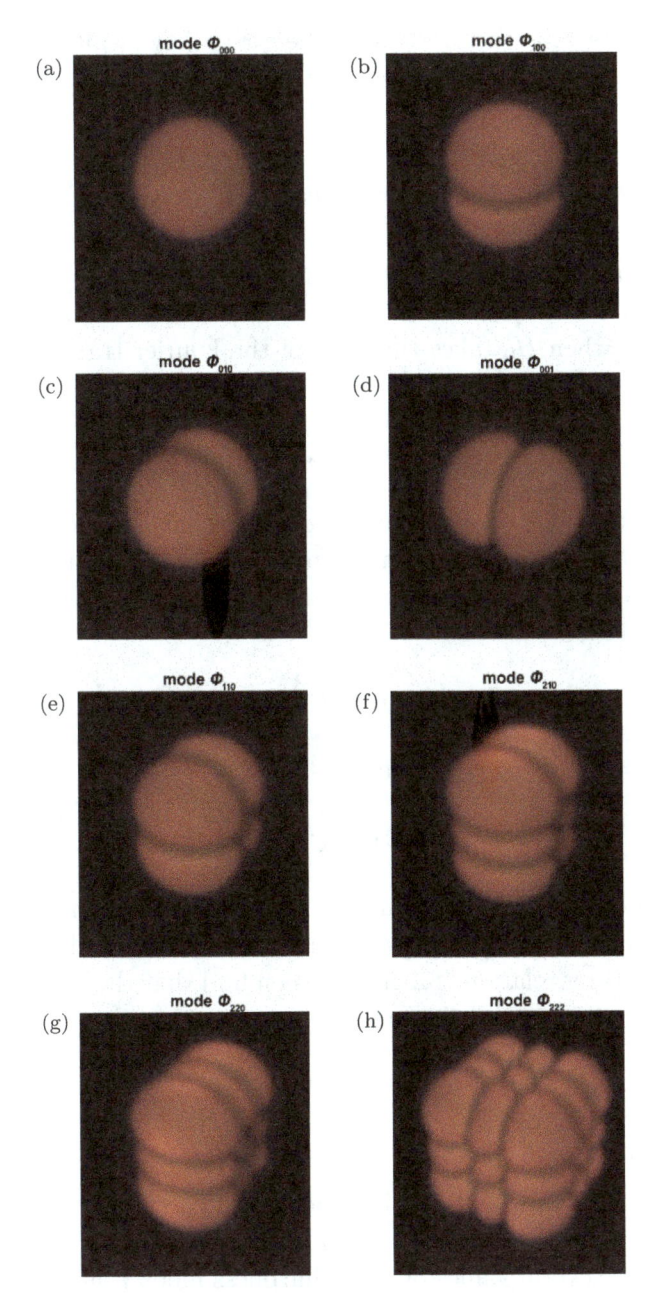

Light Scattering from Three-Dimensional Media 243

Fig. 8.11 Coherent modes of the Gaussian Schell-model medium. Reprinted with permission from Li and Korotkova (2017). © The Optical Society.

According to the Bochner theorem, any legitimate three-dimensional correlation function can be represented via the integral (Korotkova, 2015):

$$C_F(\mathbf{r}_1, \mathbf{r}_2, \omega) = \int_V H_0^*(\mathbf{v}, \mathbf{r}_1, \omega) p(\mathbf{v}, \omega) H_0(\mathbf{v}, \mathbf{r}_2, \omega) d^3 v, \quad (8.139)$$

where $\mathbf{v}$ is a three-dimensional vector, H_0 is any complex-valued function which defines correlation class, while p must be non-negative. In particular, when H_0 takes the form of the Fourier transform kernel, i.e.,

$$H_0(\mathbf{v}, \mathbf{r}, \omega) = a_F(\mathbf{r}, \omega) \exp[-2\pi i \mathbf{v} \cdot \mathbf{r}], \quad (8.140)$$

the scatterer belongs to Schell-model class, i.e., the one in which correlations are translation-invariant. In this case, function p is in the three-dimensional Fourier transform relation (denoted by tilde) with correlation μ_F

$$\mu_F(\mathbf{r}_d, \omega) = FT[p(\mathbf{v}, \omega)], \quad (8.141)$$

where $r_d = |\mathbf{r}_d|$, $\mathbf{r}_d = \mathbf{r}_1 - \mathbf{r}_2$ and

$$I_F(\mathbf{r}, \omega) = \sqrt{a_F(\mathbf{r}, \omega)}. \quad (8.142)$$

Suppose we would like to model a Schell-like medium which produces a circularly symmetric a flat intensity profile and a Gaussian-like edge. Using relation (8.141), we then find that the corresponding degree of spatial correlation must be

$$\mu_F^{(MG)}(\mathbf{r}_d, \omega) = \frac{1}{C_L} \sum_{l=1}^{L} \frac{(-1)^{l-1}}{l^{3/2}} \binom{L}{l} \exp\left[-\frac{r_d^2}{2l\delta_s^2}\right], \quad (8.143)$$

where δ_s is the r.m.s. correlation width and C_L is the normalization factor. Figure 8.12 shows $\mu_F^{(MG)}$ (g_F in the figure notations) varying with $r_d k$ for several values of edge sharpness index L and correlation width δ_s. Then, the far-field spectral density distribution of a plane wave scattered from a Gaussian Schell-model medium with μ_F in

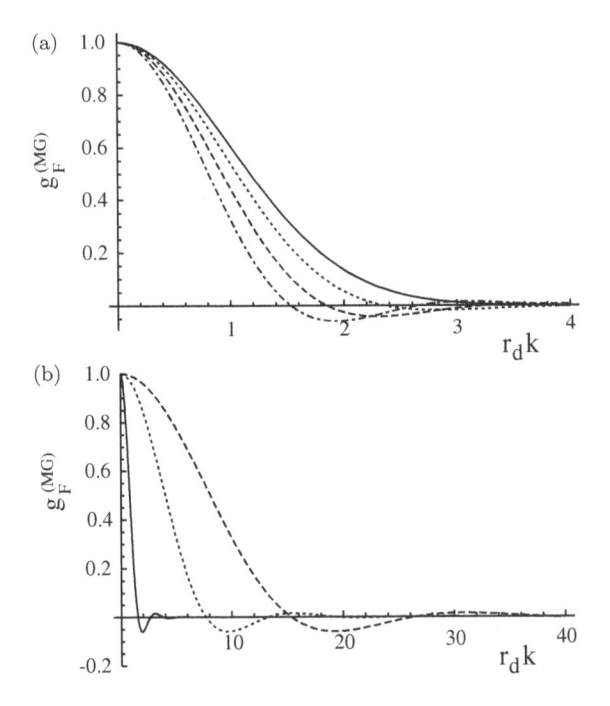

Light Scattering from Three-Dimensional Media 245

Fig. 8.12 Degree of correlation μ_F in Eq. (8.143) (g_F in figure notations) and (a) $k\delta_s = 1$, $L = 1$ (solid curve), $L = 3$ (dotted curve), $L = 10$ (dashed curve) and $L = 40$ (dash-dotted curve); (b) $L = 40$; $k\delta_s = 1$ (solid curve), $k\delta_s = 5$ (dotted curve), $k\delta_s = 10$ (dashed curve). Reprinted with permission from Korotkova (2015). © The Optical Society.

Eq. (8.143) and I_F in Eq. (8.130) for $\sigma_s \gg \delta_s$ can be approximated as

$$S^{(s)}_{(\infty)}(r\mathbf{u},\omega) = \frac{a^2(\omega)}{C_L^2 r^2} \sum_{l=1}^{L} (-1)^{l-1} \binom{L}{l} \exp\left[-\frac{lk^2\delta_s^2}{2}|\mathbf{u}-\mathbf{u}_0|^2\right]. \tag{8.144}$$

Figure 8.13 shows $S^{(s)}_{(\infty)}$ corresponding to μ_F in Fig. 8.12. Thus, structuring of the random medium's correlation function results in shaping of the far-field spectral density of the scattered field. The pure Fourier-like relation between these quantities is only valid in the regime when I_F is much flatter than μ_F (quasi-homogeneous limit); as I_F becomes narrower, its shape starts affecting that of the scattered spectral density as well.

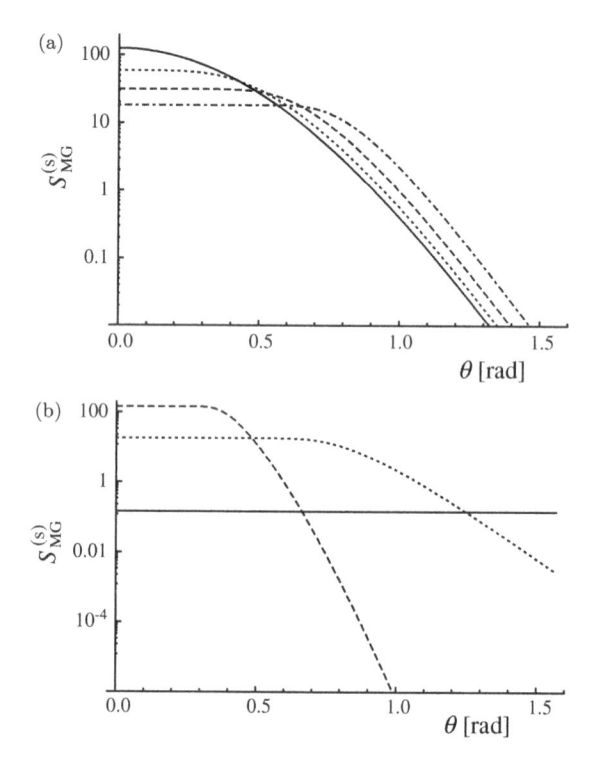

Fig. 8.13 Far-field spectral density distribution (8.144) normalized by $S^{(i)}(\omega)\pi^3\delta^3\sigma^3 B/(r^2 C_L)$ vs. azimuthal angle θ with (a) $k\delta_s = 5$; $L = 1$ (solid curve), $L = 3$ (dotted curve), $L = 10$ (dashed curve) and $L = 40$ (dash-dotted curve); (b) $L = 40$; $k\delta_s = 1$ (solid curve), $k\delta_s = 5$ (dotted curve), $k\delta_s = 10$ (dashed curve). Reprinted with permission from Korotkova (2015). © The Optical Society.

8.5.3 *Random collections of scatterers*

Consider now a random collection of identical particles. The correlation function of its scattering potential (see Eq. (8.117)) at two positions yields

$$C_F^{(\text{col})}(\mathbf{r}_1, \mathbf{r}_2, \omega) = \sum_{m=1}^{M}\sum_{n=1}^{N} \langle F^*(\mathbf{r} - \mathbf{r}_{1m}, \omega) F(\mathbf{r} - \mathbf{r}_{2n}, \omega)\rangle_m. \quad (8.145)$$

If the particles are not correlated with each other, the mixed terms in Eq. (8.145) vanish and the double sum reduces to the single

sum
$$C_F^{(\text{col})}(\mathbf{r}_1, \mathbf{r}_2, \omega) = \sum_{m=1}^{M} \langle F^*(\mathbf{r} - \mathbf{r}_{1m}, \omega) F(\mathbf{r} - \mathbf{r}_{2m}, \omega) \rangle_m. \quad (8.146)$$

Further, there are several possibilities for particles to form random collections: (A) the particles have deterministic potentials but random positions; (B) the particles have deterministic positions but their potentials are random; (C) both the positions and the potentails are random; (D) the particles can form subcollections (clusters).

Suppose first that we are interested in the spectral density of a plane wave scattered from a collection of type (A). It then suffices to find the Fourier transform of its C_F:

$$\mathfrak{C}_F^{(\text{col})}(-\mathbf{K}, \mathbf{K}, \omega) = \langle |\mathfrak{F}(\mathbf{K}, \omega)|^2 \rangle_m$$

$$= |\mathfrak{F}(\mathbf{K}, \omega)|^2 \left\langle \left| \sum_{m=1}^{M} \exp(-i\mathbf{K} \cdot \mathbf{r}_m) \right|^2 \right\rangle_m$$

$$= |\mathfrak{F}(\mathbf{K}, \omega)|^2 \left\langle \sum_{m=1}^{M} \sum_{n=1}^{N} \exp[-i\mathbf{K} \cdot (\mathbf{r}_m - \mathbf{r}_n)] \right\rangle_m. \quad (8.147)$$

The term in the angular brackets in the last line of Eq. (8.147) is known as the *structure factor*

$$\mathfrak{D}(\mathbf{K}, \omega) = \left\langle \sum_{m=1}^{M} \sum_{n=1}^{N} \exp[-i\mathbf{K} \cdot (\mathbf{r}_m - \mathbf{r}_n)] \right\rangle_m. \quad (8.148)$$

It describes the correlation properties of the collection based only on particles' locations while depending on one incident and one scattered direction. Sometimes it is defined with normalization factor $1/N$. For the two limiting cases of completely uncorrelated scatterers (ideal gas) and completely correlated scatterers (crystals), one has

$$\mathfrak{D}(\mathbf{K}, \omega) = 1, \quad (8.149)$$

and

$$\mathfrak{D}(\mathbf{K}, \omega) = \left| \sum_{n=1}^{N} \exp[-i\mathbf{K} \cdot \mathbf{r}_n] \right|^2. \quad (8.150)$$

For more general media, the structure factor can be expressed as

$$\mathfrak{D}(\mathbf{K},\omega) = 1 + \left\langle \sum_{n \neq m}^{N} \exp[-i\mathbf{K} \cdot (\mathbf{r}_m - \mathbf{r}_n)] \right\rangle_m. \tag{8.151}$$

For isotropic random media, i.e., those in which the correlations are orientation-independent, the structure factor may be represented in the form

$$\mathfrak{D}(\mathbf{K},\omega) = 1 + \rho_p \int_V \mu_F(r_d) \exp[-i\mathbf{K} \cdot \mathbf{r}] d^3 r_d, \tag{8.152}$$

where ρ_p is the average particle density in the volume V and $\mu_F(r_d)$ is a correlation function depending on separation r_d.

For fractal-like aggregates being single monomers and the combinations of N_m monomers, the (normalized) structure factor is given by the Fischer–Burford model (Fisher and Burford, 1967)

$$\mathfrak{D}(\mathbf{K},\omega) = \left(1 + \frac{2K^2 R_g}{3D}\right)^{D/2}, \tag{8.153}$$

and

$$\mathfrak{D}(\mathbf{K},\omega) = 1 + \frac{D\Gamma(D-1)[1 + (KR_g)^{-2}]^{(1-D)/2}}{(Ka)^D} \sin\left(\frac{D-1}{\tan(Da)}\right). \tag{8.154}$$

In these expressions, a is the radius of the particle, R_g is the radius of gyration, D is the fractal constant, $R_g = aN_m^{-D}$ and Γ is the Gamma function.

Another important model for structure factor has been determined for collections of highly packed spheres via the Percus–Yevick approximation (Percus and Yevick, 1958)

$$\mathfrak{D}(\mathbf{K},\omega) = \frac{1}{1 - NC(\mathbf{K})}, \tag{8.155}$$

with

$$NC(\mathbf{K}) = -24V_f \left\{ \lambda_1 \frac{\sin(K\bar{d}) - (K\bar{d})\cos(K\bar{d})}{(K\bar{d})^3} \right.$$
$$- 6V_f\lambda_2 \frac{(K\bar{d})^2\cos(K\bar{d}) - 2(K\bar{d})\sin(K\bar{d}) - 2\cos(K\bar{d}) + 2}{(K\bar{d})^4}$$
$$- V_f \frac{\lambda_1}{2}[(K\bar{d})^4\cos(K\bar{d}) - 4(K\bar{d})^3\sin(K\bar{d})$$
$$- 12(K\bar{d})^2\cos(K\bar{d}) + 24K\bar{d}\sin(K\bar{d})$$
$$\left. + 24\cos(K\bar{d}) - 24]/(K\bar{d})^6 \right\}, \qquad (8.156)$$

where

$$\lambda_1 = \frac{(1+2V_f)^2}{(1-V_f)^4}, \quad \lambda_2 = \frac{-(1+V_f/2)^2}{(1-V_f)^4}, \quad V_f = \frac{\pi(\bar{N}\bar{d})^3}{6} \qquad (8.157)$$

while V_f, $\bar{N}$ and $\bar{d}$ are the volume fraction, the density, and the effective particle diameter, respectively (see Fig. 8.14).

In situations when incident/scattered light (or both) are to be characterized for two directions (for instance for the degree of coherence evaluation), the knowledge of the structure factor is not sufficent. Indeed,

$$\mathfrak{C}_F^{(\text{col})}(-\mathbf{K}_1, \mathbf{K}_2, \omega) = \langle \mathfrak{F}^{(\text{col})*}(\mathbf{K}_1, \omega)\mathfrak{F}^{(\text{col})}(\mathbf{K}_2, \omega)\rangle_m$$
$$= \mathfrak{F}^*(\mathbf{K}_1, \omega)\mathfrak{F}(\mathbf{K}_2, \omega) \left\langle \sum_{m=1}^{M}\sum_{n=1}^{N} \right.$$
$$\left. \times \exp[-i[\mathbf{K}_1 \cdot \mathbf{r}_m - \mathbf{K}_2 \cdot \mathbf{r}_n)] \right\rangle_m. \qquad (8.158)$$

The factor in the angular brackets normalized by the number of particles can be regarded as the *pair-structure factor* (Sahin and

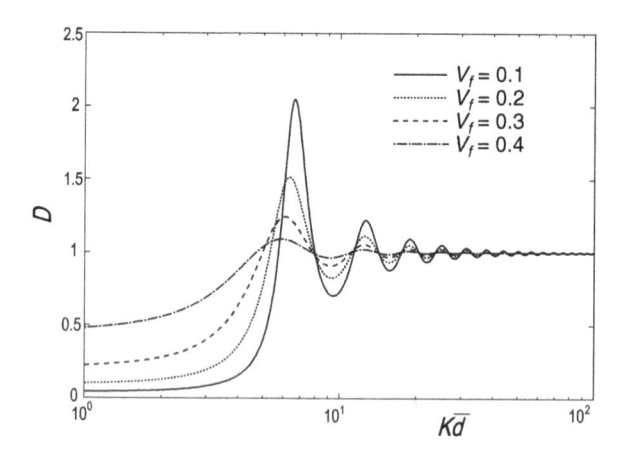

Fig. 8.14 Structure factor $\mathfrak{D}(K\bar{d})$ of hard spheres in Percus–Yevick approximation, for several values of volume fraction V_f.

Korotkova, 2009)

$$\mathfrak{Q}(\mathbf{K}_1, \mathbf{K}_2, \omega) = \left\langle \sum_{m=1}^{M} \sum_{n=1}^{N} \exp[-i[\mathbf{K}_1 \cdot \mathbf{r}_m - \mathbf{K} \cdot \mathbf{r}_n]] \right\rangle_m, \quad (8.159)$$

reducing to the structure factor for $\mathbf{K}_1 = \mathbf{K}_2 = \mathbf{K}$:

$$\mathfrak{Q}(\mathbf{K}, \mathbf{K}, \omega) = \mathfrak{D}(\mathbf{K}, \omega). \quad (8.160)$$

Therefore, for modeling of the pair-structure factor it seems plausible to employ the same type of models that one does for continuous random media, such as quasi-homogeneous, see Eq. (8.129), or Schell-model, see Eq. (8.128). In these cases,

$$\mathfrak{Q}(\mathbf{K}_1, \mathbf{K}_2, \omega) = \mathfrak{D}\left(\frac{\mathbf{K}_1 + \mathbf{K}_2}{2}, \omega\right) \mathfrak{Q}_d(\mathbf{K}_2 - \mathbf{K}_1, \omega), \quad (8.161)$$

and

$$\mathfrak{Q}(\mathbf{K}_1, \mathbf{K}_2, \omega) = \sqrt{\mathfrak{D}(\mathbf{K}_1, \omega)} \sqrt{\mathfrak{D}(\mathbf{K}_2, \omega)} \mathfrak{Q}_d(\mathbf{K}_2 - \mathbf{K}_1, \omega), \quad (8.162)$$

respectively, where $\mathfrak{Q}_d(\mathbf{K}_1, \mathbf{K}_2, \omega)$ can be regarded as a degree of correlation for the entire collection, being limited to one by magnitude.

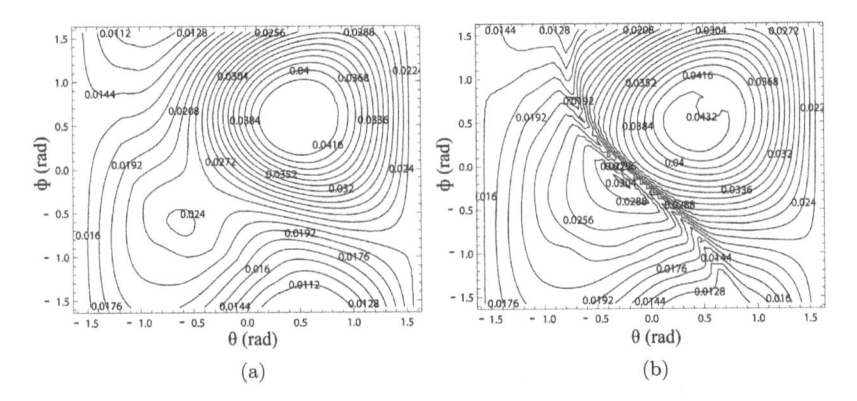

Fig. 8.15 Effects of the pair-structure factor on the far-field scattered spectral density for the incident field consisting of two partially correlated plane waves. (a) $\delta_q = 10^{-7}$ m; (b) $\delta_q = 10^{-6}$ m. Reprinted with permission from Sahin and Korotkova (2009). © The Optical Society.

Figure 8.15 illustrates the effect of the pair-structure factor with $\mathfrak{D}(\mathbf{K}, \omega)$ as in Eq. (8.155) and

$$\mathfrak{Q}_d(|\mathbf{K}_2 - \mathbf{K}_1|, \omega) = \exp\left[-\frac{|\mathbf{K}_2 - \mathbf{K}_1|^2}{k^2 \delta_q^2}\right], \quad (8.163)$$

on the spectral density of the far field produced on scattering of a pair of plane waves incident at different directions.

In cases when the random collection consists of several types of scatterers, the correlation function of the scattering potential (8.118) may be written as (Tong and Korotkova, 2011)

$$C_F^{(\text{col})}(\mathbf{r}_1, \mathbf{r}_2, \omega) = \sum_{i=1}^{L}\sum_{j=1}^{L}\sum_{m=1}^{M_i}\sum_{n=1}^{M_j} \langle F_i^*(\mathbf{r} - \mathbf{r}_{im}, \omega)$$
$$\times F_j(\mathbf{r} - \mathbf{r}_{jn}, \omega)\rangle_m. \quad (8.164)$$

Such a four-folded sum can be alternatively expressed as correlation matrix

$$\overleftrightarrow{C}_F^{(\text{col})}(\mathbf{r}_1, \mathbf{r}_2, \omega)$$
$$= \begin{bmatrix} \langle F_1^*(\mathbf{r} - \mathbf{r}_1, \omega)F_1(\mathbf{r} - \mathbf{r}_2, \omega)\rangle_m & \cdots & \langle F_1^*(\mathbf{r} - \mathbf{r}_1, \omega)F_L(\mathbf{r} - \mathbf{r}_2, \omega)\rangle_m \\ \cdots & \cdots & \cdots \\ \langle F_L^*(\mathbf{r} - \mathbf{r}_1, \omega)F_1(\mathbf{r} - \mathbf{r}_2, \omega)\rangle_m & \cdots & \langle F_L^*(\mathbf{r} - \mathbf{r}_1, \omega)F_L(\mathbf{r} - \mathbf{r}_2, \omega)\rangle_m \end{bmatrix}.$$
$$(8.165)$$

The diagonal elements of this matrix characterize the correlation properties of particles within the same type, while the off-diagonal elements address joint correlations of particles of different types.

On assuming that the particles of all types have deterministic potentials and taking the Fourier transform of matrix $\overleftrightarrow{C}_F^{(\text{col})}(\mathbf{r}_1, \mathbf{r}_2, \omega)$, one gets

$$\overleftrightarrow{\mathfrak{C}}_F^{(\text{col})}(\mathbf{K}_1, \mathbf{K}_2, \omega) = \vec{\mathfrak{F}}^\dagger(\mathbf{K}_1, \omega) \overleftrightarrow{\mathfrak{Q}}(\mathbf{K}_1, \mathbf{K}_2, \omega) \vec{\mathfrak{F}}(\mathbf{K}_2, \omega), \quad (8.166)$$

where † stands for Hermitian adjoint, $\vec{F} = (F_1, \ldots, F_L)$ is a vector of potentials and $\overleftrightarrow{\mathfrak{Q}}(\mathbf{K}_1, \mathbf{K}_2, \omega)$ is the *pair-structure matrix*, being the matrix generalization of the scalar pair-structure factor.

Chapter 9

Light Interaction with Turbulence

A large portion of the studies in statistical optics is concerned with light propagation in natural media exhibiting *optical turbulence*, i.e., continuous, relatively mild spatially and relatively rapid changes in the index of refraction. The three most explored examples of such media are atmosphere, ocean waters and soft biological tissues. Depending on the medium, optical turbulence is caused by fluctuations in advected scalars (diffusers) that can be either thermodynamic parameters or concentration of chemical compounds, or both. Intuitively, classic turbulence can be regarded as a random collection of eddies having a continuum of scales that can be characterized by power spectrum with a specific power law $\kappa^{-11/3}$, where spatial frequency κ is the Fourier variable of the eddy size. Classic turbulence is based on the assumptions of homogeneity and isotropy, the conditions that are in fact rarely met in practice. In this chapter, we will have a detailed discussion about the power spectra of classic and non-classic turbulence in cases when it is based on a single diffuser, such as temperature in the atmosphere or a double-diffuser such as temperature and salt concentration in the ocean. We will also briefly discuss the experimentally measured and fitted power spectra of the refractive index of bio-tissues for which the direct derivation is not known. Later in the chapter, several methods for characterization of the changes in statistics of light fields propagating in turbulence are overviewed and the major effects of turbulence on light are summarized.

9.1 Phenomenon of Optical Turbulence

9.1.1 *Classic and non-classic turbulence*

Most human activity is confined to the Earth's boundary layer consisting of the atmosphere and various water basins. The chemical composition and the thermodynamical state of air and water are constantly changing due to the motion of the Earth as well as the solar cycles and other celestial activity. Large masses of air and water carry heat from the equator to the Arctic and Antarctic producing global mixing of fluids with different parameters, such as temperature, pressure, humidity, etc. and with different chemical compositions. On smaller scales, local topologies and climates limit such global variations in the natural processes to much narrower ranges, but the mixing continues until very small scales are reached, at which dissipation into heat starts occurring.

In large and open volumes, the *classic turbulence*, i.e., the well-developed, homogeneous and isotropic turbulence, can be formed. The turbulent homogenuity can be understood as indendence of its statistics from specific positions, while turbulent anisotropy is associated with independence from directions. The velocity field that is caused by winds or currents produces mixing of the regions within the medium with different thermodynamic parameters or chemical concentrations. The continuum of spatial scales at which the mixing process is maintained is known as the *turbulent cascade*, a concept coined in Richardson (1922). In the classic cascade, the energy is ejected by the external velocity field at the *outer scale* (largest scale) and heat dissipation occurs at the *inner scale* (smallest scale). An inverse cascade is also possible for which the energy is ejected at an intermediate scale and is then distributed to both smaller and larger scales (Alexakis and Biferale, 2018). In particular, such inverse cascades are responsible for generation of large-scale vortex structures, and may factor into formation of tornadoes and hurricanes. In the vicinity of a boundary, for instance, the ground surface, a wall of a building, a side of a ship, or within the externally forced stratification, the structure of turbulence is strongly affected by frictional forces resulting in anisotropic statistics, constant temperature gradients and various coherent structures, e.g., Bernard cells, jets, Hairpin vortices, etc. (Dennis, 2015). Any deviation from the regime of homogeneity and isotropy is known as *non-classic turbulence*. In some cases, the non-classic turbulence can still be characterized by means of power spectra.

9.1.2 Major turbulence parameters

The differentiation between laminar (mixing-free) and turbulent flows is made with the help of the *Reynolds number* (dimensionless) defined as (Reynolds, 1894)

$$\text{Re} = \frac{l_f v_f}{\nu}, \qquad (9.1)$$

where l_f (m) is the *characteristic length* of the fluid, v_f (m/s) is its *velocity* and ν (m^2/s) is its *kinematic viscosity* (also known as eddy viscosity or momentum diffusivity). Conventionally, for Re < 2300, the flows are characterized as laminar and for Re > 2700, they are regarded as turbulent.

Depending on the medium type, one or several quantities may be advected by the velocity field. For instance, in the air, the optical turbulence is primarily caused by temperature fluctuations, while in the turbulent oceanic water, both temperature and salinity (primarily sodium chloride) concentration fluctuations are the dominant factors. Then for each advected quantity a parameter being a ratio of kinematic to frictional forces can be defined. For temperature advection, the *Prandtl number* (dimensionless) is given by the formula

$$\text{Pr}_T = \text{Pr} = \frac{\nu}{\alpha_T}. \qquad (9.2)$$

Here, kinematic viscosity can be expressed as

$$\nu = \frac{\mu_\nu}{\rho_m}, \qquad (9.3)$$

where μ_ν is *dynamic viscosity* (N s/m^2) and ρ_m (kg/m^3) is *density*. Also, α_T (m^2/s) in Eq. (9.2) is *thermal diffusivity*

$$\alpha_T = \frac{\sigma_T}{\rho_m c_p}, \qquad (9.4)$$

with σ_T (W/(m K)) and c_p (J/(kg K)) being *thermal conductivity* and *specific heat*. Hence,

$$\Pr_T = \frac{c_p \mu_\nu}{\sigma_T}. \tag{9.5}$$

For advection of chemical compounds, the *Schmidt number* is defined as

$$\Pr_S = Sc = \frac{\nu}{\alpha_S}, \tag{9.6}$$

where α_S (m^2/s) is *mass diffusivity*. Using the definition of kinematic viscosity, we may also express the Schmidt number as

$$\Pr_S = \frac{\mu_\nu}{\rho_m \alpha_S}. \tag{9.7}$$

Further, α_S obeys the Stokes–Einstein law (Poisson and Papaud, 1983)

$$\frac{\alpha_S \mu_\nu}{\langle T \rangle} = \text{constant} \approx 5.954 \times 10^{-15}, \tag{9.8}$$

where $\langle T \rangle$ is the average temperature of the medium. Thus,

$$\Pr_S \approx \frac{\mu_\nu^2}{5.954 \times 10^{-15} \langle T \rangle \rho_m}. \tag{9.9}$$

Parameters c_p, μ_ν, σ_T and ρ_m can also be directly expressed via the average temperature $\langle T \rangle$ and the average concentration $\langle S \rangle$ (see appendix) (Yao et al., 2020). The definitions of Pr and Sc required for turbulence power spectra are actually defined for laminar flows, the reader must not confuse these with Prandtl and Schmidt numbers defined elsewere for the turbulent flows by different expressions.

While the value of Pr in the air is quite small, 0.72, and stable, Pr and Sc in the ocean water are often given as 7 and 700, respectively, but vary substantially depending on average temperature and salinity (e.g., Yao et al., 2019). The values of the Prandtl and Schimdt numbers are essential in determining specific turbulent regimes for a given advected quanity.

The *Péclec numbers* (dimensionless) for temperature and concentration are defined as a ratio of rate of advection to rate of diffusion

$$\mathrm{Pe}_T = \frac{l_f v_f}{\alpha_T} = \mathrm{Re}\, Pr, \quad \mathrm{Pe}_C = \frac{l_f v_f}{\alpha_S} = \mathrm{Re}\, Sc. \qquad (9.10)$$

The *Mach number* (dimensionless) is a characteristic of the fluid velocity v_f relative to the speed of sound v_s in the medium

$$\mathrm{Ma} = \frac{v_f}{v_s}. \qquad (9.11)$$

For $\mathrm{Ma} > 1.35$ and $\mathrm{Ma} < 0.65$, the flow is characterized as *supersonic* and *subsonic*, respectively.

9.1.3 Obukhov–Corrsin power spectra

The refractive index fluctuations in natural turbulent media are conventionally treated as a random process with stationary increments (see Chapter 1). Such an assumption is based on the fact that the average value of the refractive index is not constant but varies slowly with time. Hence, such fluctuations can be characterized by the two moments, the average value and the autocovariance function:

$$n_o(\mathbf{r}) = \langle n(\mathbf{r}) \rangle_m, \qquad (9.12)$$

$$B(\mathbf{r}_1, \mathbf{r}_2) = \langle [n(\mathbf{r}_1) - n_o(\mathbf{r}_1)][n(\mathbf{r}_2) - n_o(\mathbf{r}_2)] \rangle_m, \qquad (9.13)$$

where $\mathbf{r}$ is the three-dimensional spatial position vector and subscript m denotes statistical average over medium realizations (or sufficiently long time average). In the case of homogeneous and isotropic turbulence, its spatial frequency content can be characterized by the power spectrum, on the basis of the Wiener–Khintchin theorem (see Chapter 1)

$$\mathfrak{B}(\boldsymbol{\kappa}) = \frac{1}{(2\pi)^3} \iiint B(\mathbf{r}_d) \exp[+i\boldsymbol{\kappa} \cdot \mathbf{r}_d] d^3 r_d, \qquad (9.14)$$

where $\mathbf{r}_d = \mathbf{r}_1 - \mathbf{r}_2$ and $\boldsymbol{\kappa} = \kappa_x \hat{\mathbf{x}} + \kappa_y \hat{\mathbf{y}} + \kappa_z \hat{\mathbf{z}}$ is the three-dimensional vector of spatial frequencies. Inversely,

$$B(\mathbf{r}_d) = \iiint \mathfrak{B}(\boldsymbol{\kappa}) \exp[-i\boldsymbol{\kappa} \cdot \mathbf{r}_d] d^3 \kappa. \qquad (9.15)$$

In cases when the turbulence is isotropic, its three-dimensional spectrum can be written as $\mathfrak{B}(\kappa)$, with $\kappa = |\boldsymbol{\kappa}|$ (see Chapter 1) and related to several frequently-used versions of the corresponding one-dimensional spectrum termed here as *one-sided spectrum* $\mathfrak{B}^{(1)}(\kappa)$, *two-sided spectrum* $\mathfrak{B}^{(2)}(\kappa)$, and *spherical shell spectrum* $\mathfrak{B}^{(3)}(\kappa)$:

$$\mathfrak{B}(\kappa) = -\frac{1}{2\pi\kappa}\frac{d\mathfrak{B}^{(1)}(\kappa)}{d\kappa}, \quad \mathfrak{B}^{(2)}(\kappa) = 2\mathfrak{B}^{(1)}(\kappa),$$
$$\mathfrak{B}^{(3)}(\kappa) = 4\pi\kappa^2 \mathfrak{B}(\boldsymbol{\kappa}). \tag{9.16}$$

Moreover, the three-dimensional isotropic spectrum and the spherical shell spectrum can be related to the autocovariance function as

$$B(0) = 4\pi \int_0^\infty \kappa^2 \mathfrak{B}(\boldsymbol{\kappa}) d\kappa, \quad B(0) = \int_0^\infty \mathfrak{B}^{(3)}(\boldsymbol{\kappa}) d\kappa. \tag{9.17}$$

Due to the unprecedented range and complexity of phenomena that can be generally manifested by turbulence, its complete analytic characterization by means of statistical analysis remains impossible. In most of the regimes, only experiments and computer simulations provide necessary insight into the mixing process. The only exception constitutes three-dimensional, boundary-free, homogeneous and isotropic turbulence for which Re can reach values on the order 10^8–10^9, Péclec numbers must be sufficiently high and Mach numbers must be low.

The isotropic three-dimensional power spectrum of the velocity field in such a turbulent flow was developed by Kolmogorov using dimensional analysis and was shown to obey the power law $\kappa^{-11/3}$ (Kolmogorov, 1941a,b,c, 1942). Based on this theory, the power spectrum for a scalar field advected by the turbulent velocity field was later developed (Obukhov, 1949; Corrsin, 1951). According to the Obukhov–Corrsin theory, there are three equilibrium regimes of turbulence: inertial-convective, viscous-convective and viscous-diffusive. The inertial-convective regime is pertinent to scales smaller than outer scale L_0 (the largest eddy size) but larger than parameters

$$\eta_K = \left(\frac{\nu^3}{\varepsilon}\right)^{1/4}, \quad \eta_B = \left(\frac{\alpha^3}{\varepsilon}\right)^{1/4}, \tag{9.18}$$

known as the *Kolmogorov scale* (m) and the *Batchelor scale* (m), where ε (m^2 s^{-3}) is the *turbulent kinetic energy dissipation rate* (per unit mass) and $\alpha = \alpha_T, \alpha_S$, where T and S refer to temperature and concentration. For smaller scales, the viscous-convective regime occurs, and for even smaller ones, the energy rapidly dissipates within the viscous-diffusive regime. For very small Prandtl or Schmidt numbers, the inertial-convective regime can be directly followed by the viscous-diffusive regime.

The general form of the three-dimensional power spectrum is

$$\mathcal{B}_j(\kappa) = \frac{C_0}{4\pi}\varepsilon^{-1/3}\chi_j \kappa^{-11/3} g_j(\kappa\eta_K), \qquad (9.19)$$

where $C_0 = 0.72$ is the Obukhov–Corrsin constant, χ_j is the *variance dissipation rate* whose units depend on the advected quantity: (K^{2}s^{-1}) for temperature, (g^{2}s^{-1}) for concentration and (s^{-1}) for the refractive index. The typical behavior of function g_j (j stands for tempareature or concentration) is the following: in the inertial-convective range function g_j is constant, and, hence the spectrum has the classic $-11/3$ power law, then it increases to a maximum in viscous-convective regime and decreases to zero in the viscous-diffusive regime. Depending on the advected quantity g_j also parametrically depends on Pr or Sc.

9.2 Atmospheric Turbulence

9.2.1 *Classic model*

As already stated, the Prandtl number of the atmospheric turbulence has a stable value of 0.72 for a wide range of thermodynamic conditions. In the inertial-convective regime, the power spectrum is described by Kolmogorov's power law (Andrews and Phillips, 2005; Tatarskii, 1961)

$$\mathcal{B}_T(\kappa) = 0.033 C_T^2 \kappa^{-11/3}, \qquad (9.20)$$

where C_T^2 (m$^{-2/3}$) is the temperature structure parameter

$$C_T^2 = \frac{9\Gamma(1/3)}{10}C_0 \chi_T \varepsilon^{-1/3}, \qquad (9.21)$$

χ_T being the temperature dissipation rate and Γ standing for Gamma function. For higher spatial frequencies, the spectrum involves function g_T:

$$\mathfrak{B}_T(\kappa) = 0.033 C_T^2 \kappa^{-11/3} g_T(\kappa \eta_K). \tag{9.22}$$

It is a convention in atmopsheric propagation studies to use inner scale l_0 instead of η_K. The two quantities are related as

$$l_0 = \left(\frac{27\Gamma(1/3) C_0}{5 Pr} \right)^{3/4} \eta_K, \tag{9.23}$$

reducing to $l_0 = 7.42 \eta_K$ for classic atmospheric case. The widely used models for function $g_l(\kappa l_0) = g_T(\kappa \eta_K)$ were suggested in Tatarskii (1971)

$$g_l(\kappa l_0) = \exp\left[-\left(\frac{\kappa l_0}{5.92} \right)^2 \right], \tag{9.24}$$

and in Frehlich (1992)

$$g_l(\kappa l_0) = \exp(1.1090 l_0 \kappa)[1 + 0.70937 l_0 \kappa + 2.8235(l_0 \kappa)^2$$
$$- 0.28086(l_0 \kappa)^3 + 0.08277(l_0 \kappa)^4]. \tag{9.25}$$

While Tatarskii's model gives a smooth decay, the Frehlich model predicts the occurrence of a bump as spatial frequencies transition from inertial-convective to viscous-diffusing regime (see Fig. 9.1). The two models are consistent with the first principles of theormodynamics (Muschinski, 2015).

The major effect of classic turbulence on a light field is the gradual randomization of its phase, and, hence, intensity, along the path, the rate of randomization being the same along the two directions transverse to direction of propagation.

9.2.2 Non-classic extensions

Any deviation from the classic turbulent regime, such as anisotropy, non-Kolmogorov power law, the presence of refractive index stable gradients and other coherent structures can be referred to as *non-classic turbulence*. All aforementioned phenomena have been measured in the close proximity of a boundary (up to several meters)

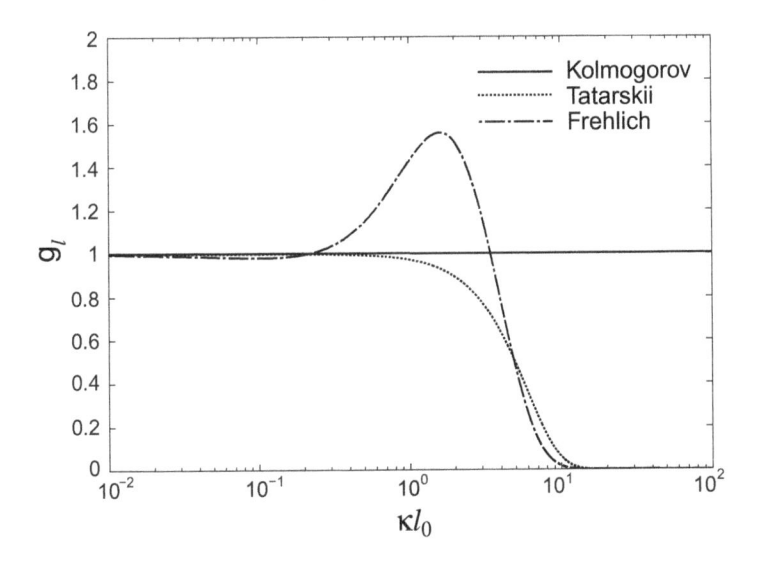

Fig. 9.1 Function g_l depending on κl_0.

and some of them, such as anisotropy and the non-Kolmogorov power law, were also reported at high altitudes. The anisotropy of natural atmospheric turbulence was measured for the first time in Consortini *et al.* (1970) using the beam wander (dancing) analysis of a pair of closely launched laser beams, set along a horizontal path. It was revealed in that and later measurements that sufficiently close to the ground the turbulent eddies are not spherically symmetric (on average) as in the case of the classic turbulence, but rather resemble horizontally stretched "pancakes". Such an eddy profile implies that the refractive index spatial correlations are higher in (any) horizontal direction than those in the vertical. Hence, on passing through anisotropic turbulence the light beam's behavior depends on its orientation.

Two important special cases have been analyzed in detail: vertical and horizontal propagation paths (see Fig. 9.2). When the beam propagates along the vertical direction, the anisotropy only acts as a scaling factor. However, on propagation in horizontal direction, the anisotropy affects the spatial profiles of the beam statistics. For example, the beam's phase decorrelates faster along the vertical

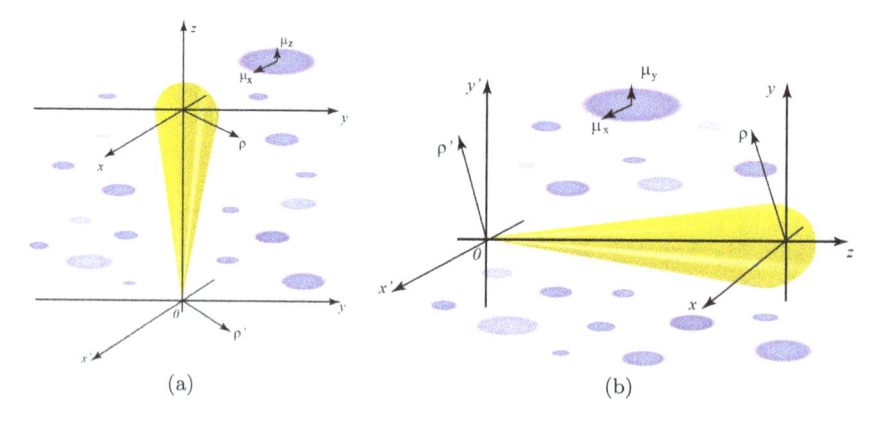

Fig. 9.2 Two cases of optical beam propagation through anisotropic turbulence: vertical path (a) and horizontal path (b). Reprinted with permission from Wang and Korotkova (2016b). © The Optical Society.

direction, hence, resulting in higher diffraction at larger propagation distances. An initially circular beam intensity would then gradually acquire a vertically stretched elliptical shape.

The deviation of the power law from the Kolmogorov value 11/3 to any value in the general interval [3, 4] is also regarded to be a consequence of stratification and anisotropy (Belenkii et al., 1999). For values of power law smaller/larger than 11/3, the turbulent energy is distributed more towards smaller/larger scales and results in apprearance of finer/cruder structures. The power spectrum model accounting for the turbulent eddy symmetry axis placed along the optical axis, say, z (vertical path) and non-Kolmogorov power law α has the form (Toselli et al., 2011)

$$\mathfrak{B}_T(\kappa, \alpha, \mu_z) = \frac{\Gamma(\alpha - 1)\cos(\alpha\pi/2)\hat{C}_T^2\mu_z^2}{4\pi^2[\mu_z^2\kappa_z^2 + (\kappa_x^2 + \kappa_y^2)]^{\alpha/2}}, \qquad (9.26)$$

where Γ stands for the Gamma function and $\hat{C}_T^2$ is the generalized refractive index structure parameter with units $m^{3-\alpha}$. The power spectrum describing the axis of eddy symmetry being orthogonal to the optical axis z (horizontal path) is given as

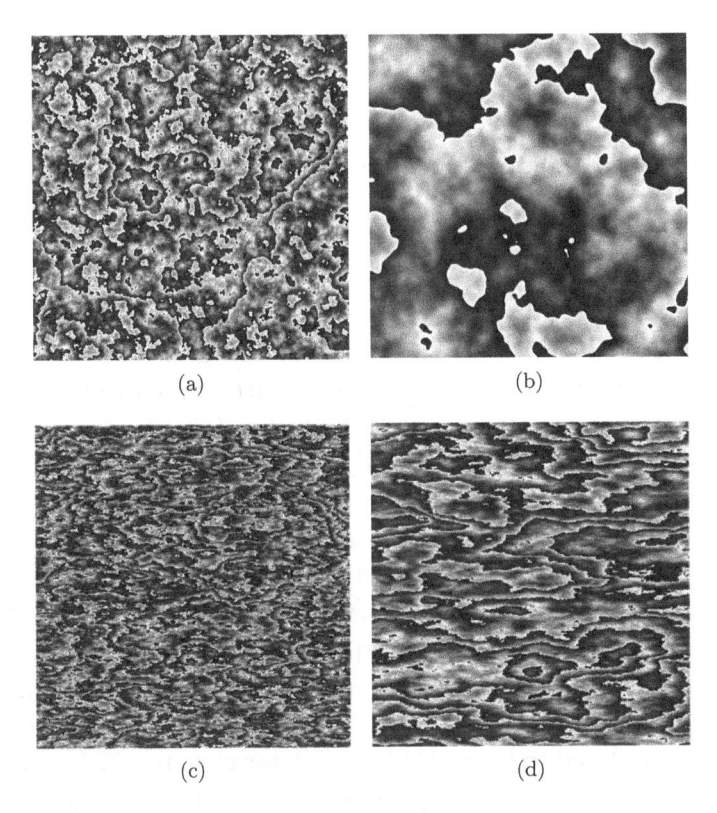

Fig. 9.3 Computer realizations of turbulent phase screens with 2π wrapped phase. Vertical turbulence with $\alpha = 3.2$ (a) $\mu_z = 1$, (b) $\mu_z = 5$. Horizontal turbulence with $\mu_x = 3$, $\mu_y = 1$, (c) $\alpha = 3.2$, (d) $\alpha = 11/3$. Reprinted with permission from Toselli et al. (2015) and Xiao et al. (2016). © The Optical Society.

(Andrews et al., 2014)

$$\mathfrak{B}_T(\kappa_x, \kappa_y, \alpha, \mu_x, \mu_y) = \frac{\Gamma(\alpha-1)\cos(\alpha\pi/2)\hat{C}_T^2 \mu_x \mu_y}{4\pi^2[\kappa_z^2 + \mu_x^2\kappa_x^2 + \mu_y^2\kappa_y^2]^{\alpha/2}}. \qquad (9.27)$$

In these spectra, the inner scale and the bump at high spatial frequencies are not included but can also be modeled in. Figure 9.3 presents typical two-dimensional turbulent phase distributions wrapped by 2π. The modeling of non-Kolmogorov isotropic power spectrum was considered in Zilberman et al. (2008) and Toselli et al. (2008).

The experimental results also suggest that in close proximity of the ground, the orientation of the semi-major axis of the ellipse associated with a typical shape of a turbulent eddy might deviate from a horizontal direction (Wang *et al.*, 2017).

9.3 Oceanic Turbulence

9.3.1 *Classic model*

Oceanic turbulence is a much more complex environment as compared with its atmospheric counterpart. First of all, the average refractive index of water is much higher (about 1.33) compared to that of air (about 1.0003) already introducing a number of diffculties for optical system's operation. The second important contradistinction concerns molecular absorption, being low in air and very high in water, the latter reducing the actual light beam travel paths underwater to several tens of meters, even at high initial power levels. Moreover, refractive and absorptive properties of a water column are very sensitive to its average temperature and average salinity as well as light wavelength. The smallest absorption coefficient (for fresh water) is measured to be 0.00442 m^{-1} at wavelength 417.5 nm, which explains the blue color of large, pure water volumes and suggests the optimal optical transmission window (Pope and Fry, 1997).

Oceanic optical turbulence is driven by two simultaneous and coupled random processes: temperature and salinity (sodium chloride) concentration fluctuations (Hill, 1978c). The pertinent Prandtl and Schmidt numbers may exceed unity by three orders of magnitude, which greatly influences the shape of functions $g_j(\kappa\eta_K)$.

Four approximate models for g_j have been obtained in Hill (1978a,b,c) two of which became particularly wide-spread. The first of the Hill's models, H1, gives

$$g_j^{H1}(\kappa\eta_K) = [1 + C_1(\kappa\eta_K)^{2/3}]\exp[-b_j\delta_H], \quad (j = T, S), \quad (9.28)$$

where $b_j = C_0 C_1^{-2}/Pr_j$, $\delta_H = 1.5 C_1^2(\kappa\eta_K)^{4/3} + C_1^3(\kappa\eta_K)^2$, $C_0 = 0.72$ and $C_1 = 2.35$ (a data-fitted constant).

The combination of the two separate (temperature and salinity) spectra has been first proposed in Nikishov and Nikishov (2000) as a linearized approximation of a more general high-power polynomial expansion developed in Ruddick and Schirtcliffe (1979). It was assumed, however, that $\mathrm{Pr}_T \approx 7$ and $\mathrm{Pr}_S \approx 700$, which corresponds to $\langle T \rangle \approx 20°C$ and $\langle S \rangle \approx 35$ppt. The fluctuating portion of the refractive index was expressed as a linear combination

$$n' = -a_T T' + a_S S', \quad (9.29)$$

with thermal expansion coefficient $a_T = 2.6 \times 10^{-4}$ (L/degree) and saline contraction coefficient $a_S = 1.7 \times 10^{-4}$ (L/g). After taking correlation of the refractive index and applying to it the Fourier transform, the resulting power spectrum can be shown to have the form

$$\mathfrak{B}(\kappa) = a_T^2 \mathfrak{B}_T(\kappa) + a_S^2 \mathfrak{B}_S(\kappa) + 2a_T a_S \mathfrak{B}_{TS}(\kappa), \quad (9.30)$$

where $\mathfrak{B}_T, \mathfrak{B}_S, \mathfrak{B}_{TS}$ are the temperature spectrum, the salinity spectrum, their co-spectrum, respectively, and a_T and a_S are the linear coefficients. More explicitly,

$$\mathfrak{B}(\kappa) = \frac{C_0}{4\pi} \epsilon^{-1/3} \kappa^{-11/3} [1 + C_1(\kappa \eta_K)^{2/3}]$$
$$\times [G_T(\kappa) + G_S(\kappa) - 2G_{TS}(\kappa)], \quad (9.31)$$

with

$$G_T(\kappa) = a_T^2 \chi_T \exp[-b_T \delta_H], \quad G_S(\kappa) = a_S^2 \chi_S \exp[-b_S \delta_H],$$
$$G_{TS}(\kappa) = a_T a_S \chi_{TS} \exp[-b_{TS} \delta_H], \quad (9.32)$$

where $b_{TS} = 0.5 C_0 C_1^{-2}(1/\mathrm{Pr}_T + 1/\mathrm{Pr}_S)$, while δ_H and a_j $(j = T, S)$ were defined above. Further, using the relation among the dissipation rates pertinent to homogeneous and isotropic turbulence

$$\chi_n = a_T^2 \chi_T + a_S^2 \chi_S - 2a_T a_S \chi_{TS}, \quad (9.33)$$

and expressing the individual dissipation rates as

$$\chi_T = K_T \left(\frac{d\langle T \rangle}{dz}\right)^2, \quad \chi_S = K_S \left(\frac{d\langle S \rangle}{dz}\right)^2,$$
$$\chi_{TS} = 0.5(K_T + K_S)\left(\frac{d\langle T \rangle}{dz}\right)\left(\frac{d\langle S \rangle}{dz}\right), \quad (9.34)$$

with K_T and K_S being the coefficiens of eddy thermal diffusivity and diffusion of salt, and $\left(\frac{d\langle T\rangle}{dz}\right)$, $\left(\frac{d\langle S\rangle}{dz}\right)$ being the gradients of temperature and salinity, respectively, we find that

$$\mathfrak{B}(\kappa) = \frac{C_0}{4\pi}\chi_T \varepsilon^{-1/3}\kappa^{-11/3}[1+C_1(\kappa\eta_K)^{2/3}]$$
$$\times \frac{\omega_b^2 d_r \exp[-b_T\delta_H] + \exp[-b_S\delta_H] - \omega_b(1+d_r)\exp[-b_{TS}\delta_H]}{\omega_b^2 d_r + 1 - \omega_b(1+d_r)}.$$
(9.35)

We also set

$$\chi_n = \frac{a_T^2 \chi_T(\omega_b-1)^2}{\omega_b^2}, \quad \chi_S = \frac{a_T^2}{\omega_b^2 a_S^2}\chi_T, \quad \chi_{TS} = \sqrt{\chi_T \chi_S}, \quad (9.36)$$

and define the *temperature–salinity balance parameter* (dimensionless)

$$\omega_b = a_T\frac{d\langle T\rangle}{dz}\bigg/ a_S\frac{d\langle S\rangle}{dz}, \qquad (9.37)$$

varying between values -5 for temperature-dominated fluctuations and 0 for salinity-dominated fluctuations, and $d_r = K_T/K_S$ for *eddy diffusivity ratio* reducing to unity under stable stratification. Then spectrum in Eq. (9.35) reduces, with $d_r = 1$, to

$$\mathfrak{B}(\kappa) = \frac{C_0}{4\pi}\chi_T a_T^2 \varepsilon^{-1/3}\kappa^{-11/3}[1+C_1(\kappa\eta_K)^{2/3}]$$
$$\times [\omega_b^2 \exp(-b_T\delta_H) + \exp(-b_S\delta_H)$$
$$- 2\omega_b \exp(-b_{TS}\delta_H)]. \qquad (9.38)$$

The values of the two major oceanic turbulence parameters ε and χ_T span several orders of magnitude: ε varies from $10^{-2}\,\mathrm{m^2\,s^{-3}}$ near the ocean surface to $10^{-8}\,\mathrm{m^2\,s^{-3}}$ at larger depths; χ_T varies from $10^{-4}\,\mathrm{K^2\,s^{-1}}$ close to the surface to $10^{-10}\,\mathrm{K^2\,s^{-1}}$ in deeper layers (Thorpe, 2012).

9.3.2 Natural Earth's water turbulence

The previous section offered a relatively simple power spectrum model, but it is not applicable if average temperature $\langle T\rangle$ and average salinity $\langle S \rangle$ of water substantially devaite from the fixed values, and it is also wavelength insensitive. We will now discuss the power spectrum that has the same structure as that in Eq. (9.30) but such that its contributions, including the temperature spectrum $\mathfrak{B}_T$, the salinity spectrum $\mathfrak{B}_S$ and their co-spectrum $\mathfrak{B}_{TS}$, as well as coefficients a_T and a_S may depend on $\langle T \rangle$, $\langle S \rangle$ and λ as

$$\mathfrak{B}(\kappa, \langle T \rangle, \langle S \rangle, \lambda)$$
$$= A_T^2(\langle T \rangle, \langle S \rangle, \lambda) \mathfrak{B}_T(\kappa, \langle T \rangle, \langle S \rangle)$$
$$+ A_S^2(\langle T \rangle, \langle S \rangle, \lambda) \mathfrak{B}_S(\kappa, \langle T \rangle, \langle S \rangle)$$
$$+ 2A_T(\langle T \rangle, \langle S \rangle, \lambda) A_S(\langle T \rangle, \langle S \rangle, \lambda) \mathfrak{B}_{TS}(\kappa, \langle T \rangle, \langle S \rangle). \quad (9.39)$$

Here, the individual power spectra $\mathfrak{B}_j$, $(j = T, S, TS)$ defined in Eq. (9.19) can be accurately predicted by the fourth of Hill's models, H4, giving g_j^{H4} as the solution of non-linear differential equation

$$\frac{d}{d\xi}\left\{(\xi^{2b}+1)^{-1/3b}\left[-\frac{11}{3}f(\xi)+\xi\frac{df(\xi)}{d\xi}\right]\right\}$$
$$= \frac{22}{3}c_j\xi^{1/3}f(\xi), \quad (9.40)$$

where

$$f(\xi) = f(a\xi) = f(\kappa \eta_K), \quad (9.41)$$

and $\xi = \kappa \eta_K / a$, $c_j = C_0 a^{4/3}/\mathrm{Pr}_j$, $a = 0.072$, $b = 1.9$.

For characterization of the interaction between light and water having $\langle T \rangle \in [0°C, 30°C]$ and $\langle S \rangle \in [0\mathrm{ppt}, 40\mathrm{ppt}]$, which we will term *natural water*, Eq. (9.40) has been numerically fitted in Yao *et al.* (2019) in the interval $[3, 3000]$ for Pr or Sc to the function:

$$g_j(\kappa, \langle T \rangle, \langle S \rangle) = [1 + 21.61(\kappa \eta_K)^{0.61} c_j^{0.02} - 18.18(\kappa \eta_K)^{0.55} c_j^{0.04}]$$
$$\times \exp[-174.90(\kappa \eta_K)^2 c_j^{0.96}], \quad (9.42)$$

where $(j = T, S, TS)$. Interval $[3, 3000]$ is dictated by Pr and Sc varying with natural water's temperature and salinity (see Fig. 9.4). For dependence of all quantities used in definition of Pr and Sc on

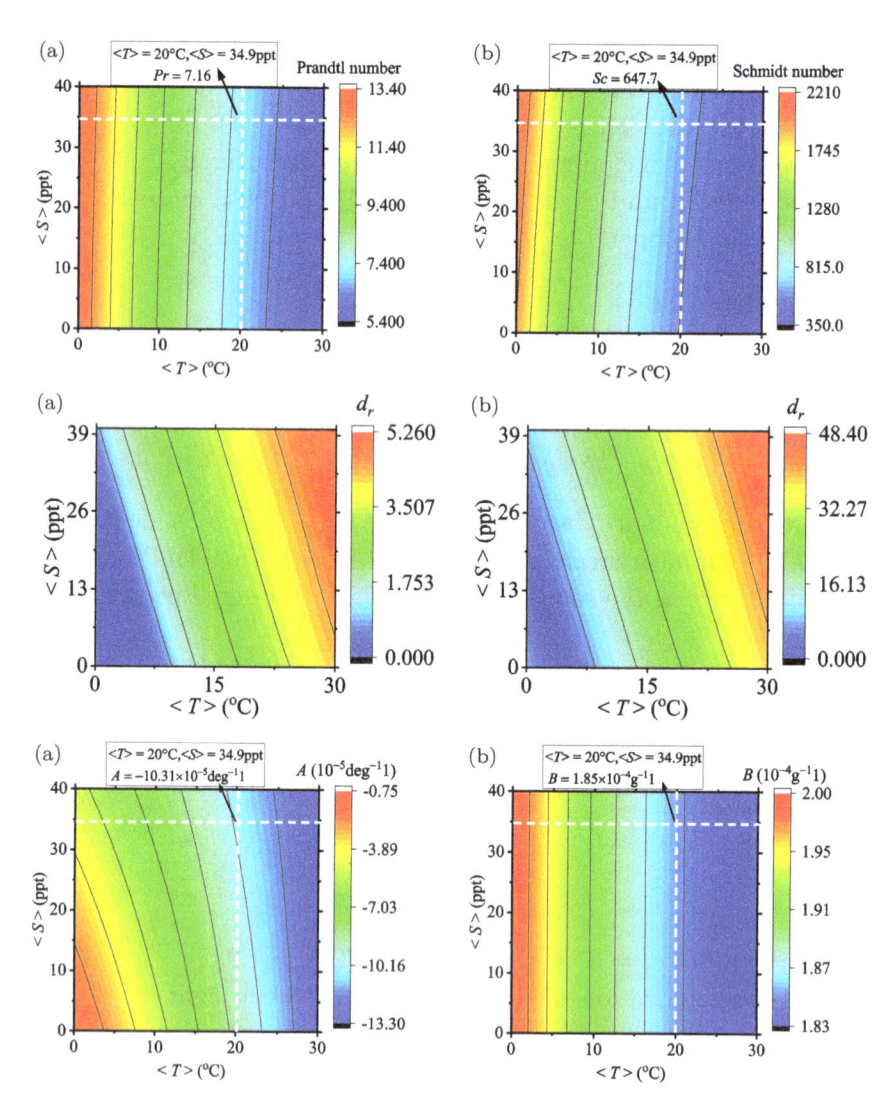

Fig. 9.4 Top: Pr and Sc versus $\langle T \rangle$ and $\langle S \rangle$. Middle: d_r varying with $\langle T \rangle$ and $\langle S \rangle$ at two values of H_g. (a) $H_g = -10°\text{C} \cdot \text{ppt}^{-1}$; (b) $H_g = -100°\text{C} \cdot \text{ppt}^{-1}$. Bottom: coefficients A_T and A_S (A and B in figure notations) varying with $\langle T \rangle$ and $\langle S \rangle$ at $\lambda = 532$ nm. Reprinted with permission from Yao *et al.* (2020). © The Optical Society.

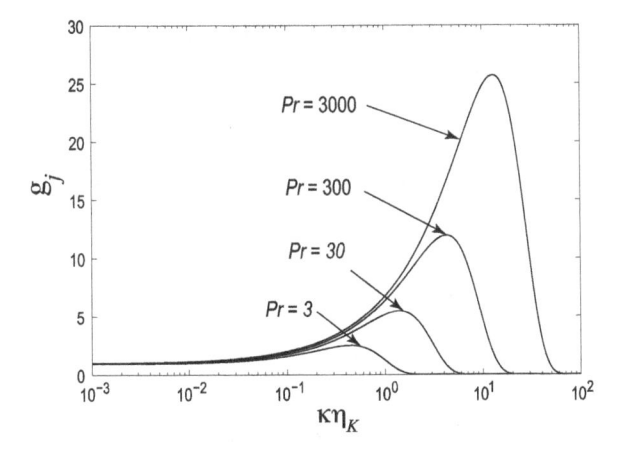

Fig. 9.5 Function g_j approximated in Eq. (9.42), for several values of Pr or Sc.

$\langle T \rangle$ and $\langle S \rangle$, see Appendix A. Thus, for each of these three spectra, we write

$$\mathfrak{B}_j(\kappa, \langle T \rangle, \langle S \rangle) = \frac{C_0}{4\pi} \varepsilon^{-\frac{1}{3}} \kappa^{-\frac{11}{3}} \chi_j \exp[-174.90(\kappa\eta_K)^2 c_j^{0.96}]$$
$$\times [1 + 21.61(\kappa\eta_K)^{0.61} c_j^{0.02}$$
$$- 18.18(\kappa\eta_K)^{0.55} c_j^{0.04}], \tag{9.43}$$

where $(j = T, S, TS)$. Figure 9.5 shows g_j for various Pr (or Sc) numbers. In Eqs. (9.42), and (9.43), c_j (dimensionless) are

$$c_T = a^{4/3} C_0 \mathrm{Pr}^{-1}(\langle T \rangle, \langle S \rangle), \quad c_S = a^{4/3} C_0 \mathrm{Sc}^{-1}(\langle T \rangle, \langle S \rangle),$$
$$c_{TS} = a^{4/3} C_0 \frac{\mathrm{Pr}(\langle T \rangle, \langle S \rangle) + \mathrm{Sc}(\langle T \rangle, \langle S \rangle)}{2\mathrm{Pr}(\langle T \rangle, \langle S \rangle)\mathrm{Sc}(\langle T \rangle, \langle S \rangle)}, \tag{9.44}$$

where c_{TS} is based on the coupling between Pr and Sc and

$$\eta_K = \left[\frac{\mu(\langle T \rangle, \langle S \rangle)}{\rho(\langle T \rangle, \langle S \rangle)}\right]^{3/4} \varepsilon^{-1/4}. \tag{9.45}$$

Hence, c_j and η_K also depend on $\langle T \rangle$ and $\langle S \rangle$.

We will now express χ_j ($j = S, TS$) through χ_T. Let us write eddy diffusivity ratio d_r as

$$d_r = K_S/K_T \approx \begin{cases} R_\rho + R_\rho^{0.5}(R_\rho - 1)^{0.5}, & R_\rho \geq 1, \\ 1.85 R_\rho - 0.85, & 0.5 \leq R_\rho < 1, \\ 0.15 R_\rho, & R_\rho < 0.5, \end{cases} \quad (9.46)$$

where R_ρ is *density ratio* (dimensionless) given by the expression

$$R_\rho = \frac{a_T}{a_S} |H_g|, \quad (9.47)$$

with a_T and a_S being the thermal expansion coefficient and the saline contraction coefficient, respectively, and

$$H_g = \frac{d\langle T \rangle}{dz} \bigg/ \frac{d\langle S \rangle}{dz} \quad (9.48)$$

is the *temperature–salinity gradient ratio* [°C ppt^{-1}]. See Appendix A for dependence of a_T and a_S on $\langle T \rangle$ and $\langle S \rangle$.

Under the assumption of atmospheric pressure

$$a_T(\langle T \rangle, \langle S \rangle) = \frac{1}{V_s} \frac{\partial V_s}{\partial T}\bigg|_S \quad \text{and} \quad a_S(\langle T \rangle, \langle S \rangle) = \frac{1}{V_s} \frac{\partial V_s}{\partial S}\bigg|_T. \quad (9.49)$$

Here, V_s is given in the Appendix (see also Yao et al., 2020 for details). Based on Eqs. (9.46)–(9.49), $d_r = d_r(\langle T \rangle, \langle S \rangle, H_g)$ (see Fig. 9.4).

The ensemble-averaged variance dissipation rates χ_j were given in Eq. (9.34). Combining Eq. (9.34) with Eqs. (9.46)–(9.49) finally gives

$$\chi_S(\langle T \rangle, \langle S \rangle, H_g, \chi_T) = \frac{d_r(\langle T \rangle, \langle S \rangle, H_g)}{H_g^2} \chi_T,$$
$$\chi_{TS}(\langle T \rangle, \langle S \rangle, H_g, \chi_T) = \frac{1 + d_r(\langle T \rangle, \langle S \rangle, H_g)}{2H_g} \chi_T. \quad (9.50)$$

In order to find the coefficients A_T and A_S in Eq. (9.39), the formula for the still natural water refractive index obtained in Quan and Fry (1997) on the basis of experimental data can be used:

$$n(T, S, \lambda) = a_0 + (a_1 + a_2 T + a_3 T^2)S + a_4 T^2$$
$$+ \frac{a_5 + a_6 S + a_7 T}{\lambda} + \frac{a_8}{\lambda^2} + \frac{a_9}{\lambda^3}, \quad (9.51)$$

with constants fitted as

$$a_0 = 1.31405, \quad a_1 = 1.779 \times 10^{-4}, \quad a_2 = -1.05 \times 10^{-6},$$
$$a_3 = 1.6 \times 10^{-8}, \quad a_4 = -2.02 \times 10^{-6},$$
$$a_5 = 15.868, \quad a_6 = 0.01155, \quad (9.52)$$
$$a_7 = -0.00423, \quad a_8 = -4382, \quad a_9 = 1.1455 \times 10^6.$$

Expression (9.51) is valid for $0°C \leq T \leq 30°C$ and $0\,\text{ppt} \leq S \leq 40\,\text{ppt}$ but also depends on wavelength in the visible range $400\,\text{nm} \leq \lambda \leq 700\,\text{nm}$. Refractive index n can be expressed as a sum of its average value n_0 and relative fluctuation n':

$$n = n_0(\langle T \rangle, \langle S \rangle, \lambda) + n'. \quad (9.53)$$

Here, the first quantity is simply given by Eq. (9.51), but with T and S substituted by $\langle T \rangle$ and $\langle S \rangle$, respectively. The linearization

$$n' \approx A_T(\langle T \rangle, \langle S \rangle, \lambda)T' + A_S(\langle T \rangle, \langle S \rangle, \lambda)S', \quad (9.54)$$

can then be employed, with T' and S' being the fluctuating components of the temperature and salinity concentration distributions, respectively. Unlike in the model (Nikishov and Nikishov, 2000), A_T and A_S also depend on $\langle T \rangle, \langle S \rangle$ and λ. Indeed, taking the differential

$$dn(T, S, \lambda) = \frac{\partial n(T, S, \lambda)}{\partial T} dT + \frac{\partial n(T, S, \lambda)}{\partial S} dS + \frac{\partial n(T, S, \lambda)}{\partial \lambda} d\lambda, \quad (9.55)$$

and setting $d\lambda = 0$ yields

$$A_T(\langle T \rangle, \langle S \rangle, \lambda) = \left. \frac{\partial n(T, S, \lambda)}{\partial T} \right|_{T=\langle T \rangle, S=\langle S \rangle}$$
$$= a_2 \langle S \rangle + 2a_3 \langle T \rangle \langle S \rangle + 2a_4 \langle T \rangle + \frac{a_7}{\lambda}, \quad (9.56)$$

and

$$A_S(\langle T\rangle, \langle S\rangle, \lambda) = \left.\frac{\partial n(T, S, \lambda)}{\partial S}\right|_{T=\langle T\rangle, S=\langle S\rangle}$$
$$= a_1 + a_2\langle T\rangle + a_3\langle T\rangle^2 + \frac{a_6}{\lambda}. \quad (9.57)$$

Thus, the power spectrum model (9.39) describes the natural water turbulence for the majority of conditions existing in the Earth's water basins. It combines the two scalar spectra and their co-spectrum with the Pr/Sc, d_r, A_T and A_S depending on $\langle T\rangle$ and $\langle S\rangle$; A_T and A_S depending on λ and d_r depending on H_g. To summarize the dependence on $\langle T\rangle$ and $\langle S\rangle$, we again refer the reader to Fig. 9.4 showing such variation for (top row) Pr and Sc; (bottom row) A_T and A_S at $\lambda = 532\,\text{nm}$; (middle row) of d_r for two values of H_g. Figure 9.6 shows the natural turbulent water power spectrum varying with $\eta\kappa$ $\eta_K\kappa$ for selected values of thremodynamic and source parameters.

The effects of oceanic turbulence on propagating light is similar to that of the atmospheric turbulence, but depending on its parameters, locally it can be several orders of magnitude stronger. Also, with increasing average temperature and average salinity, the turbulence effects on light become stronger.

9.4 Bio-tissue Turbulence

Some optically soft, relatively homogeneous and isotropic biological tissues with typical refractive index of about 1.35 may also be viewed as distributions of inhomogeneites obeying the turbulent cascade with a negative power law. The seminal measurement of the spatial correlation functions and power spectra of human and mouse epidermis and liver was obtained in Schmidt and Kumar (1996). The results for the power spectrum inferred from the microscopy images demonstarted that there is a well-pronounced inertial range of scales, with the power law that may deviate from the Kolmogorov law depending on the tissue type. The outer scale of the measured tissues were shown to be on the order of several micron. The inner scale was not detected, but it can be argued that it is on the order of 0.1–0.3 μm,

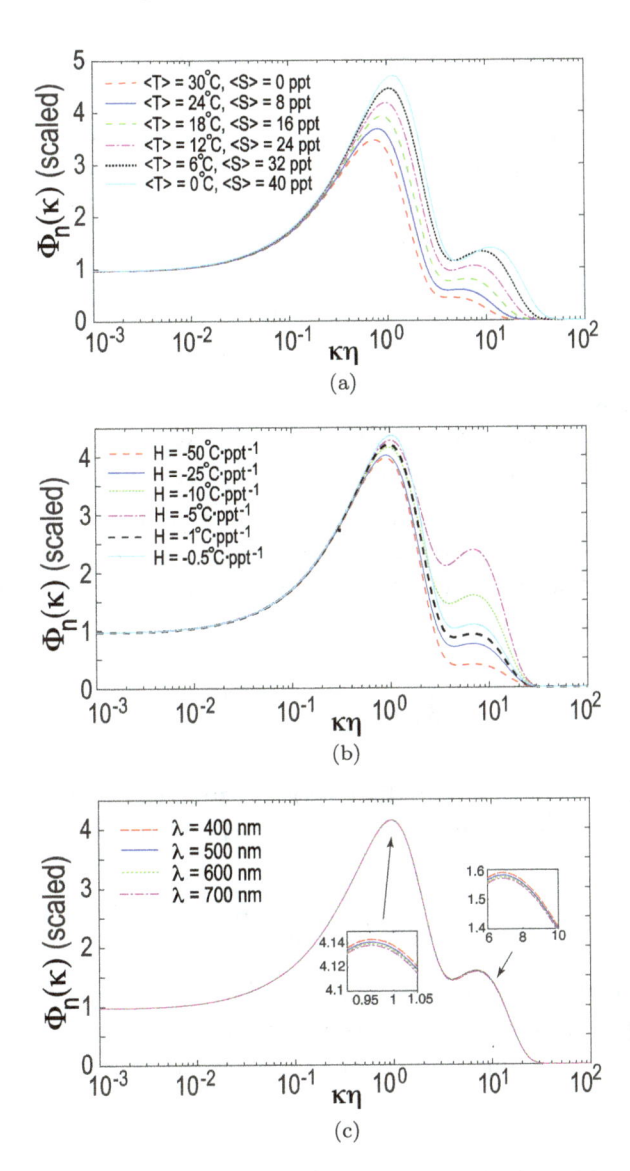

Fig. 9.6 The natural turbulent water power spectrum varying with average temperature, salinity, gradient ratio and source wavelength. Reprinted with permission from Ata and Korotkova (2021) @ the Optical Society.

being comparable to the size of an organelle. The typical values of the refractive index variance σ_n^2 are on the order of 10^{-3}. The inertial range extent and the σ_n^2 values make it a much stronger scattering medium as compared to atmospeheric and oceanic turbulence. The original, image-based, two-dimensional power spectrum model suggested in Schmidt and Kumar (1996) uses the van Karman form

$$\mathfrak{B}_n(\kappa) = \frac{4\pi\sigma_n^2 L_0^2(\alpha'-1)}{(1+\kappa^2 L_0^2)^{\alpha'}}, \qquad (9.58)$$

where L_0 is the outer scale and $\alpha' = \alpha/2$ (α was introduced for non-classic atmopsheric turbulence) is the exponent varying in the interval from 1.28 to 1.41. The relation between the two-dimensional and the three-dimensional power spectra required for light propagation problems may be obtained first getting the one-dimensional spectrum using the projection-slice theorem (Sheppard, 1996) and then using the relation

$$\mathfrak{B}_n(\kappa) = -\frac{d\mathfrak{B}_n^{(1)}(\kappa)}{2\pi\kappa d\kappa}, \qquad (9.59)$$

where $\mathfrak{B}_n^{(1)}$ is the one-dimensional spectrum. The resulting three-dimensional spectrum is (Chen et al., 2020)

$$\mathfrak{B}_n(\kappa) = \frac{\sigma_n^2 L_0^3 \Gamma(\alpha')}{\pi^{3/2}\Gamma(\alpha'-3/2)(1+\kappa^2 L_0^2)^{\alpha'}}, \qquad (9.60)$$

where Γ denotes the Gamma function. The dependence of the tissue type on power law α' affects the statistics of light scattered from it, which can be employed in early diagnostics of cancerous tissues (Hunter et al., 2006).

9.5 Methods for Light–turbulence Interaction

9.5.1 Extended Huygens–Fresnel method

For the mildly-fluctuating media, i.e., those in which $\delta n/n_0 \ll 1$, where δn is the varying part, and n_0 is the average value of the refractive index, the terms in Maxwell's equations (see Chapter 2) that are coupling different electric (magnetic) field components vanish. Then each field component obeys the wave equation:

$$\nabla^2 \mathcal{U}(\mathbf{r},t) + k^2 n^2(\mathbf{r},t) \frac{\partial^2 \mathcal{U}(\mathbf{r},t)}{\partial t^2} = 0, \qquad (9.61)$$

in the space–time representation or the Helmholtz equation

$$\nabla^2 U(\mathbf{r},\omega) + k^2 n^2(\mathbf{r},\omega) U(\mathbf{r},\omega) = 0, \qquad (9.62)$$

in the space–frequency representation. The methods for solving these equations largely depend on the type of statistical properties of light to be studied and the particular regime of turbulence. In this section, we will introduce a method widely used in dealing with the light–turbulence interaction, known as the *Extended Huygens–Fresnel* method.

The solution of Eq. (9.62) can be found for the field $U(\mathbf{r})$ (ω is omitted) propagating at distance z from the source plane in the form

$$U(\mathbf{r}) = -\frac{ik \exp(ikz)}{2\pi z} \int U(\boldsymbol{\rho}') \exp\left[ik\frac{(\boldsymbol{\rho}' - \boldsymbol{\rho})^2}{2z}\right]$$
$$\times \exp[\Psi(\boldsymbol{\rho}', \mathbf{r})] d^2 \rho', \qquad (9.63)$$

where we recognize the Huygens–Fresnel integral derived in Chapter 4 adjusted by the complex phase perturbation of a spherical wave $\Psi(\boldsymbol{\rho}', \mathbf{r})$ propagating from source point $\boldsymbol{\rho}'$ to field point $\mathbf{r}$.

If the cross-spectral density function of the optical field in the source plane, at points $\boldsymbol{\rho}_1'$ and $\boldsymbol{\rho}_2'$, is

$$W^{(0)}(\boldsymbol{\rho}_1', \boldsymbol{\rho}_2') = \langle U^*(\boldsymbol{\rho}_1') U(\boldsymbol{\rho}_2') \rangle, \qquad (9.64)$$

where the average is taken over the ensemble of monochromatic realizations and that after propagation at distance z from the source is

$$W(\mathbf{r}_1, \mathbf{r}_2) = \langle U^*(\mathbf{r}_1) U(\mathbf{r}_2) \rangle, \qquad (9.65)$$

then on substituting from Eq. (9.63) into Eq. (9.65) we find that

$$W(\mathbf{r}_1, \mathbf{r}_2) = \int \int W^{(0)}(\boldsymbol{\rho}_1', \boldsymbol{\rho}_2') K(\mathbf{r}_1, \mathbf{r}_2, \boldsymbol{\rho}_1', \boldsymbol{\rho}_2') d^2 \rho_1' d^2 \rho_2'. \qquad (9.66)$$

Here, propagator K has the form

$$K(\mathbf{r}_1, \mathbf{r}_2, \boldsymbol{\rho}_1', \boldsymbol{\rho}_2') = \left(\frac{k}{2\pi z}\right)^2 \exp\left[-ik\frac{(\boldsymbol{\rho}_1' - \boldsymbol{\rho}_1)^2 - (\boldsymbol{\rho}_2' - \boldsymbol{\rho}_2)^2}{2z}\right]$$

$$\times \langle \exp[\Psi^*(\boldsymbol{\rho}_1', \mathbf{r}_1) + \Psi(\boldsymbol{\rho}_2', \mathbf{r}_2)]\rangle_m, \qquad (9.67)$$

where wave perturbation induced by random fluctuations in the medium is accounted by correlation function of the spherical wave, shown in the angular brackets with subscript m, denoting the statistical average over the medium realizations.

In the general case of anisotropic fluctuations, the spherical wave correlation function has the form (Yura, 1972)

$$\langle \exp[\Psi^*(\boldsymbol{\rho}_1', \mathbf{r}_1) + \Psi(\boldsymbol{\rho}_2', \mathbf{r}_2)]\rangle_m$$

$$= \exp\left[-2\pi k^2 z \int_0^1 d\xi \int_0^\infty d^2\kappa\right.$$

$$\left.\times \mathfrak{B}(\boldsymbol{\kappa})\left[1 - \exp\left[\xi\boldsymbol{\rho}_d \cdot \boldsymbol{\kappa} + (1-\xi)\boldsymbol{\rho}_d' \cdot \boldsymbol{\kappa}_\perp\right]\right]\right], \qquad (9.68)$$

where $\boldsymbol{\rho}_d' = \boldsymbol{\rho}_1' - \boldsymbol{\rho}_2'$, $\boldsymbol{\rho}_d = \boldsymbol{\rho}_1 - \boldsymbol{\rho}_2$ are difference vectors in transverse planes and $\boldsymbol{\kappa}_\perp = (\kappa_x, \kappa_y, 0)$ is the transverse spatial-frequency vector. For derivation of Eq. (9.68), Markov approximation was employed under which the fluctuations in the refractive index are assumed to be delta-correlated at any pair of points in the direction of propagation z.

In isotropic turbulence, the power spectrum is a function of $\kappa_\perp$ and, hence, expression (9.68) reduces to

$$\langle \exp[\Psi^*(\boldsymbol{\rho}_1', \mathbf{r}_1) + \Psi(\boldsymbol{\rho}_2', \mathbf{r}_2)]\rangle_m$$

$$= \exp\left[-4\pi^2 k^2 z \int_0^1 d\xi \int_0^\infty d\kappa \mathfrak{B}(\kappa)\right.$$

$$\left.\times [1 - J_0[|(1-\xi)(\mathbf{r}_1 - \mathbf{r}_2) + \xi(\boldsymbol{\rho}_1 - \boldsymbol{\rho}_2)|\kappa_\perp]]\right], \qquad (9.69)$$

where $J_0(x)$ is the 0th order Bessel function of the first kind. For field points $\mathbf{r}_1$ and $\mathbf{r}_2$ located in the region sufficiently close to the beam axis, two-term approximation of the Bessel function $J_0(x) \approx 1 - x^2/4$

is frequently used. Hence, expression (9.69) reduces to the form (Gbur and Wolf, 2002; Lu et al., 2007)

$$\langle \exp[\Psi^*(\boldsymbol{\rho}'_1, \mathbf{r}_1) + \Psi(\boldsymbol{\rho}'_2, \mathbf{r}_2)]\rangle_m$$

$$= \exp\left[-\frac{\pi^2 k^2 z}{3}[(\boldsymbol{\rho}_1 - \boldsymbol{\rho}_2)^2 + (\boldsymbol{\rho}_1 - \boldsymbol{\rho}_2)(\boldsymbol{\rho}'_1 - \boldsymbol{\rho}'_2)\right.$$

$$\left. + (\boldsymbol{\rho}'_1 - \boldsymbol{\rho}'_2)^2]\int_0^\infty \kappa^3 \mathfrak{B}(\kappa)d\kappa\right]. \tag{9.70}$$

For evaluation of the spherical wave correlation function for an anisotropic power spectrum with two different anisotropy coefficients μ_x and μ_y in the two transverse directions to the beam axis, the change of variables $\kappa'_x = \mu_x \kappa_x$ and $\kappa'_y = \mu_y \kappa_y$ can be made. Under this transformation, the power spectrum can be shown to depend on scalar $\kappa' = \sqrt{\kappa'^2_x + \kappa'^2_y}$ (Wang and Korotkova, 2016b).

9.5.2 Convolution method

The Extended Huygens–Fresnel approach has a very simple and intuitive form if applied to Schell-model sources and isotropic random media for calculation of the spectral density of an optical beam. Namely, the spectral density can be represented as a convolution of three factors, relating to the source spectral density, the source degree of coherence and the refractive index correlation function of the medium (Wang and Korotkova, 2016c). Indeed, in this case the spherical wave structure function (9.69) calculated for $\mathbf{r}_1 = \mathbf{r}_2 = \mathbf{r}$, can be written as

$$\langle \exp[\Psi^*(\boldsymbol{\rho}'_1, \mathbf{r}) + \Psi(\boldsymbol{\rho}'_2, \mathbf{r})]\rangle_m$$

$$= \exp\left[\frac{1}{3}\pi^2 k^2 z(\boldsymbol{\rho}'_2 - \boldsymbol{\rho}'_1)^2 \int_0^\infty \kappa^3 \mathfrak{B}(\kappa)d\kappa\right]$$

$$= \int p_2(\mathbf{v}) \exp\left[ik(\boldsymbol{\rho}'_2 - \boldsymbol{\rho}'_1) \cdot \mathbf{v}\right] d^2v, \tag{9.71}$$

where

$$p_2(\mathbf{v}) = \frac{1}{2}k^2\rho_0^2 \exp[-k^2\rho_0^2 v^2], \quad \rho_0^2 = \frac{3}{2}\pi^2 k^2 z \int_0^\infty \kappa^3 \mathfrak{B}(\kappa)d\kappa, \tag{9.72}$$

ρ_0 being the spherical wave coherence radius, i.e., the characteristic transverse separation distance at which it becomes spatially incoherent. As the next step, we write the Schell-model source cross-spectral density (Chapter 5) as

$$W^{(0)}(\boldsymbol{\rho}'_1, \boldsymbol{\rho}'_2) = a^*(\boldsymbol{\rho}'_1)a(\boldsymbol{\rho}'_2)FT[p_1(\boldsymbol{\rho}'_2 - \boldsymbol{\rho}'_1)], \qquad (9.73)$$

where $FT[p_1] = \mu$ is the source degree of coherence. Then, introducing function

$$A(\boldsymbol{\rho}') = a(\boldsymbol{\rho}') \exp\left[\frac{ik\rho'^2}{2z}\right], \qquad (9.74)$$

and employing variables $\boldsymbol{\rho}'_d = \boldsymbol{\rho}'_1 - \boldsymbol{\rho}'_2$ and $\boldsymbol{\rho}'_s = (\boldsymbol{\rho}'_1 + \boldsymbol{\rho}'_2)/2$ we write:

$$A^*(\boldsymbol{\rho}'_s - \boldsymbol{\rho}'_d/2) = \frac{k^2}{4\pi^2} \int FT[A^*(\mathbf{u}_1)] \exp[-ik\mathbf{u}_1 \cdot (\boldsymbol{\rho}'_s - \boldsymbol{\rho}'_d/2)]d^2u_1 \qquad (9.75)$$

and

$$A(\boldsymbol{\rho}'_s + \boldsymbol{\rho}'_d/2) = \frac{k^2}{4\pi^2} \int FT[A(\mathbf{u}_2)] \exp[-ik\mathbf{u}_2 \cdot (\boldsymbol{\rho}'_s + \boldsymbol{\rho}'_d/2)]d^2u_2. \qquad (9.76)$$

Then it follows from Eqs. (9.66), (9.67), and (9.71)–(9.76) that

$$S(\mathbf{r}) = \frac{1}{\lambda^2 z^2} \iint W^{(0)}(\boldsymbol{\rho}'_1, \boldsymbol{\rho}'_2) \exp\left[\frac{ik(\rho'^2_2 - \rho'^2_1)}{2z} - \frac{ik(\boldsymbol{\rho}'_2 - \boldsymbol{\rho}'_1) \cdot \boldsymbol{\rho}}{z}\right]$$
$$\times \langle \exp[\Psi^*(\boldsymbol{\rho}'_1, \mathbf{r}) + \Psi(\boldsymbol{\rho}'_2, \mathbf{r})]\rangle_M d^2\rho'_1 d^2\rho'_2 \qquad (9.77)$$

and, hence, it can be expressed as

$$S(\mathbf{r}) = \frac{1}{\lambda^2 z^2} \iiiint A^*(\boldsymbol{\rho}'_s - \boldsymbol{\rho}'_d/2) A(\boldsymbol{\rho}'_s + \boldsymbol{\rho}'_d/2)$$
$$\times \exp[-ik\boldsymbol{\rho}'_d \cdot (\mathbf{v}_1 + \mathbf{v}_2)] p_1(\mathbf{v}_1) p_2(\mathbf{v}_2)$$
$$\times \exp\left[\frac{-ik}{z}\boldsymbol{\rho}'_d \cdot \boldsymbol{\rho}\right] d^2\rho'_d d^2\rho'_s d^2v_1 d^2v_2. \qquad (9.78)$$

Integration over $\boldsymbol{\rho}'_s$, $\boldsymbol{\rho}'_d$, $\mathbf{u}_1$ and $\mathbf{u}_2$ (see Eqs. (9.75) and (9.76)) results in triple convolution

$$S(\mathbf{r}) = S_f(\mathbf{u}) \circledast p_1(\mathbf{u}) \circledast p_2(\mathbf{u}), \qquad (9.79)$$

where

$$S_f(\mathbf{u}) = \frac{1}{\lambda^2 z^2}|FT[A(-\mathbf{u})]|^2, \qquad (9.80)$$

and $\mathbf{u} = -\boldsymbol{\rho}/z$. Thus, Eq. (9.79) makes it possible to express the spectral density of a random beam generated by a Schell-model source and propagating in a linear random isotropic medium as a convolution of three functions: $S_f(\mathbf{u})$, $p_1(\mathbf{u})$ and $p_2(\mathbf{u})$ all depending on the direction $\mathbf{u}$.

The first term S_f is associated with the transverse distribution of the spectral density of the coherent portion of the beam propagating in vacuum at distance z from the source; the second term, p_1, is the Fourier transform of the source degree of coherence; the last function, p_2, is the Fourier transform of the spherical wave correlation function propagating through the random medium. Due to the simplicity of this result, straighforward analytic calculations may be performed for numerous combinations of the source and the medium properties. Moreover, the Fast Fourier Transform [FFT] method can be readily implemented for rapid numerical calculations with the help of the alternative formula

$$S(\mathbf{r}) = FT^{-1}[FT[S_f(\mathbf{f})]FT[p_1(\mathbf{f})]FT[p_2(\mathbf{f})]](\mathbf{u})], \qquad (9.81)$$

where $\mathbf{f}$ is a two-dimensional vector in the Fourier domain. To obtain Eq. (9.81), the two-dimensional convolution theorem was used.

To illustrate the versatility and ease of the convolution approach, we show in Fig. 9.7 the propagating average intensity of a beam radiated by a coherent Bessel–Gauss source with

$$a(\boldsymbol{\rho}) = J_1(\beta\rho)\exp\left(-\frac{\rho^2}{2\sigma_0^2}\right)\exp(i\theta_\rho), \qquad (9.82)$$

where $\boldsymbol{\rho} = (\rho, \theta_\rho)$, J_1 being the first-kind Bessel function of order 1, and β being its transverse wave number, propagating in free space (top) and oceanic turbulence (bottom). In Fig. 9.8, a random beam

280 *Theoretical Statistical Optics*

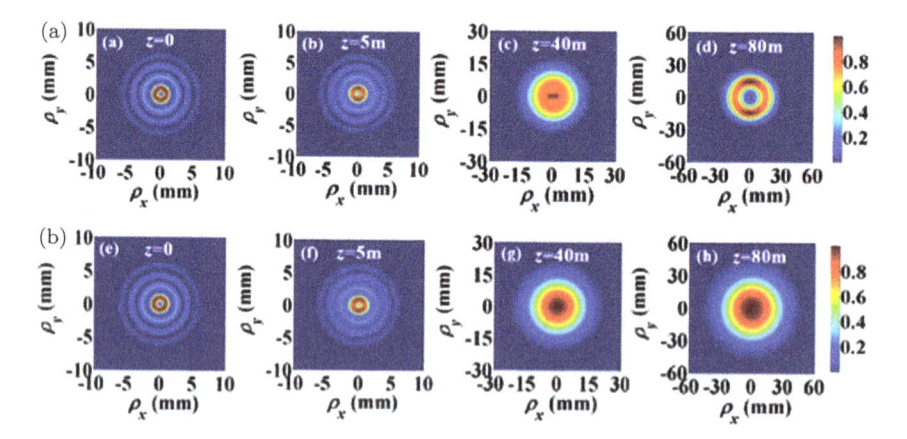

Fig. 9.7 Bessel–Gauss beam in free space (a); oceanic turbulence (b). Reprinted with permission from Wang and Korotkova (2016c). © The Optical Society.

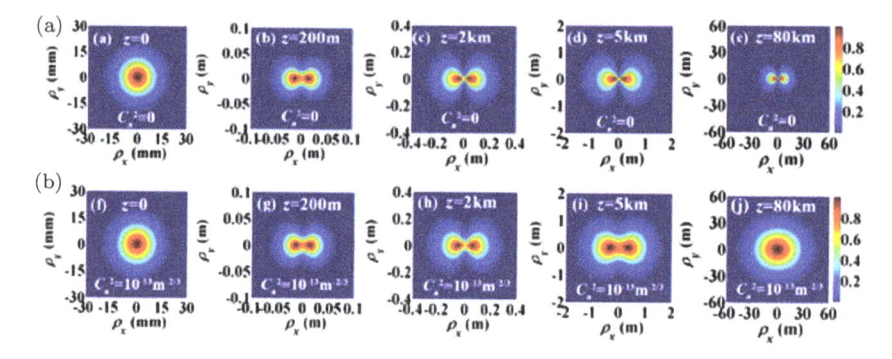

Fig. 9.8 Random beam with structured coherence in free space (a); atmopsheric turbulence (b). Reprinted with permission from Wang and Korotkova (2016c). © The Optical Society.

with

$$a(\boldsymbol{\rho}) = \exp\left(-\frac{\rho^2}{2\sigma_0^2}\right),$$

$$p_1(\mathbf{v}) = \frac{k^2\delta_0^2}{\pi} \exp\left(-\frac{k^2\delta_0^2 v^2}{2}\right) \cos^2\theta_v, \qquad (9.83)$$

where $\mathbf{v} = (v, \theta_v)$, propagating in free space (top row) and atmospheric turbulence (bottom row) is presented. In both cases, free space propagation leads to formation of the far-field average intensity being the Fourier transform of the source degree of coherence, while propagation at sufficiently long distances in turbulence results in a Gaussian-like profile.

9.5.3 Other methods

The *Rytov method* makes it possible to predict the statistical moments of the fluctuating optical fields of any order but is based on small perturbations and, hence, is limited to weak turbulence regime, i.e., relatively small propagation distances and sufficiently weak local turbulence. The key assumption of the Rytov method is the multiplicative nature of the optical field modulation by the random medium (Born and Wolf, 1999)

$$U(\mathbf{r}) = U_0(\mathbf{r}) \exp[\Psi(\mathbf{r})]. \tag{9.84}$$

Here, $U(\mathbf{r})$ is the optical field in the presence of random medium, $U_0(\mathbf{r})$ is field propagating to the same position but in the absence of the medium and $\Psi(\mathbf{r})$ is the complex phase perturbation caused by the refractive index fluctuations in the medium. It is frequently approximated by the two first terms, $\Psi(\mathbf{r}) \approx \Psi_1(\mathbf{r}) + \Psi_2(\mathbf{r})$, which is sufficient for obtaining accurate enough results for the second and fourth-order field moments. In particular, the *angular spectrum method* introduced in Gbur and Korotkova (2007) is based on the Rytov method developed for a pair of arbitrarily tilted plane waves. Then the correlation functions of any order for the field propagating in a random medium can be found by superposing all such pairs.

The *parabolic equation method* is, on the other hand, a very general approach which is applicable to a wide variety of turbulent regimes (Rytov et al., 1989). It uses the field in the form $U(\mathbf{r}) = \exp[ikz]U^{(p)}(\mathbf{r})$ in Eq. (9.62), which, after neglecting $U_{zz}^{(p)}(\mathbf{r})$ (subscript denotes partial derivative) and confirming that $\lambda < l_0$, reduces to parabolic equation

$$2ikU_z^{(p)}(\mathbf{r}) + \nabla_\perp^2 U^{(p)}(\mathbf{r}) + 2k^2\bar{n}(\mathbf{r})U^{(p)}(\mathbf{r}) = 0, \tag{9.85}$$

where $\bar{n}$ is obtained as the weak medium fluctuation condition $n^2(\mathbf{r}) \approx 1 + 2\bar{n}(\mathbf{r})$ is imposed. Then using the Markov approximation for correlation:

$$\langle \bar{n}(\mathbf{r}_1)\bar{n}(\mathbf{r}_2)\rangle_m = N^\perp(\boldsymbol{\rho}_1 - \boldsymbol{\rho}_2, z)\delta(z_1 - z_2), \tag{9.86}$$

one can find the field correlation functions. For example, for the cross-spectral density, one gets

$$4kW_z(\boldsymbol{\rho}_s, \boldsymbol{\rho}_d, z) - 4i\nabla_{r_s} \cdot \nabla_{r_d} W(\boldsymbol{\rho}_s, \boldsymbol{\rho}_d, z)$$
$$+ \pi k^3 D(\boldsymbol{\rho}_d) W(\boldsymbol{\rho}_s, \boldsymbol{\rho}_d, z) = 0, \tag{9.87}$$

where subindices denote partial derivatives, $\boldsymbol{\rho}_s = (\boldsymbol{\rho}_1 + \boldsymbol{\rho}_2)/2$, $\boldsymbol{\rho}_d = (\boldsymbol{\rho}_1 - \boldsymbol{\rho}_2)/2$ and

$$D(\boldsymbol{\rho}_d) = 8 \iint \mathfrak{B}_T(\kappa_x, \kappa_y, \kappa_z = 0)[1 - \exp(i\boldsymbol{\kappa}_\perp \cdot \boldsymbol{\rho}_d)]d^2\kappa. \tag{9.88}$$

The solution of this equation can be shown to precisely coincide with that developed by the extended Huygens–Fresnel method. The true power of the parabolic method is in accurate predictions of the light statistics of any order in any turbulent regime, however, requiring the use of the numerical methods.

9.6 Behavior of Light Beams in Turbulence

9.6.1 *General phenomena*

The aim of this section is to briefly overview the general results of various theoretical outcomes regarding the effects of classic turbulence on the propagating light beams radiated by scalar, coherent sources. Deviation from these assumptions results in a broad range of other possibilities that are out of scope of this text.

- *Beam wander.* In sufficiently weak turbulence, the light beam wanders, i.e., exhibits a snake-like motion, being caused by refraction of the beam by large turbulent eddies. However, after some propagation distance the beam gradually breaks down into speckles which gradually surpresses wandering. If the beam is tracked and short-exposure images are recorded, its centroid changes position with respect to the origin. The temporal rate of turbulence depends on viscosity of the substance. For instance, the atmopsheric turbulence turnover is on the order of a fraction of a millisecond, the oceanic and tissue turbulence are orders of magnitude slower. This defines the temporal scales of the detector needed to resolve the beam wander.

- *Turbulence-induced diffraction.* If the light beam is not tracked and the recorded intensity is averaged over a sufficiently long time (much longer than the turbulence turnover time), then the intensity averages to a profile larger than that of the short-term centroid.
- *Spectral density reshaping to Gaussian-like.* Regardless of the spectral density of the beam-like field in the source plane the turbulent media gradually convert it to a Gaussian-like profile with increasing propagation distance.
- *Loss of coherence.* The initially spatially coherent light gradually looses coherence becoming nearly incoherent at large propagation distances.
- *Changes in topological charge distribution.* The phase discrepancies accumulated in the light field propagating over a sufficiently long path may exceed values of 2π. In this case, generation of local vortex structures in the field by turbulence is possible.
- *Intensity fluctuations.* The phase discrepancies resulting in constructive and distructive interference events along the path result in fluctuation of the instantaneous intensity of the light field at a fixed point, called *scintillations*, and at two points, known as *intensity correlations*. Since the turbulent perturbations are not obeying Gaussian statistics, the high-order moments of intensity cannot be related to the average intensity (spectral density) by the Gaussian moment theorem.
- *Phase conjugation.* In monostatic links (double-pass links with coinciding transmitter and receiver) with a retro-reflector, the phases accumulated along the two paths cancel each other leading to the mechanism of random phase conjugation, and, hence, annihilating the effects of turbulence in a region close to the optical axis (see Chapter 4).

9.6.2 Probability density functions of intensity

Since the fluctuations in the parameters of a turbulent medium are not governed by Gaussian statistics, the moments of the fluctuating phase and intensity of the optical field higher than the second, and, more generally, their probability density functions (PDFs), become of particular interest. In this section, we will overview the intensity PDFs of a typical field radiated by a laser source widely used in studies relating to light propagation in turbulent air and water. Most of these models are heuristic and are based on experimental fits. Some models may be applied to turbulent paths of any strength and others are only suitable to particular regimes. The strength of turbulence along the propagation path is conventionally characterized by the *scintillation index*, generally defined as the normalized variance (Andrews and Phillips, 2005):

$$\sigma_I^2(\mathbf{r}) = \frac{\langle \mathcal{I}^2(\mathbf{r}) \rangle}{\langle \mathcal{I}(\mathbf{r}) \rangle^2} - 1, \tag{9.89}$$

of fluctuating intensity $\mathcal{I}$, of a plane wave, resulting for the Obukhov–Corrsin spectrum in the expression

$$\sigma_{I,pl}^2 = 1.23 C_T^2 k^{7/6} z^{11/6}, \tag{9.90}$$

where C_T^2 is the structure parameter, k is the wave number and z is the propagation length. Parameter $\sigma_{I,pl}^2$ is also known as the *Rytov variance*. For $\sigma_{I,pl}^2 \ll 1$, ≈ 1, and $\gg 1$, the turbulence is characterized as weak, moderate and strong.

We will now outline the PDFs of the intensity normalized by their average values $\langle \mathcal{I} \rangle$, for the reasons of comparison among various families and also because of the fact that the important fade statistics in lasercom channels use this normalization (Jamali et al., 2018).

The most popular one-parametric PDF model is *log-normal*, accurately describing the fluctuating intensity in the weak turbulence regime:

$$p_I^{(L)}(\mathcal{I}) = \frac{1}{2\mathcal{I}\sqrt{2\pi\sigma_L^2}} \exp\left[-\frac{(\ln \mathcal{I} - 2\mu_L)^2}{8\sigma_L^2}\right], \tag{9.91}$$

where μ_L, σ_L^2 are the mean and the variance of the corresponding Gaussian distribution (lognormal PDF is derived for random variable $\mathcal{X}$ from relation $\mathcal{X} = \ln \mathcal{I}/2$, where $\mathcal{X}$ is normally distributed). It immediately follows from $\langle \mathcal{I} \rangle = 1$ that $\mu_L = -\sigma_L^2$, hence producing one-parametric family. The scintillation index can then be expressed as $\sigma_I^2 = \exp(4\sigma_L^2) - 1$.

Another one-parametric family is described by *Gamma PDF* model:

$$p_I^{(G)}(\mathcal{I}) = \frac{\mathcal{I}^{k_G-1} \exp(-\mathcal{I}/\theta_G)}{\Gamma(k_G)(\theta_G)^{k_G}}, \qquad (9.92)$$

where k_G and θ_G are shape and scale parameters. Gamma PDF can be applied to various turbulent regimes. Condition $\langle \mathcal{I} \rangle = 1$ implies that $k_G \theta_G = 1$. Also, since $\langle \mathcal{I} \rangle^2 = k_G(k_G + 1)\theta_G^2$, the scintillation index becomes $\sigma_I^2 = \theta_G = 1/k_G$.

The *K-distribution* is an accurate model in the strong turbulent regime:

$$p_I^{(K)}(\mathcal{I}) = \frac{2\alpha_K}{\Gamma(\alpha_K)} (\alpha_K \mathcal{I})^{(\alpha_K-1)/2} K_{\alpha_K-1}(2\sqrt{\alpha_K \mathcal{I}}), \qquad (9.93)$$

where K_m is the mth order Bessel function of the second kind. With $\langle \mathcal{I} \rangle = 1$, we get the scintillation index $\sigma_I^2 = 2/\alpha_K + 1$.

The *Weibull* PDF is a one parametric family that can be applied to a wide range of turbulent strengths

$$p_I^{(W)}(\mathcal{I}) = \frac{\beta_W}{\eta_W} \left(\frac{\mathcal{I}}{\eta_W}\right)^{\beta_W-1} \exp\left[\left(-\frac{\mathcal{I}}{\eta_W}\right)^{\beta_W}\right]. \qquad (9.94)$$

For this distribution, $\langle \mathcal{I} \rangle = \eta_W \Gamma(1+1/\beta_W)$, $\langle \mathcal{I} \rangle^2 = \eta_W^2 \Gamma(1+2/\beta_W)$. Since $\eta_W = 1/\Gamma(1+1/\beta_W)$, the scintillation index becomes $\sigma_I^2 = \Gamma(1+2/\beta_W)/\Gamma^2(1+1/\beta_W) - 1$.

The two-parametric *Gamma–Gamma PDF* model became very popular for characterizing the effects of weak-to-strong atmospehric turbulence on laser beam fluctuating intensity (Al-Habash et al., 2001). It was derived from the assumption that the effects of small and large turbulent eddies are different. The Gamma–Gamma distribution has the form

$$p_I^{(GG)}(\mathcal{I}) = \frac{2\alpha_G \beta_G}{\Gamma(\alpha_G)\Gamma(\beta_G)} \mathcal{I}^{(\alpha_G+\beta_G)/2-1} K_{\alpha_G-\beta_G}(2\sqrt{\alpha_G \beta_G \mathcal{I}}),$$
(9.95)

where α_G and β_G are the variances of the fluctuating intensity associated with the large and the small turbulence scales, respectively, K_m is the second-kind Bessel function of order m. Further, parameters α_G and β_G are related to the normalized variance of the fluctuating intensity σ_W^2 in weak turbulence as

$$\alpha_G = \left[\exp\left[\frac{0.49\sigma_W^2}{(1+1.11\sigma_W^{12/5})^{7/6}}\right] - 1\right]^{-1},$$

$$\beta_G = \left[\exp\left[\frac{0.51\sigma_W^2}{(1+0.69\sigma_W^{12/5})^{5/6}}\right] - 1\right]^{-1}.$$
(9.96)

The scintillation index is expressed as $\sigma_I^2 = 1/\alpha_G + 1/\beta_G + 1/(\alpha_G \beta_G)$.

The *exponential Weibul* is a three-parametric PDF model applicable to all turbulent strengths

$$p_I^{(E)}(\mathcal{I}) = \frac{\alpha_E \beta_E}{\eta_E}\left(\frac{\mathcal{I}}{\eta_E}\right)^{\beta_E-1} \exp\left[-\left(\frac{\mathcal{I}}{\eta_E}\right)^{\beta_E}\right]$$

$$\times \left[1 - \exp\left[\left(-\frac{\mathcal{I}}{\eta_E}\right)^{\beta_E}\right]\right]^{\alpha_E-1}.$$
(9.97)

Generally, the moments of this PDF are defined by fast-convergent series

$$\langle \mathcal{I} \rangle^{(n)} = \alpha_E \eta_E^n \Gamma(1+n/\beta_E) g_n(\alpha_E, \beta_E),$$
(9.98)

where

$$g_n(\alpha_E, \beta_E) = \sum_{i=0}^{\infty} \frac{(-1)^i \Gamma(\alpha_E)}{i!(i+1)^{1+n/\beta_E}\Gamma(\alpha_E-i)}.$$
(9.99)

Normalization $\langle \mathcal{I} \rangle = 1$ leads to the scintillation index

$$\sigma_I^2 = \frac{\Gamma(1+2/\beta_E) g_2(\alpha_E, \beta_E)}{\alpha_E[\Gamma(1+1/\beta_E) g_1(\alpha_E, \beta_E)]^2} - 1.$$
(9.100)

Other distributions, such as Rician, Malaga, Rayleigh, are also used.

Chapter 10

Non-stationary Pulse Ensembles

In the preceding chapters, our discussion was concerned with the statistically stationary optical fields in which different spectral components are completely uncorrelated. On the other hand, in a deterministic pulse the spectral components are perfectly correlated. It appears possible to introduce a variety of non-stationary fields in which spectral components have partial correlation and would reduce to stationary radiation and to deterministic pulses in the two limiting cases. Partial correlation of the field at frequency pairs can be translated to that at time moment pairs, on applying the time–frequency Fourier transform, in analogy to the duality of correlations in direct spatial variables and spatial frequencies, for stationary radiation. The aim of this chapter is to introduce the major definitions and concepts relating to non-stationary pulse ensembles, discuss modeling of the sources radiating them and illustrate their behavior in dispersive media.

10.1 Theory of Quasi-stationary Pulses

The non-stationary fields must be characterized by statistical moments at a set of frequencies and not at a single frequency as is the case with the stationary fields. Hence, in time domain the characterization must also be made for a set of time instants (Lajunen et al., 2005). On using the generalized Wiener–Khintchin theorem (see Chapter 1), one can establish the Fourier transform relation between the mutual coherence function at moments t_1 and t_2 and the cross-spectral density function, at frequencies ω_1 and ω_2:

$$W(\mathbf{r}_1, \mathbf{r}_2, \omega_1, \omega_2) = \frac{1}{(2\pi)^2} \int_{-\infty}^{\infty} \int_{-\infty}^{\infty} \mathcal{W}(\mathbf{r}_1, \mathbf{r}_2, t_1, t_2)$$
$$\times \exp[-i(\omega_1 t_1 - \omega_2 t_2)] dt_1 dt_2, \quad (10.1)$$

$$\mathcal{W}(\mathbf{r}_1, \mathbf{r}_2, t_1, t_2) = \int_0^{\infty} \int_0^{\infty} W(\mathbf{r}_1, \mathbf{r}_2, \omega_1, \omega_2)$$
$$\times \exp[i(\omega_1 t_1 - \omega_2 t_2)] d\omega_1 d\omega_2. \quad (10.2)$$

The expectation values of the intensity envelope and the spectral density of a non-stationary field can be deduced from expressions

$$\mathcal{I}(\mathbf{r}, t) = \mathcal{W}(\mathbf{r}, \mathbf{r}, t, t), \quad (10.3)$$

$$S(\mathbf{r}, \omega) = W(\mathbf{r}, \mathbf{r}, \omega, \omega), \quad (10.4)$$

respectively. Further, the degrees of coherence in space–time and space–frequency domains generalize to expressions

$$\gamma(\mathbf{r}_1, \mathbf{r}_2, t_1, t_2) = \frac{\mathcal{W}(\mathbf{r}_1, \mathbf{r}_2, t_1, t_2)}{\sqrt{\mathcal{I}(\mathbf{r}_1, t_1)} \sqrt{\mathcal{I}(\mathbf{r}_2, t_2)}}, \quad (10.5)$$

$$\mu(\mathbf{r}_1, \mathbf{r}_2, \omega_1, \omega_2) = \frac{W(\mathbf{r}_1, \mathbf{r}_2, \omega_1, \omega_2)}{\sqrt{S(\mathbf{r}_1, \omega_1)} \sqrt{S(\mathbf{r}_2, \omega_2)}}. \quad (10.6)$$

The free-space propagation laws of a non-stationary field in the space–time domain become a pair of wave equations (Cairns and Wolf, 1987)

$$\nabla_j^2 \mathcal{W}(\mathbf{r}_1, \mathbf{r}_2, t_1, t_2) = \frac{1}{c^2} \frac{\partial^2}{\partial t_j^2} \mathcal{W}(\mathbf{r}_1, \mathbf{r}_2, t_1, t_2), \quad (10.7)$$

where differentiation is performed with respect to a coordinate with subindex ($j = 1, 2$). Taking the Fourier transform of Eqs. (10.7) yields the corresponding pair of the Helmholtz equations for W:

$$\nabla_j^2 W(\mathbf{r}_1, \mathbf{r}_2, \omega_1, \omega_2) + \frac{\omega_j^2}{c^2} W(\mathbf{r}_1, \mathbf{r}_2, \omega_1, \omega_2) = 0. \quad (10.8)$$

Other properties (non-negative definiteness, quasi-Hermiticity, square-integrability) and their consequences (e.g., coherent mode decomposition) can also be readily established (Lajunen et al., 2005).

10.2 Mathematical Models

10.2.1 *Laser-based examples*

The most common example of an optical field with perfectly correlated spectral components is a monochromatic Gaussian pulse:

$$\mathcal{U}(\mathbf{r},t) = U(\mathbf{r})\exp\left[-\frac{t^2}{2\sigma_t^2}\right]\exp(i\omega_0 t). \tag{10.9}$$

Here, $U(\mathbf{r})$ is the time-independent complex amplitude of the pulse while σ_t and ω_0 are its width and central frequency. The average pulse intensity

$$\begin{aligned}\mathcal{I}(\mathbf{r},t) &= \langle\mathcal{U}^*(\mathbf{r},t)\mathcal{U}(\mathbf{r},t)\rangle \\ &= |U(\mathbf{r})|^2\exp\left[-\frac{t^2}{\sigma_t^2}\right]\end{aligned} \tag{10.10}$$

shows the explicit time dependence.

Consider now a laser source radiating fully correlated monochromatic fields at frequencies ω_{01} and ω_{02} (Bertolotti et al., 1995)

$$\mathcal{U}(\mathbf{r},t) = U(\mathbf{r})[\alpha_1\exp(i\omega_{01}t) + \alpha_2\exp(i\omega_{02}t)]. \tag{10.11}$$

Here, α_1 and α_2 are the complex amplitudes of these components and $U(\mathbf{r})$ is the common spatial source aperture profile. While the correlation between the spectral components of this source, say, ν_c, is unity by absolute value

$$|\nu_c| = \frac{|\langle\alpha_1^*\alpha_2\rangle|}{\sqrt{\langle|\alpha_1|^2\rangle}\sqrt{\langle|\alpha_2|^2\rangle}}, \tag{10.12}$$

its intensity has the form

$$\begin{aligned}\mathcal{I}(\mathbf{r}) &= \langle\mathcal{U}^*(\mathbf{r},t)\mathcal{U}(\mathbf{r},t)\rangle \\ &= |U(\mathbf{r})|^2\{I_1 + I_2 + 2\sqrt{I_1}\sqrt{I_2} \\ &\quad \times \cos[(\omega_{01}-\omega_{02})t]\},\end{aligned} \tag{10.13}$$

with $I_j = |\alpha_j|^2$, $(j = 1,2)$ being the individual intensities. Equation (10.13) implies that the intensity is strictly periodic. The partial correlation of two signals can be obtained if one sets $0 < |\nu_c| < 1$ with the limiting case of complete uncorrelation if $\nu_c = 0$.

10.2.2 Gaussian Schell-model pulses

Let us assume that in the source plane the mutual coherence function factorizes into spatial and temporal parts (Christov, 1986)

$$\mathcal{W}(\mathbf{r}_1, \mathbf{r}_2, t_1, t_2) = \mathcal{W}_0 \mathcal{W}_R(\mathbf{r}_1, \mathbf{r}_2) \mathcal{W}_T(t_1, t_2), \tag{10.14}$$

and that the temporal part $\mathcal{W}_T$ is of the Gaussian Schell-model form

$$\begin{aligned}\mathcal{W}_T(t_1, t_2) &= \exp[-\mathcal{T}(t_1, t_2)] \\ &= \exp\left[-\frac{t_1^2 + t_2^2}{2\sigma_t^2}\right] \exp\left[-\frac{(t_1 - t_2)^2}{2\delta_t^2}\right] \\ &\quad \times \exp[i\omega_0(t_1 - t_2)],\end{aligned} \tag{10.15}$$

where, as before, ω_0 is the central frequency, while σ_t and δ_t are the root-mean-squared values of the pulse width and the pulse coherence width, respectively. On taking the Fourier transform of the cross-spectral density $\mathcal{W}(\mathbf{r}_1, \mathbf{r}_2, t_1, t_2)$ with respect to t_1 and t_2, we arrive, for its two-frequency portion, at the formula

$$\begin{aligned}W_S(\omega_1, \omega_2) &= W_0 \exp[-\mathcal{S}(\omega_1, \omega_2)] \\ &= W_0 \exp\left[-\frac{(\omega_1 - \omega_0)^2 + (\omega_2 - \omega_0)^2}{2\sigma_\omega^2}\right] \\ &\quad \times \exp\left[-\frac{(\omega_1 - \omega_2)^2}{2\delta_\omega^2}\right],\end{aligned} \tag{10.16}$$

with

$$\sigma_\omega^2 = \frac{1}{\sigma_t^2} + \frac{2}{\delta_t^2}, \quad \delta_\omega = \frac{\delta_t \sigma_\omega}{\sigma_t}, \quad W_0 = \frac{\mathcal{W}_0 \sigma_t}{2\pi \sigma_\omega}. \tag{10.17}$$

When $\delta_\omega \to \infty$, the conventional Gaussian pulse is deduced and as $\delta_\omega \to 0$, the frequency components become uncorrelated approaching in the limit a stationary polychromatic source. This implies that parameter δ_ω sets the connection between stationary fields and spectrally coherent (conventional) pulses (Päläkkönen et al., 2002). We also note that the spatial part $\mathcal{W}_R(\mathbf{r}_1, \mathbf{r}_2)$ is not affected by the Fourier transform and, hence, in the source plane the cross-spectral density is separable either in the space–time or the space–frequency domain.

Spatial part $\mathcal{W}_R$ may be a polychromatic plane wave, a Gaussian Schell-model source or can be structured, with methods similar to those discussed in Chapter 5). Let us write it as

$$\mathcal{W}_R(\mathbf{r}_1, \mathbf{r}_2) = \exp[-\mathcal{R}(\mathbf{r}_1, \mathbf{r}_2)], \qquad (10.18)$$

and assume that it is in the Gaussian Schell-model form

$$\mathcal{W}_R(\mathbf{r}_1, \mathbf{r}_2) = \exp\left[-\frac{x_1^2 + y_1^2 + x_2^2 + y_2^2}{\sigma^2}\right]$$

$$\times \exp\left[-\frac{(x_1 - x_2)^2 + (y_1 - y_2)^2}{2\delta^2}\right]. \qquad (10.19)$$

The version of the extended Huygens–Fresnel integral suitable for the analysis of non-stationary (paraxial) fields in the space–frequency domain following from Eq. (10.8) has the form (Lajunen et al., 2005)

$$W(\mathbf{r}_1, \mathbf{r}_2, \omega_1, \omega_2)$$
$$= \frac{\omega_1 \omega_2}{4z_1 z_2 \pi^2 c^2} \exp[i(\omega_2 z_2 - \omega_1 z_1)/c]$$
$$\times \int_{-\infty}^{\infty}\int_{-\infty}^{\infty}\int_{-\infty}^{\infty}\int_{-\infty}^{\infty} W(x_1', y_1', 0, x_2', y_2', 0, \omega_1, \omega_2)$$
$$\times \exp\left[\frac{i\omega_2(x_2 - x_2')^2}{2cz_2} - \frac{i\omega_1(x_1 - x_1')^2}{2cz_1}\right]$$
$$\times \exp\left[\frac{i\omega_2(y_2 - y_2')^2}{2cz_2} - \frac{i\omega_1(y_1 - y_1')^2}{2cz_1}\right]$$
$$\times dx_1' dx_2' dy_1' dy_2'. \qquad (10.20)$$

On applying this source model in the Huygens–Fresnel formula for non-stationary fields in Eq. (10.20), the following expression for the cross-spectral density of the propagating pulse can be obtained:

$$W(\mathbf{r}_1, \mathbf{r}_2, \omega_1, \omega_2)$$
$$= \frac{W_0 a_1 a_2}{\sqrt{A_2(a_1, a_2)}} \exp[-\mathcal{S}(\omega_1, \omega_2)]$$

$$\times \exp\left\{-\frac{1}{A_2(a_1,a_2)}\left[a_1^2 a_2^2 \mathcal{R}(x_1,y_1,x_2,y_2) + \frac{a_3^2}{\omega_0^2} A_1(a_1,a_2)\right]\right\}$$

$$\times \exp\left[i\left(A_5(a_1,a_2) + \frac{a_4 a_1 a_2 (a_1 - a_2)}{A_2(a_1,a_2)}\mathcal{R}(x_1,y_1,x_2,y_2)\right.\right.$$

$$\left.\left. -\frac{A_3(a_1,a_2)}{A_2(a_1,a_2)}A_4(a_1,a_2,a_3) + \frac{\omega_2 z_2 - \omega_1 z_1}{c}\right)\right]. \tag{10.21}$$

Here, functions A_j, $j = 1, \ldots, 5$ are defined as

$$A_1(a_1,a_2) = \frac{a_1^2(x_1^2 + y_1^2) + a_2^2(x_2^2 + y_2^2)}{\sigma^2}$$

$$+ \frac{(a_1 x_1 - a_2 x_2)^2 + (a_1 y_1 - a_2 y_2)^2}{2\delta^2},$$

$$A_2(a_1,a_2) = A_3(a_1,a_1)A_3(a_2,a_2) + \frac{(a_1-a_2)^2}{4\delta^4},$$

$$A_3(a_1,a_2) = \frac{a_3^2}{\sigma^2} + a_1 a_2, \tag{10.22}$$

$$A_4(a_1,a_2,a_3) = a_3^2 \frac{a_1(x_1^2 + y_1^2) - a_2(x_2^2 + y_2^2)}{\sigma^2},$$

$$A_5(a_1,a_2) = \arctan\left[\frac{a_4(a_1-a_2)}{A_3(a_1,a_2)}\right].$$

Further, parameters a_j, $j = 1, \ldots, 4$ are given by expressions

$$a_j = \frac{\omega_j}{2cz_j}, \quad (j=1,2), \quad a_3 = \left(\frac{1}{\sigma^2} + \frac{1}{\delta^2}\right)^{-1}, \quad a_4 = \frac{1}{\sigma^2} + \frac{1}{2\delta^2}. \tag{10.23}$$

The pulse's spectral density then reduces to a particularly simple expression

$$S(\mathbf{r},\omega) = \frac{W_0 \sigma^2}{\sigma^2(z,\omega)} \exp\left[-\frac{2(x^2+y^2)}{\sigma^2(z,\omega)}\right] \exp\left[-\frac{(\omega-\omega_0)^2}{\sigma_\omega^2}\right], \tag{10.24}$$

where $\sigma(z,\omega)$ is the effective spectral expansion coefficient:

$$\sigma(z,\omega) = \sigma^2 \sqrt{1 + \left(\frac{2cza_3}{\sigma\omega}\right)}. \tag{10.25}$$

Equation (10.24) implies that the spectral and the spatial contributions to the pulse ensemble that are separable at the source plane become mixed on propagation.

10.3 Propagation of Pulse Ensembles in Dispersive Media

We will now ignore the transverse spatial properties of the field and focus entirely on evolution of its temporal correlations on propagation in dispersive media along z direction. Let the field realization at time t and position z have the form

$$\mathcal{U}(t,z) = \mathcal{V}(t,z) \exp[i(\omega_0 t - k_0 z)], \qquad (10.26)$$

where $\mathcal{V}(t,z)$ is the slowly varying, complex-valued envelope, $k_0 = \omega_0/c$ is the wave number corresponding to central frequency ω_0. In a medium with dispersive properties up to the second order, $\mathcal{V}(t,z)$ obeys equation (Born and Wolf, 1999)

$$2\mathcal{V}_z(t,z) + 2\beta' \mathcal{V}_t(t,z) - i\beta'' \mathcal{V}_{tt}(t,z) = 0, \qquad (10.27)$$

where β' and β'' are the *inverse group velocity* and the *group velocity dispersion coefficient*, respectively. Making changes in variables

$$\tau = t - \beta' z, \quad \zeta = \omega_0 \beta'' z, \qquad (10.28)$$

where τ is the time lag with respect to a frame moving with group velocity $1/\beta'$ and ζ is a scaled distance, in Eq. (10.29) results in the equation

$$\mathcal{V}_{\tau\tau}(\tau,\zeta) + 2i\omega_0 \mathcal{V}_\zeta(\tau,\zeta) = 0. \qquad (10.29)$$

This equation was solved in Zhang and Fan (1992) for dispersive media specified by the temporal $ABCD$ matrices that can be regarded as one-dimensional, temporal analog of the spatial $ABCD$ matrices introduced in Chapter 7, but describes the effects of dispersion rather than diffraction. For example, for the second-order dispersive medium and an ideal chirper:

$$\overleftrightarrow{\Upsilon}_d = \begin{bmatrix} 1 & \omega_0 \beta'' z \\ 0 & 1 \end{bmatrix}, \quad \overleftrightarrow{\Upsilon}_c = \begin{bmatrix} 1 & 0 \\ s_c/\omega_0 & 1 \end{bmatrix}, \quad (10.30)$$

where s_c is the chirp coefficient. The resulting integral

$$\mathcal{V}(\tau, \zeta) = \sqrt{\frac{i\omega_0}{B}} \int \mathcal{V}(\tau', \zeta') \\ \times \exp\left[-\frac{i\omega_0}{2B}(A\tau'^2 + D\tau^2 - 2\tau'\tau)\right] d\tau' \quad (10.31)$$

resembles the Collins integral for the spatial domain (see Chapter 7). Its substitution into the mutual coherence function of the partially coherent pulse ensemble

$$\mathcal{W}(\zeta_1, \zeta_2, \tau_1, \tau_2) = \langle \mathcal{V}^*(\zeta_1, \tau_1,)\mathcal{V}(\zeta_2, \tau_2) \rangle \quad (10.32)$$

yields the propagation law (Lin et al., 2003)

$$\mathcal{W}(\zeta_1, \zeta_2, \tau_1, \tau_2) = \frac{\omega_0}{2\pi B} \iint \exp\left[\frac{i\omega_0}{2B}(A\tau_1'^2 + D\tau_1^2 - 2\tau_1'\tau_1)\right] \\ \times \exp\left[-\frac{i\omega_0}{2B}(A\tau_2'^2 + D\tau_2^2 - 2\tau_2'\tau_2)\right] \\ \times \mathcal{W}(\zeta_1', \zeta_2', \tau_1', \tau_2') d\tau_1' d\tau_2'. \quad (10.33)$$

Assume that the pulse ensemble is radiated by the temporal Gaussian Schell-model source with unit maximum intensity (see Eq. (10.15)) and its mutual coherence function is expressed in variables τ_1', τ_2' as

$$W_\tau(0, 0, \tau_1', \tau_2') = \exp\left[-\frac{\tau_1'^2 + \tau_2'^2}{2\sigma_\tau^2}\right] \exp\left[-\frac{(\tau_1' - \tau_2')^2}{2\delta_\tau^2}\right]. \quad (10.34)$$

Then using this equation in the integral (10.33) yields

$$W_\tau(\tau_1, \tau_2) = \frac{1}{\Delta_\tau} \exp\left[-\frac{(\tau_1 + \tau_2)^2}{4\sigma_\tau^2 \Delta_\tau^2}\right] \exp\left[-\frac{(\tau_1 - \tau_2)^2}{2\alpha_\tau^2 \Delta_\tau^2}\right] \\ \times \exp[i\phi_\tau(\tau_1^2 - \tau_2^2)/2], \quad (10.35)$$

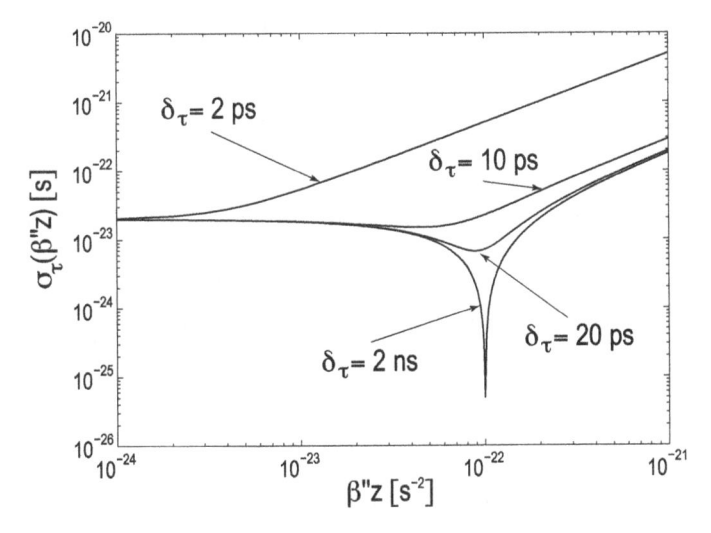

Fig. 10.1 A Gaussian Schell-model pulse propagating in a dispersive medium with a chirper.

where $\Delta_\tau = \Delta_\tau(z)$ is the temporal pulse broadening rate and $\phi_\tau = \phi_\tau(z)$ is the frequency sweep rate, given by expressions

$$\Delta_\tau = \sqrt{A^2 + \frac{2B^2}{\omega_0^2 \sigma_\tau^2}\alpha_\tau^2}, \quad \phi_\tau = \frac{\omega_0}{\Delta_\tau^2}\left[AC + \frac{2BD}{\omega_0^2 \sigma_\tau^2}\alpha_\tau^2\right],$$
$$\frac{2}{\alpha_\tau^2} = \frac{1}{\sigma_\tau^2} + \frac{2}{\delta_\tau^2}.$$
(10.36)

Setting $\mathcal{I}(\tau) = \mathcal{W}(\tau,\tau)$ implies that the evolving pulse width is $\sigma_\tau(z) = 2\sigma_\tau \Delta_\tau(z)$. Figure 10.1 shows an example of the evolution of the pulse width in dispersive medium with a chirp, specified by matrix $\overleftrightarrow{\Upsilon} = \overleftrightarrow{\Upsilon}_d \cdot \overleftrightarrow{\Upsilon}_c$, with $\sigma_\tau = 20ps$, $s_c = -1 \times 10^{22} s^{-2}$, for different source coherence values δ_τ.

10.4 Structured Pulse Coherence

10.4.1 *Bochner's theorem method*

Since the mutual coherence function $\mathcal{W}_\tau$ must be quasi-Hermitian and non-negative definite with respect to its temporal arguments, according to the Bochner theorem (see Chapter 5) it is sufficient for it to be genuine if representation

$$\mathcal{W}_\tau^{(0)}(\tau_1, \tau_2) = \int H_0^*(\tau_1, \nu) p(\nu) H_0(\tau_2, \nu) d\nu \qquad (10.37)$$

holds. Here, just like in the spatial case, $H_0(\tau, \nu)$ is an arbitrary kernel and $p(\nu)$ is a real, non-negative function. As in Chapter 5, all the choices for H and p that are intrinsically one-dimensional can be employed. In particular, H_0 can be either of the uniformly-correlated (Schell-model) class or of the non-uniformly correlated class. For the pulse ensembles of the Schell-class, we choose H_0 as a Fourier-like kernel

$$H_0(\tau, \nu) = a(\tau) \exp[-i\tau\nu], \qquad (10.38)$$

and obtain a mutual coherence function of the form

$$\mathcal{W}_\tau^{(0)}(\tau_1, \tau_2) = a^*(\tau_1) a(\tau_2) \gamma(\tau_d), \qquad (10.39)$$

where $\tau_d = \tau_1 - \tau_2$, $a(\tau)$ is the complex amplitude profile, $|a(\tau)| = \sqrt{\mathcal{I}(\tau)}$, $\mathcal{I}(\tau)$ is the average intensity of the pulse train, and $\gamma(\tau_d)$ is the temporal (complex) degree of coherence, such that $FT[\gamma(\tau_d)] = p(v)$. One can obtain a variety of models for $\mathcal{W}_\tau^{(0)}$. For example, pulse ensembles with the cosine-Gaussian distribution (Ding et al., 2014) can be obtained on setting

$$p(\nu) = \frac{\delta_\tau}{\sqrt{2\pi}} \cosh(n\sqrt{2\pi}\delta_\tau \nu) \exp(-\delta_\tau^2 \nu^2 / 2 + n^2 \pi), \qquad (10.40)$$

and

$$H_0(\tau, \nu) = \exp\left[-\frac{\tau^2}{2\sigma_\tau^2}\right] \exp[-i\tau\nu]. \qquad (10.41)$$

Then the mutual coherence function takes the form

$$\mathcal{W}_\tau^0(\tau_1, \tau_2) = \exp\left[-\frac{\tau_1^2 + \tau_2^2}{2\sigma_\tau^2}\right] \cos\left[\frac{n\sqrt{2\pi}(\tau_1 - \tau_2)}{\delta_\tau}\right]$$

$$\exp\left[-\frac{(\tau_1 - \tau_2)^2}{2\delta_\tau^2}\right]. \qquad (10.42)$$

Figure 10.2 shows the average intensity $\mathcal{I}(\tau, z)$ of the pulse obtained on substituting from Eq. (10.42) into Eq. (10.33), then calculating the result at $\tau_1 = \tau_2 = \tau$ and assuming that the disperisve medium is given by the ABCD matrix $\overleftrightarrow{\Upsilon}_d$ in Eq. (10.30).

Non-stationary Pulse Ensembles 297

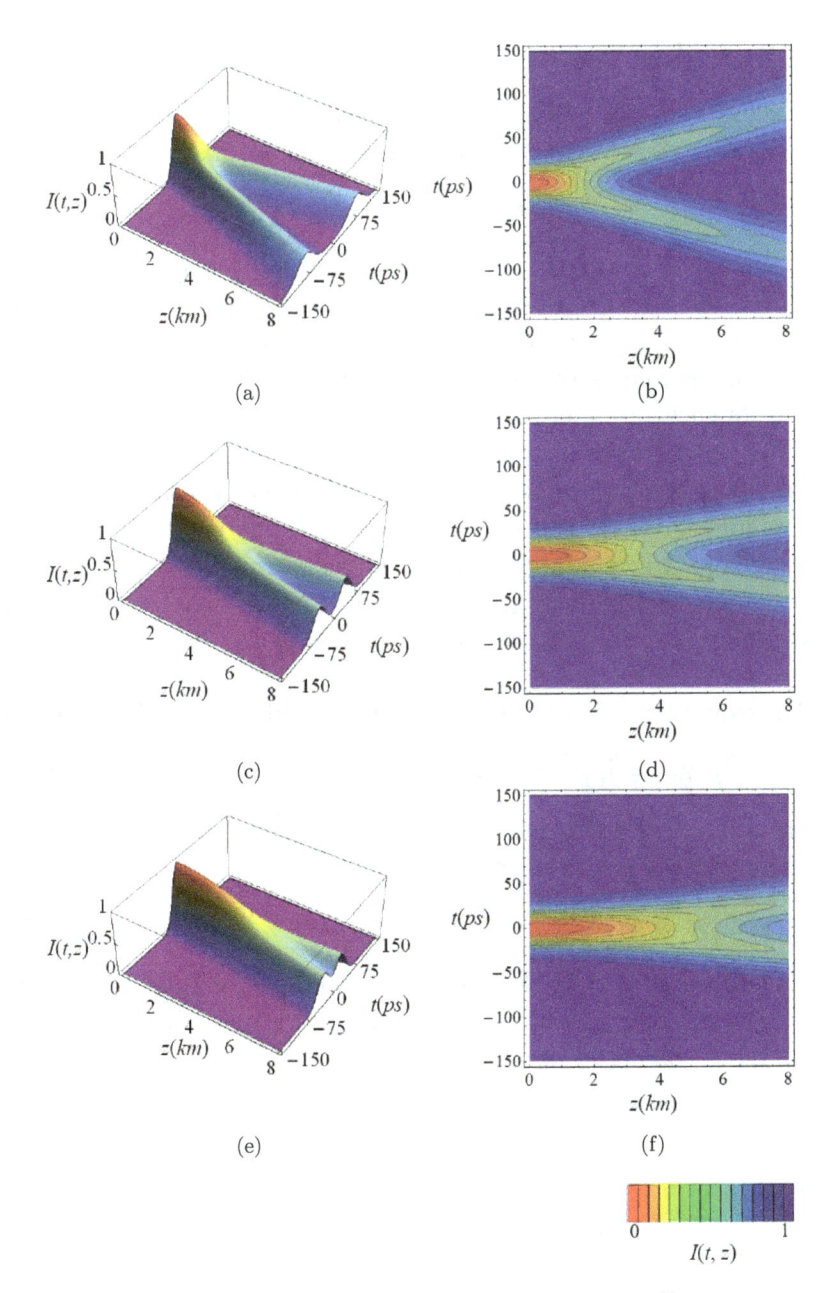

Fig. 10.2 Cos-Gaussian Schell-model pulse ($n = 2$, $\sigma_\tau = 15\sqrt{2}\,\text{ps}$) propagating in a dispersive medium with $\beta'' = 20\,\text{ps}^2\,\text{km}^{-1}$. (a), (d) $\delta_\tau = 10\,\text{ps}$, (b), (e) $\delta_\tau = 20\,\text{ps}$, and (c), (f) $\delta_\tau = 10\,\text{ps}$. Reprinted with permission from Ding et al. (2014). © The Optical Society.

10.4.2 Sliding function method

In analogy with the sliding function method of Chapter 5, $\gamma(\tau_d)$ can be expressed as self-convolution (Zhang et al., 2020a):

$$\gamma(\tau_d) = g(\tau_d) \circledast g(\tau_d), \qquad (10.43)$$

for some Hermitian $g(\tau_d)$. Moreover, $\gamma(0) = 1$ implies $g(0) = 1$. Further, we require that $\mathcal{I}(\tau)$ and $|\gamma(\tau_d)|$ must be such that $\iint |\mathcal{W}(\tau_1, \tau_2)| d\tau_1 d\tau_2 < \infty$. Since the choice of the phase of $g(\tau_d)$, which carries some structure over to $\gamma(\tau_d)$ is quite arbitrary, a large variety of such functions can be introduced.

The concept of a coherence curve considered in Chapter 5 can be carried over to the temporal domain. The *temporal coherence curve* is a planar curve in the complex plane parametrized by the time difference τ_d, whose x and y components coincide with the real and the imaginary parts of $\gamma(\tau_d)$, respectively. It must be unity at $\tau_d = 0$, be contained within the unit circle of the complex plane, and converge to zero as $\tau_d \to \infty$.

For example, a sliding function having a cubic phase dependence:

$$g(\tau_d) = \delta_\tau^{-1/2} \pi^{-1/4} \exp\left(-\tau_d^2/2\delta_\tau^2 + ib\tau_d^3\right), \qquad (10.44)$$

where b is a real constant, can be used in Eqs. (10.43) and (10.44) to yield γ in the closed form

$$\gamma(\tau_d) = (1 - i3\delta_\tau^2 b\tau_d)^{-1/2} \exp\left(-\tau_d^2/4\delta_\tau^2 + ib\tau_d^3/4\right). \qquad (10.45)$$

The average intensity of the propagating pulse ensemble can be then shown to gradually acquire Airy-like patterns for sufficiently low δ_τ and large z (Zhang et al., 2020b).

Appendix A

Natural Water Parameters Varying with $\langle T \rangle$ and $\langle S \rangle$

Specific heat $c_p (\text{J} \cdot \text{kg}^{-1} \cdot \text{K}^{-1})$:

$$c_p = 1000 \times (a_{11} + a_{12}\langle T \rangle + a_{13}\langle T \rangle^2 + a_{14}\langle T \rangle^2),$$
$$a_{11} = 5.328 - 9.76 \times 10^{-2}\langle S \rangle + 4.04 \times 10^{-4}\langle S \rangle^2,$$
$$a_{12} = -6.913 \times 10^{-3} + 7.351 \times 10^{-4}\langle S \rangle - 3.15 \times 10^{-6}\langle S \rangle^2, \quad \text{(A.1)}$$
$$a_{13} = 9.6 \times 10^{-6} - 1.927 \times 10^{-6}\langle S \rangle + 8.23 \times 10^{-9}\langle S \rangle^2,$$
$$a_{14} = 2.5 \times 10^{-9} + 1.666 \times 10^{-9}\langle S \rangle - 7.125 \times 10^{-12}\langle S \rangle^2.$$

Thermal conductivity σ_T $(\text{W} \cdot \text{m}^{-1}\text{K}^{-1})$:

$$\log(\sigma_T) = 0.434 \times \left(2.3 - \frac{343.5 + 0.037\langle S_h \rangle}{\langle T_h \rangle + 273.15}\right)$$
$$\times \left[1 - \frac{\langle T_h \rangle + 273.15}{647.3 + 0.03\langle S_h \rangle}\right]^{1/3}$$
$$+ \log(240 + 0.0002\langle S_h \rangle) - 3, \quad \langle T_h \rangle = 1.00024\langle T \rangle,$$
$$\langle S_h \rangle = \langle S \rangle / 1.00472. \quad \text{(A.2)}$$

Dynamic viscosity μ_v $(\text{N} \cdot \text{s} \cdot \text{m}^2)$ at atmospheric pressure:

$$\mu_v = \mu_0(a_{21}\langle s\rangle + a_{22}\langle s\rangle^2), \quad \langle s\rangle = \langle S\rangle \times 10^{-3},$$
$$a_{21} = 1.5409136040 + 1.9981117208 \times 10^{-2}\langle T\rangle$$
$$\quad - 9.5203865864 \times 10^{-5}\langle T\rangle^2,$$
$$a_{22} = 7.9739318223 - 7.5614568881 \times 10^{-2}\langle T\rangle$$
$$\quad + 4.7237011074 \times 10^{-4}\langle T\rangle^2,$$
$$\mu_0 = \left[0.15700386464 \times (\langle T\rangle + 64.992620050)^2 - 91.296496657\right]^{-1}$$
$$\quad + 4.2844324477 \times 10^{-5}. \tag{A.3}$$

Density of water ρ_m (kg $\cdot$ m^3) at atmospheric pressure is:

$$\rho_m = \rho_T + \rho_S,$$
$$\rho_T = 9.9992293295 \times 10^2 + 2.0341179217 \times 10^{-2}\langle T\rangle$$
$$\quad - 6.1624591598 \times 10^{-3}\langle T\rangle^2 + 2.2614664708 \times 10^{-5}\langle T\rangle^3$$
$$\quad - 4.6570659168 \times 10^{-8}\langle T\rangle^4, \tag{A.4}$$
$$\rho_{TS} = \langle s\rangle[8.0200240891 \times 10^2 - 2.0005183488\langle T\rangle$$
$$\quad + 1.6771024982 \times 10^{-2}\langle T\rangle^2 - 3.0600536746 \times 10^{-5}\langle T\rangle^3$$
$$\quad - 1.6132224742 \times 10^{-5}\langle T\rangle^2\langle s\rangle],$$

Thermal expansion a_T (L/degree) and saline contraction a_S [L/g] coefficients:

$$a_T = 0.025VT/V, \quad a_S = -VS/(16(35.16504/7))/(VX),$$
$$X = [(35.16504 \times 40/35)^{-1}\langle S\rangle + 0.5971840214030754]^{1/2}, \tag{A.5}$$
$$Y = 0.025\langle T\rangle, \quad Z = \text{Pressure} \times 10^{-4},$$

$$VT = a000 + X(a100 + X(a200 + X(a300 + X(a400 + a500X))))$$
$$\quad + Y(a010 + X(a110 + X(a210 + X(a310 + a410X))))$$
$$\quad + Y(a020 + X(a120 + X(a220 + a320X)) + Y(a030$$
$$\quad + X(a130 + a230X) + Y(a040 + a140X + a050Y))))$$

$$+ Z(a001 + X(a101 + X(a201 + X(a301 + a401X)))$$
$$+ Y(a011 + X(a111 + X(a211 + a311X)) + Y(a021$$
$$+ X(a121 + a221X) + Y(a031 + a131X + a041Y)))$$
$$+ Z(a002 + X(a102 + X(a202 + a302X)) + Y(a012$$
$$+ X(a112 + a212X) + Y(a022 + a122X + a032Y))$$
$$+ Z(a003 + a103X + a013Y + a004Z))), \tag{A.6}$$

$$VS = b000 + X(b100 + X(b200 + X(b300 + X(b400 + b500X))))$$
$$+ Y(b010 + X(b110 + X(b210 + X(b310 + b410X))) + Y(b020$$
$$+ X(b120 + X(b220 + b320X)) + Y(b030 + X(b130 + b230X)$$
$$+ Y(b040 + b140X + b050Y)))) + Z(b001 + X(b101 + X(b201$$
$$+ X(b301 + b401X))) + Y(b011 + X(b111 + X(b211$$
$$+ b311X)) + Y(b021 + X(b121 + b221X) + Y(b031 + b131X$$
$$+ b041Y))) + Z(b002 + X(b102 + X(b202 + b302X))$$
$$+ Y(b012 + X(b112 + b212X) + Y(b022 + b122X + b032Y))$$
$$+ Z(b003 + b103X + b013Y + b004Z))), \tag{A.7}$$

$$V = v000 + X(v100 + X(v200 + X(v300 + X(v400 + X(v500$$
$$+ v600X))))) + Y(v010 + X(v110 + X(v210 + X(v310$$
$$+ X(v410 + v510X)))) + Y(v020 + X(v120 + X(v220$$
$$+ X(v320 + v420X))) + Y(v030 + X(v130 + X(v230$$
$$+ v330X)) + Y(v040 + X(v140 + v240 * X) + Y(v050 + v150X$$
$$+ v060Y))))) + Z(v001 + X(v101 + X(v201 + X(v301$$
$$+ X(v401 + v501X)))) + Y(v011 + X(v111 + X(v211$$
$$+ X(v311 + v411X))) + Y(v021 + X(v121 + X(v221$$
$$+ v321X)) + Y(v031 + X(v131 + v231X) + Y(v041 + v141X$$

$$+ v051Y)))) + Z(v002 + X(v102 + X(v202 + X(v302$$
$$+ v402X))) + Y(v012 + X(v112 + X(v212 + v312X))$$
$$+ Y(v022 + X(v122 + v222X) + Y(v032 + v132X + v042Y)))$$
$$+ Z(v003 + X(v103 + v203X) + Y(v013 + v113X + v023Y)$$
$$+ Z(v004 + v104X + v014Y + Z(v005 + v006Z)))))), \quad (A.8)$$

Here, the sets of coefficients $vXXX$, $aXXX$ and $bXXX$ are as follows:

$v000 = 1.0769995862e - 3,\quad v100 = -3.1038981976e - 4,\quad v220 = 3.5907822760e - 5,$
$v001 = -6.0799143809e - 5,\quad v101 = 2.4262468747e - 5,\quad v221 = 2.9283346295e - 6,$
$v002 = 9.9856169219e - 6,\quad v102 = -5.8484432984e - 7,\quad v222 = -6.5731104067e - 7,$
$v003 = -1.1309361437e - 6,\quad v103 = 3.6310188515e - 7,\quad v230 = -1.4353633048e - 5,$
$v004 = 1.0531153080e - 7,\quad v104 = -1.1147125423e - 7,\quad v231 = 3.1655306078e - 7,$
$v005 = -1.2647261286e - 8,\quad v110 = 3.5009599764e - 5,\quad v240 = 4.3703680598e - 6,$
$v006 = 1.9613503930e - 9,\quad v111 = -9.5677088156e - 6,\quad v300 = -8.5047933937e - 4,$
$v010 = -1.5649734675e - 5,\quad v112 = -5.5699154557e - 6,\quad v301 = 3.7470777305e - 5,$
$v011 = 1.8505765429e - 5,\quad v113 = -2.7295696237e - 7,\quad v302 = 4.9263106998e - 6,$
$v012 = -1.1736386731e - 6,\quad v120 = -3.7435842344e - 5,\quad v310 = 3.4532461828e - 5,$
$v013 = -3.6527006553e - 7,\quad v121 = -2.3678308361e - 7,\quad v311 = -9.8447117844e - 6,$
$v014 = 3.1454099902e - 7,\quad v122 = 3.9137387080e - 7,\quad v312 = -1.3544185627e - 6,$
$v020 = 2.7762106484e - 5,\quad v130 = 2.4141479483e - 5,\quad v320 = -1.8698584187e - 5,$
$v021 = -1.1716606853e - 5,\quad v131 = -3.4558773655e - 6,\quad v321 = -4.8826139200e - 7,$
$v022 = 2.1305028740e - 6,\quad v132 = 7.7618888092e - 9,\quad v330 = 2.2863324556e - 6,$
$v023 = 2.8695905159e - 7,\quad v140 = -8.7595873154e - 6,\quad v400 = 5.8086069943e - 4,$
$v030 = -1.6521159259e - 5,\quad v141 = 1.2956717783e - 6,\quad v401 = -1.7322218612e - 5,$
$v031 = 7.9279656173e - 6,\quad v150 = -3.3052758900e - 7,\quad v402 = -1.7811974727e - 6,$
$v032 = -4.6132540037e - 7,\quad v200 = 6.6928067038e - 4,\quad v410 = -1.1959409788e - 5,$
$v040 = 6.9111322702e - 6,\quad v201 = -3.4792460974e - 5,\quad v411 = 2.5909225260e - 6,$
$v041 = -3.4102187482e - 6,\quad v202 = -4.8122251597e - 6,\quad v420 = 3.8595339244e - 6,$
$v042 = -6.3352916514e - 8,\quad v203 = 1.6746303780e - 8,\quad v500 = -2.1092370507e - 4,$
$v050 = -8.0539615540e - 7,\quad v210 = -4.3592678561e - 5,\quad v501 = 3.0927427253e - 6,$
$v051 = 5.0736766814e - 7,\quad v211 = 1.1100834765e - 5,\quad v510 = 1.3864594581e - 6,$
$v060 = 2.0543094268e - 7,\quad v212 = 5.4620748834e - 6,\quad v600 = 3.1932457305e - 5,$

$$(A.9)$$

$a000 = -1.5649734675e - 5,\quad a050 = 1.23258565608e - 6,\quad a210 = 7.1815645520e - 5,$
$a001 = 1.8505765429e - 5,\quad a100 = 3.5009599764e - 5,\quad a211 = 5.8566692590e - 6,$
$a002 = -1.1736386731e - 6,\quad a101 = -9.5677088156e - 6,\quad a212 = -1.31462208134e - 6,$
$a003 = -3.6527006553e - 7,\quad a102 = -5.5699154557e - 6,\quad a220 = -4.3060899144e - 5,$
$a004 = 3.1454099902e - 7,\quad a103 = -2.7295696237e - 7, a221 = 9.4965918234e - 7,$
$a010 = 5.5524212968e - 5,\quad a110 = -7.4871684688e - 5,\quad a230 = 1.74814722392e - 5,$
$a011 = -2.3433213706e - 5,\quad a111 = -4.7356616722e - 7,\quad a300 = 3.4532461828e - 5,$
$a012 = 4.2610057480e - 6,\quad a112 = 7.8274774160e - 7,\quad a301 = -9.8447117844e - 6,$
$a013 = 5.7391810318e - 7,\quad a120 = 7.2424438449e - 5,\quad a302 = -1.3544185627e - 6,$
$a020 = -4.9563477777e - 5,\quad a121 = -1.03676320965e - 5, a310 = -3.7397168374e - 5,$
$a021 = 2.37838968519e - 5,\quad a122 = 2.32856664276e - 8,\quad a311 = -9.7652278400e - 7,$
$a022 = -1.38397620111e - 6, a130 = -3.50383492616e - 5, a320 = 6.8589973668e - 6,$
$a030 = 2.76445290808e - 5,\quad a131 = 5.1826871132e - 6,\quad a400 = -1.1959409788e - 5,$
$a031 = -1.36408749928e - 5, a140 = -1.6526379450e - 6,\quad a401 = 2.5909225260e - 6,$
$a032 = -2.53411666056e - 7, a200 = -4.3592678561e - 5,\quad a410 = 7.7190678488e - 6,$
$a040 = -4.0269807770e - 6,\quad a201 = 1.1100834765e - 5,\quad a500 = 1.3864594581e - 6,$
$a041 = 2.5368383407e - 6,\quad a202 = 5.4620748834e - 6,$

(A.10)

and

$b000 = -3.1038981976e - 4,\quad b050 = -3.3052758900e - 7,\quad b210 = 1.03597385484e - 4,$
$b001 = 2.4262468747e - 5, b100 = 1.33856134076e - 3,\quad b211 = -2.95341353532e - 5,$
$b002 = -5.8484432984e - 7,\quad b101 = -6.9584921948e - 5,\quad b212 = -4.0632556881e - 6,$
$b003 = 3.6310188515e - 7,\quad b102 = -9.62445031940e - 6, b220 = -5.6095752561e - 5,$
$b004 = -1.1147125423e - 7,\quad b103 = 3.3492607560e - 8,\quad b221 = -1.4647841760e - 6,$
$b010 = 3.5009599764e - 5,\quad b110 = -8.7185357122e - 5,\quad b230 = 6.8589973668e - 6,$
$b011 = -9.5677088156e - 6,\quad b111 = 2.2201669530e - 5,\quad b300 = 2.32344279772e - 3,$
$b012 = -5.5699154557e - 6, b112 = 1.09241497668e - 5,\quad b301 = -6.9288874448e - 5,$
$b013 = -2.7295696237e - 7,\quad b120 = 7.1815645520e - 5,\quad b302 = -7.1247898908e - 6,$
$b020 = -3.7435842344e - 5,\quad b121 = 5.8566692590e - 6,\quad b310 = -4.7837639152e - 5,$
$b021 = -2.3678308361e - 7,\quad b122 = -1.31462208134e - 6, b311 = 1.0363690104e - 5,$
$b022 = 3.9137387080e - 7,\quad b130 = -2.8707266096e - 5,\quad b320 = 1.54381356976e - 5,$
$b030 = 2.4141479483e - 5,\quad b131 = 6.3310612156e - 7,\quad b400 = -1.05461852535e - 3,$
$b031 = -3.4558773655e - 6,\quad b140 = 8.7407361196e - 6,\quad b401 = 1.54637136265e - 5,$
$b032 = 7.7618888092e - 9,\quad b200 = -2.55143801811e - 3, b410 = 6.9322972905e - 6,$
$b040 = -8.7595873154e - 6,\quad b201 = 1.12412331915e - 4,\quad b500 = 1.9159474383e - 4,$
$b041 = 1.2956717783e - 6,\quad b202 = 1.47789320994e - 5.$

(A.11)

Bibliography

Alexakis, A. and Biferale, L. (2018). Cascades and transitions in turbulent flows, *Phys. Rep.*, Vol. 767–769, pp. 1–101.

Al-Habash, M. A., Andrews, L. C. and Phillips, R. L. (2001). Mathematical model for the irradiance PDF of a laser beam propagating through turbulent media, *Opt. Eng.*, Vol. 40, pp. 1554–1562.

Alonso, M. A., Korotkova, O. and Wolf, E. (2006). Propagation of the electric correlation matrix and the Cittert–Zernike theorem for random electromagnetic fields, *J. Mod. Opt.*, Vol. 53, pp. 969–978.

Andrews, L. C. (1992). An analytical model for the refractive index power spectrum and its application to optical scintillations in the atmosphere, *J. Mod. Opt.*, Vol. 39, pp. 1849–1853.

Andrews, L. C. and Phillips, R. L. (2005). *Laser Beam Propagation in Random Media*, 2nd edn. (SPIE Press, Bellington, WA).

Andrews, L. C., Phillips, R. L. and Crabbs, R. (2014). Propagation of a Gaussian-beam wave in general anisotropic turbulence, *Proc. SPIE*, Vol. 9224, 922402.

Arteaga, O., Ossikovski, R., Kuntman, E., Kuntman, M. A., Canillas, A. and Garcia-Caurel, E. (2017). Mueller matrix polarimetry on a Young's double-slit experiment analog, *Opt. Lett.* 42, pp. 3900–3903.

Ata, Y. and Korotkova, O. (2021). Adaptive optics correction in natural turbulent waters, *J. Opt. Soc. A*, Vol. 38, pp. 587–594.

Barabanenkov, Y. N., Kravtsov, Y. A., Ozrin, V. D. and Saichev, A. I. (1991). Enhanced backscattering in optics, in *Progress in Optics XXIX*, Wolf E. (ed.) (Elsevier, Amsterdam, Holland).

Belenkii, M. S., Barchers, J. D., Karis, S. J., Osmon, C. L., Brown, J. M. and Fugate, R. Q. (1999). Preliminary experimental evidence of anisotropy of turbulence and the effect of non-Kolmogorov turbulence on wavefront tilt statistics, *Proc. SPIE*, Vol. 3762, pp. 396–406.

Belenkii, M. S. and Mironov, V. L. (1972). Diffraction of optical radiation on a mirror disc in a turbulent atmosphere, *Quant. Electron.*, Vol. 5, pp. 38–45.

Bennink, R. S., Bentley, S. J. and Boyd, R. W. (2002). "Two-photon" coincidence imaging with a classical source, *Phys. Rev. Lett.*, Vol. 89, 113601.

Bernoulli, J. (1713). *Ars conjectandi, opus posthumum. Accedit Tractatus de seriebus infinitis, et epistola gallicé scripta de ludo pilae reticularis* (Thurneysen Brothers, Basel).

Bertolotti, M., Ferrari, A. and Sereda, L. (1995). Coherence properties of nonstationary polychromatic light sources, *J. Opt. Soc. Am. B*, Vol. 12, pp. 341–347.

Blomstedt, K., Friberg, A. T. and Setälä, T. (2017). Classical coherence of blackbody radiation, *Prog. Opt.*, Vol. 62, pp. 293–346.

Born, M. and Wolf, E. (1999). *Principles of Optics*, 7th edn. (Cambridge University Press, Amsterdam, Holland).

Brown, R. H., Davis, J. and Allen, L. R. (1967). The stellar interferometer at Narrabri Observatory, *Month. Notices Roy. Astron. Soc.*, Vol. 137, 375–392.

Brown, R. H. and Twiss, R. Q. (1954). A new type of interferometer for use in radio astronomy, *Phil. Mag.*, Vol. 45, pp. 663–682.

Brown, R. H. and Twiss, R. Q. (1956). Correlation between photons in two coherent beams of light, *Nature*, Vol. 177, pp. 27–29.

Brosseau, C. (1998). *Fundamentals of Polarized Light* (Wiley-Interscience).

Cairns, B. and Wolf, E. (1987). The instantaneous cross-spectral density of non-stationary wavefields, *Opt. Commun.*, Vol. 62, pp. 215–218.

Cai, Y. Lin, Q. and Korotkova, O. (2009). Ghost imaging with twisted Gaussian Schell-model beam, *Opt. Exp.*, Vol. 17, pp. 2453–2464.

Cai, Y. and Wang, F. (2007). Lensless imaging with partially coherent light, *Opt. Lett.*, Vol. 32, 205–207.

Cairns, B. and Foley, J. T. (1993). Contribution of the second-order Born approximation to the scattered intensity, *Opt. Commun.*, Vol. 101, pp. 144–150.

Cardano G. (1961). *The Book of Games of Chances* (Holt, Rinehart & Winston, New York) Opera Omia, Vol. 1; Lyon, (1663).

Carter, W. H. and Wolf, E. (1977). Coherence and radiometry with quasi-homogeneous sources, *J. Opt. Sos. Am.*, Vol. 67, pp. 785–796.

Chandrasekhar, S. (1960). *Radiative Transfer* (Dover Publications Inc., New York).

Chen, X. and Korotkova, O. (2018). Famous planar curves for source coherence structuring, *Opt. Lett.*, Vol. 43, pp. 2676–2679.

Chen, X. and Korotkova, O. (2019). Phase structuring of 2D complex coherence states, *Opt. Lett.*, Vol. 44, pp. 2470–2473.

Chen, X., Li, J. and Korotkova, O. (2020). Light scintillation in soft biological tissues, *Waves Rand. Compl. Med.*, Vol. 30, pp. 481–489.

Cheng, J. and Han, S. (2004). Incoherent coincidence imaging and its applications in X-ray diffraction, *Phys. Rev. Lett.*, Vol. 92, 093903.

Christov, I. P. (1986). Propagation of partially coherent light pulses, *Opt. Acta*, Vol. 33, pp. 63–72.

Collett, E. and Wolf, E. (1978). Is complete spatial coherence necessary for the generation of highly directional light beams? *Opt. Lett.*, Vol. 2, pp. 27–29.

Collins, S. A. Jr. (1970). Lens-system diffraction integral written in terms of matrix optics, *J. Opt. Soc. Am.*, Vol. 60, pp. 1168–1177.

Considine, P. S. (1966). Effects of coherence on imaging systems. *J. Opt. Soc. Am.*, Vol. 56, pp. 1001–1009.

Consortini, A., Ronchi, L. and Stefanutti, L. (1970). Investigation of atmospheric turbulence by narrow laser beams, *Appl. Opt.*, Vol. 9, pp. 2543–2547.

Corrsin, S. (1951). On the spectrum of isotropic temperature fluctuations in an isotropic turbulence, *J. Appl. Phys.*, Vol. 22, pp. 469–473.

Dačić, Z. and Wolf, E. (1988). Changes in the spectrum of a partially coherent light beam propagating in free space, *J. Opt. Soc. Am. A*, Vol. 5, pp. 1118–1126.

Davis, B. J. (2007). Observable coherence theory for statistically periodic fields, *Phys. Rev. A*, Vol. 76, 043843.

Debnath, L. and Bhatta, D. (2015). *Integral Transforms and Their Applications* (CRC Press, Boca Raton, London, New York).

Dennis, D. J. C. (2015). Coherent structures in wall bounded turbulence, *An. Acad. Bras. Cienc.*, Vol. 87, pp. 1161–1193.

de Moivre, A. (1718). The Doctrine of Chances: Or, a Method for Calculating the Probabilities of Events in Play (W. Pearson, London).

Ding, C., Korotkova, O., Zhang, Y. and Pan, L. (2014). Cosine Gaussian correlated Schell-model pulsed beams, *Opt. Exp.*, Vol. 22, pp. 931–942.

Divitt, S. and Novotny, L. (2015). Spatial coherence of sunlight and its implications for light management in photovoltaics, *Optica*, Vol. 2, pp. 95–103.

Divitt, S., Lapin, Z. J. and Novotny, L. (2014). Measuring coherence functions using non-parallel double slits, *Opt. Exp.*, Vol. 22, pp. 8277-8290.

Duffieux, P. M. (1946). *L'Integral de Fourier et ses Applications a l'Optique* (Rennes, Imprimeries Oberthur, Rennes, France).

Einstein, A. (1905). On a Heuristic Point of View about the Creation and Conversion of Light, *Annalen der Physik*, Vol. 17, 132–148.

Ellis, J., Dogariu, A., Ponomarenko, S. A. and Wolf, E. (2004a). Correlation matrix of a completely polarized, statistically stationary electromagnetic field, *Opt. Lett.*, Vol. 29, pp. 1536–1538.

Ellis, J., Dogariu, A., Ponomarenko, S. A. and Wolf, E. (2004b). Degree of polarization of statistically stationary electromagnetic fields, *Opt. Commun.*, Vol. 248, pp. 333–337.

Frehlich, R. G. (1992). Laser scintillation measurements of the temperature spectrum in the atmospheric surface layer, *J. Atmos. Sci.*, Vol. 49, pp. 1494–1509.

Fisher, M. E. and Burford, R. J. (1967). Theory of critical-point scattering and correlations. I. The Ising model., *Phys. Rev.*, Vol. 156, pp. 583–622.

Gabor, D. (1946). Theory of communication, part 1: The analysis of information, *J. Inst. Electr. Eng.* Vol. 3 93, pp. 429–457.

Gardner, W. A., Napolitano, A. and Paura, L. (2006). Cyclostationarity: Half a century of research, *Signal Process.*, Vol. 86, pp. 639–697.

Gauss, C. F. (1809). *Theoria Motus Corporum Celestium* (Hamburg, Perthes et Besser).

Gbur, G. and Wolf, E. (2002). Spreading of partially coherent beams in random media, *J. Opt. Soc. Am. A*, Vol. 19, pp. 1592–1598.

Gbur, G. and Korotkova, O. (2007). Angular spectrum representation for propagation of arbitrary coherent and partially coherent beams through atmospheric turbulence, *J. Opt. Soc. Am. A*, Vol. 24, pp. 745–752.

Gerrard, A. and Burch, J. M. (1994). *Introduction to Matrix Methods in Optics* (Courier Dover, New York).

Goodman, J. W. (2000). *Statistical Optics* (Wiley-Interscience).

Gori, F. (1980). Collett-Wolf sources and multimode lasers, *Opt. Commun.*, Vol. 34, pp. 301–305.

Gori, F. (1994). Flattened Gaussian beams, *Opt. Commun.*, Vol. 107, pp. 335–341.

Gori, F., Santarsiero, M., Vicalvi, S., Borghi, R. and Guattari, G. (1998). Beam coherence polarization matrix, *Pure Appl. Opt.: J. Eur. Opt. Soc. A*, Vol. 7, pp. 941–951.

Gori, F., Santarsiero, M., Borghi, R. and Piquero, G. (2000). Use of the van Cittert–Zernike theorem for partially polarized sources, *Opt. Lett.*, Vol. 25, pp. 1291–1293.

Gori, F., Santarsiero, M., Piquero, G., Borghi, R., Mondello, A. and Simon, R. (2001). Partially polarized Gaussian Schell model beams, *J. Opt. A: Pure and Appl. Opt.*, Vol. 3, pp. 1–9.

Gori, F. and Santarsiero, M. (2007). Devising genuine spatial correlation functions, *Opt. Lett.*, Vol. 32, pp. 3531–3533.

Gori, F. Santarsiero, M. and Borghi, R. (2007) Maximizing Young's fringe visibility through reversible optical transformations, *Opt. Lett.*, Vol. 32, pp. 588–590.

Gori, F. and Korotkova, O. (2009). Modal expansion for spherical homogeneous sources, *Opt. Commun.*, Vol. 282, pp. 3859–3861.

Gori, F. and Santarsiero, M. (2014). Difference of two Gaussian Schell-model cross-spectral densities, *Opt. Lett.*, Vol. 39, pp. 2731–2734.

Grimes, D. N. and Thompson, B. J. (1967). Two-point resolution with partially coherent light, *J. Opt. Soc. Am.*, Vol. 57, pp. 1330–1334.

Hielscher, A. H., Eick, A. A., Mourant, J. R., Shen, D., Freyer, J. P. and Bigio, I. J. (1997). Diffuse backscattering Mueller matrices of highly scattering media, *Opt. Exp.*, Vol. 1, pp. 441–453.

Hill, R. J. (1978a). Models of the scalar spectrum for turbulent advection, *J. Fluid Mech.*, Vol. 88, pp. 541–562.

Hill, R. J. (1978b). Optical propagation in turbulent water, *J. Opt. Soc. Am.*, Vol. 68, pp. 1067–1072.

Hill, R. J. (1978c). Spectra of fluctuations in refractivity, temperature, humidity and the temperature-humidity cospectrum in the inertial and dissipation ranges, *Radio Sci.*, Vol. 13, pp. 953–961.

Hopkins, H. H. (1951). The concept of partial coherence in optics, *Proc. Roy. Soc.* Vol. A208, pp. 263–277.

Hopkins, H. H. (1952). On the diffraction theory of optical images, *Proc. Roy. Soc.* Vol. A217, pp. 408–432.

Hunter, M., Backman, V., Popescu, G., Kalashnikov, M., Boone, C. W., Wax, A., Gopal, V., Badizadegan, K., Stoner, G. D. and Feld, M. S. (2006). Tissue self-affinity and polarized light scattering in the Born approximation: A new model for precancer detection. *Phys. Rev. Lett.*, Vol. 97, 138102.

Hyde, M. W. IV, Bose-Pillai, S. R. and Korotkova, O. (2018). Monte-Carlo simulations of three-dimensional electromagnetic Gaussian Schell-model sources, *Opt. Exp.* Vol. 26, pp. 2301–2313.

Jackson, J. D. (1998). *Classical Electrodynamics*, 3rd edn. (John Wiley and Sons, Berkley, CA).

Jamali, M. V., Mirani, A., Parsay, A., Abolhassani, B., Nabavi, P., Chizari, A., Khorramshahi, P., Abdollahramezani, S. and Salehi, J. A. (2018). Statistical studies of fading in underwater wireless optical channels in the presence of air bubble, temperature, and salinity random variations, *IEEE Trans. Comm.*, Vol. 66, pp. 4706–4723.

James, D. F. V. (1994). Change of polarization of light beams on propagation in free space, *J. Opt. Soc. Am. A*, Vol. 11, pp. 1641–1643.

Jannson, J., Jannson, T. and Wolf, E. (1988). Spatial coherence discrimination in scattering, *Opt. Lett.*, Vol. 13, pp. 1060–1062.

Jones, R. C. (1941). A new calculus for the treatment of optical systems. I. Description and discussion of the new calculus, *J. Opt. Soc. Am.*, Vol. 31, pp. 488–493.

Jönsson, C. (1961). Elektroneninterferenzen an mehreren künstlich hergestellten Feinspalten, *Zeitschrift fur Physik*, Vol. 161, pp. 454–474. [Translation to English: Brandt, D. and Hitschi, S. (1974). Electron diffraction at multiple slits, *Am. J. Phys.*, Vol. 42, pp. 4–11].

Kim, S. M. and Gbur, G. (2009). Momentum conservation in partially coherent wave fields, *Phys. Rev. A*, Vol. 79, 033844.

Kim, S. M. and Gbur, G. (2012). Angular momentum conservation in partially coherent wave fields, *Phys. Rev. A*, Vol. 86, 043814.

Klyshko, D. N. (1988). A simple method of preparing pure states of an optical field, of implementing the Einstein–Podolsky–Rosen experiment, and of demonstrating the complementarity principle, *Sov. Phys. Usp.*, Vol. 31, pp. 74–85.

Klyshko, D. N. (1988). Combine epr and two-slit experiments: Interference of advanced waves, *Phys. Lett. A*, Vol. 132, pp. 299–304.

Koivurova, M., Partanen, H., Lahyani, J., Cariou, N. and Turunen, J. (2019). Scanning wavefront folding interferometers, *Opt. Exp.*, Vol. 27, pp. 7738–7750.

Kolmogorov, A. N. (1941a). The local structure of turbulence in incompressible viscous fluid for very large Reynolds numbers, *Proc. USSR Acad. Sci.* (in Russian), Vol. 30, pp. 299–303.

Kolmogorov, A. N. (1941b). On degeneration (decay) of isotropic turbulence in an incompressible viscous liquid, *Proc. USSR Acad. Sci.* (in Russian), Vol. 31, pp. 538–540.

Kolmogorov, A. N. (1941c). Dissipation of energy in the locally isotropic turbulence, *Proc. USSR Acad. Sci.* (in Russian), Vol. 32, pp. 16–18.

Kolmogorov, A. N. (1942). Equations of turbulent motion of an incompressible fluid, *Izvestia Akademii Nauk USSR, Ser. Fiz.*, Vol. 6, pp. 56–58.

Korotkova, O. (2013). *Random Beams: Theory and Applications* (CRC Press, Boca Raton, FL).

Korotkova, O. (2018). Enhanced backscatter in LIDAR systems with retroreflectors operating through a turbulent ocean, *J. Opt. Soc. Am. A*, Vol. 35, pp. 1797–1804.

Korotkova, O., Sahin, S. and Shchepakina, E. (2012). Multi-Gaussian Schell-model beams, *J. Opt. Soc. Am. A*, Vol. 29, pp. 2159–2164.

Korotkova, O. (2014). Random sources for rectangular far fields, *Opt. Lett.*, Vol. 39, pp. 64–67.

Korotkova, O., Sahin, S. and Shchepakina, E. (2014). Potential scattering of light from particles of different shapes and semi-hard edges, *J. Opt. Soc. Am. A*, Vol. 31, pp. 1782–1787.

Korotkova, O. (2015). Design of weak scattering media for controllable light scattering, *Opt. Lett.*, Vol. 40, pp. 284–287.

Korotkova, O. and Mei, Z. (2015). Convolution of degrees of coherence, *Opt. Lett.*, Vol. 40, pp. 3073–3076.

Korotkova, O., Ahad, L. and Setala, T. (2017). Three-dimensional electromagnetic Gaussian Schell-model sources, *Opt. Lett.* Vol. 42, pp. 1792–1795.

Korotkova, O. and Chen, X. (2018). Phase structuring of complex degree of coherence, *Opt. Lett.*, Vol. 43, pp. 4727–4730.

Korotkova, O. and Soresi, A. (2019). Polarization signature of a mono-static double-pass system with a corner-cube reflector in the turbulent air, *Appl. Opt.*, Vol. 58, pp. 7139–7144.

Korotkova, O., Salem, M. and Wolf, E. (2004). Beam conditions for radiation generated by an electromagnetic Gaussian Schell-model source, *Opt. Lett.*, Vol. 29, pp. 1173–1175.

Korotkova, O. and Wolf, E. (2004). Spectral degree of coherence of a random three-dimensional electromagnetic field, *J. Opt. Soc. Am. A*, Vol. 21, pp. 2382–2385.

Korotkova, O. and Wolf, E. (2005a). Changes in the state of polarization of a random electromagnetic beam on propagation, *Opt. Commun.*, Vol. 246, pp. 35–43.

Korotkova, O. and Wolf, E. (2005b). Generalized Stokes parameters of random electromagnetic beams, *Opt. Lett.*, Vol. 30, pp. 198–200.

Korotkova, O. and Wolf, E. (2005c). Effects of linear non-image-forming devices on spectra and on coherence and polarization properties of stochastic electromagnetic beams: Part I: General theory, *J. Mod. Opt.*, Vol. 52, pp. 2659–2671.

Korotkova, O. and Wolf, E. (2005d). Effects of linear non-image-forming devices on spectra and on coherence and polarization properties of stochastic electromagnetic beams: Part II: Examples, *J. Mod. Opt.*, Vol. 52, pp. 2673–2685.

Korotkova, O. and Wolf E. (2007). Scattering matrix theory for stochastic scalar fields, *Phys. Rev. E*, Vol. 75, 056609.

Lajunen, H., Vahimaa, P. and Tervo, J. (2005). Theory of spatially and spectrally partially coherent pulses, *J. Opt. Soc. Am. A*, Vol. 22, pp. 1536–1545.

Lajunen, H. and Saastamoinen, T. (2011). Propagation characteristics of partially coherent beams with spatially varying correlations, *Opt. Lett.*, Vol. 36, pp. 4104–4106.

Li, Y. (2002). Light beams with flat-topped profiles, *Opt. Lett.*, Vol. 27, pp. 1007–1009.

Li, J., Wang, F. and Korotkova, O. (2016). Random sources for cusped beams, *Opt. Exp.*, Vol. 24, pp. 17779–17791.

Li, J. and Korotkova, O. (2017). Deeterministic mode representation of random stationary media for scattering problems, *J. Opt. Soc. Am. A*, Vol. 34, pp. 1021–1028.

Li, Y., Lee, H. and Wolf, E. (2003). Effect of edge rounding and sloping of sidewalls on the readout signal of the information pits, *Opt. Eng.*, Vol. 42, pp. 2707–2720.

Liang, C., Liu, X., Wang, F., Cai, Y. and Korotkova, O. (2014). Experimental generation of cosine-Gaussian-correlated Schell-model beams with rectangular symmetry, *Opt. Lett.*, Vol. 39, pp. 769–772.

Liang, C., Wu, G., Wang, F., Li, W., Cai, Y. and Ponomarenko, S. A. (2017). Overcoming the classical Rayleigh diffraction limit by controlling two-point correlations of partially coherent light sources, *Opt. Exp.*, Vol. 25, pp. 28352–28362.

Liang, C., Monfared, Liu, X., Qi, B., Wang, F., Cai, Y. and Korotkova, O. (2021). Optimizing illumination's complex coherence state for overcoming Rayleigh's resolution limit, *Chin. Opt. Lett.*, Vol. 19, 052601.

Lin, Q., Wang, L. and Zhu, S. (2003). Partially coherent light pulse and its propagation, *Opt. Commun.* Vol. 219, pp. 65–70.

Liu J. and Azzam, R. M. A. (1997). Polarization properties of corner-cube retroreflectors: Theory and experiment, *Appl. Opt.*, Vol. 36, pp. 1553–1559.

Lu, W., Liu, L., Sun, J., Yang, Q. and Zhu, Y. (2007). Change in degree of coherence of partially coherent electromagnetic beams propagating through atmospheric turbulence, *Opt. Commun.*, Vol. 271, pp. 1–8.

Luneberg, R. K. (1964). *Mathematical Theory of Optics* (University of California Press, Berkley, CA), pp. 319–320.

Ma, L. and Ponomarenko, S. A. (2014). Optical coherence gratings and lattices, *Opt. Lett.*, Vol. 39, pp. 6656–6659.

Mandel. L. and Wolf, E. (1995). *Optical Coherence and Quantum Optics* (Cambridge University Press, Cambridge, UK).

Mei, Z., (2017). Modeling for partially spatially coherent vortex beams, *IEEE Photon. J.*, Vol. 9 (IEEE Soc.) 6102306.

Mei, Z. and Korotkova, O. (2012). Random light scattering by collections of ellipsoids, *Opt. Exp.*, Vol. 20, pp. 29296–29307.

Mei, Z. and Korotkova, O. (2013). Random sources generating ring-shaped beams, *Opt. Lett.*, Vol. 38, pp. 91–93.

Mei, Z. and Korotkova, O. (2015). Alternating series of cross-spectral densities, *Opt. Lett.*, Vol. 40, (Am. Opt. Soc.), pp. 2473–2476.

Mei, Z. and Korotkova, O. (2017). Random sources for rotating spectral densities, *Opt. Lett.*, Vol. 42, pp. 255–258.

Mei, Z. and Korotkova, O. (2018a). Sources for random arrays with structured complex degree of coherence, *Opt. Lett.*, Vol. 43, pp. 2676–2679.

Mei, Z. and Korotkova, O. (2018b). Twisted EM beams with structured correlations, *Opt. Lett.* Vol. 43, pp. 3905–3908.
Mei, Z., Korotkova, O. and Shchepakina, E. (2013). Electromagnetic multi-Gaussian Schell-model beams, *J. Opt.*, Vol. 15, 025705.
Mei, Z., Korotkova O. and Mao Y. (2014). Products of Schell-model cross-spectral densities, *Opt. Lett.* Vol. 39, pp. 6879–6882.
Mei, Z., Zhao, D., Korotkova, O. and Mao, Y. (2015). Gaussian Schell-model arrays, *Opt. Lett.*, Vol. 40, pp. 5662–5665.
Mercer, J. (1909). Functions of positive and negative type an their connection with the theory of integral equations, *Phil. Trans. Roy. Soc. A*, Vol. 209, pp. 415–446.
Michelson, A. and Morley, E. (1887). On the relative motion of the Earth and the luminiferous ether, *Am. J. Sci.*, Vol. 34, pp. 333–345.
Mie, G. (1908). Contributions to the optics of cloudy media, especially of colloidal metalic solutions, *Annalen der Physik*, Vol. 330, pp. 377–445.
Millikan, R. (1916). A direct photoelectric determination of Planck's "h", *Phys. Rev.* Vol. 7, pp. 355–388.
Mishchenko, M. I., Travis, L. D. and Lacis, A. A. (2006). *Multiple Scattering of Light by Particles* (Cambridge University Press, UK).
Moreau, P. A., Toninelli, E., Gregory, T. and Padgett, M. J. (2018). Ghost imaging using optical correlations, *Las. and Photon. Rev.*, Vol. 12, 1700143.
Mueller, H. (1948). The foundation of optics, *J. Opt. Am.*, Vol. 38, pp. 661–661.
Muschinski, A. (2015). Temperature variance dissipation equation and its relevance for optical turbulence modeling, *J. Opt. Soc. Am. A*, Vol. 32, pp. 2195–2200.
Nikishov, V. V. and Nikishov, V. I. (2000). Spectrum of turbulent fluctuations of the sea-water refraction index, *Int. J. Fluid Mech. Res.*, Vol. 27, pp. 82–98.
Novotny L. and Hecht, B. (2006). *Principles of Nano-Optics* (Cambridge University, Cambridge, UK).
Nussenzveig, H. M. (1977). The theory of a rainbow, *Sci. Amer.*, Vol. 236, pp. 116–127.
Obukhov, A. M. (1949). The structure of the temperature field in a turbulent flow, *Izvestiya Akademii Nauk SSSR, Seriya Geografii i Geophysiki*, Vol. 13, pp. 58–69.
Päläkkönen, P., Turunen, J., Vahimaa, P., Friberg A. T. and Wyrowski, F. (2002). Partially coherent Gaussian pulses, *Opt. Commun.*, Vol. 204, pp. 53–58.
Partanen, H., Friberg, A. T., Setälä, T. and Turunen, J. (2019). Spectral measurement of coherence Stokes parameters of random broadband light beams, *Photon. Res.*, Vol. 7, pp. 669–677.

Padgett, M. J. and Boyd, R. W. (2017). An introduction to ghost imaging: Quantum and classical, *Phil. Trans. Roy. Soc. A*, Vol. 375, 20160233.

Papoulis, A. and Pillai, S. U. (2002). *Probability, Random Variables and Stochastic Processes*, 4th edn. (McGraw-Hill, New York, USA).

Peck, E. R. (1948a). A new principle in interferometer design, *J. Opt. Soc. Am.*, Vol. 38, pp. 66–66.

Peck, E. R. (1948b). Theory of corner-cube retro-reflector, *J. Opt. Soc. Am.*, Vol. 38, pp. 1015–1024.

Percus, J. K. and Yevick, G. J. (1958). Analysis of classical statistical mechanics by means of collective coordinates, *Phys. Rev.* Vol. 110, pp. 1–13.

Peyvasteh, M., Dubolazov, A., Popov, A., Ushenko, A., Ushenko, Y. and Meglinski, I. (2020). Two-point Stokes vector diagnostic approach for characterization of optically anisotropic biological tissues, *J. Phys. D: Appl. Phys.*, Vol. 53, (IOP), 395401.

Pittman, T. B., Shih, Y. H., Strekalov, D. V. and Sergienko, A. V. (1995). Optical imaging by means of two photon quantum entanglement, *Phys. Rev. A*, Vol. 52, R3429(R).

Plank, M. (1901). On the Law of the Energy distribution in the normal spectrum, *Annalen der Physik* Vol. 4, pp. 553–553.

Poisson, A. and Papaud, A. (1983). Diffusion coefficients of major ions in seawater, *Marine Chem.*, Vol. 13, pp. 265–280.

Pope, R. M. and Fry, E. S. (1997). Absorption spectrum (380–700 nm) of pure water. II. Integrating cavity measurements, *Appl. Opt.*, Vol. 36, pp. 8710–8723.

Quan, X. and Fry, E. S. (1995). Empirical equation for the index of refraction of seawater, *Appl. Opt.*, Vol. 34, pp. 3477–3480.

Raman, C. V. (1928). A new radiation, *Ind. J. Phys.*, Vol. 2, pp. 387–398.

Rayleigh, L. (1879). Investigations in optics, with special reference to the spectroscope, *Phil. Mag.*, Vol. 8, pp. 261–274, 403–411, 477–486.

Reynolds, O. (1894). On the dynamical theory of incompressible viscous fluids and the determination of the criterion, *Phil. Trans. A*, Vol. 186, pp. 123–164.

Richardson, L. F. (1922). *Weather prediction by Numerical Process* (Cambridge University Press, Cambridge, UK).

Roychowdhury, H. and Korotkova, O. (2005). Realizability conditions for electromagnetic Gaussian Schell-model sources, *Opt. Commun.*, Vol. 249, pp. 379–385.

Roychowdhury, H. and Wolf, E. (2005). Young's interference experiment with light of any state of coherence and of polarization, *Opt. Commun.*, Vol. 252, pp. 268–274.

Ruddick, B. R. and Shirtcliffe, T. G. L. (1979). Data for double diffusers: Physical properties of aqueous salt-sugar solutions, *Deep Sea Research*, Vol. 20A, pp. 775–787.

Ruffine, R. S. and de Wolfe, D. A. (1965). Cross-polarized electromagnetic backscatter from turbulent plasmas, *J. Geophys. Res.*, Vol. 70, pp. 4313–4321.

Ryczkowski, P., Barbier, M., Friberg, A. T., Dudley, J. M. and Genty, G. (2016). Ghost imaging in the time domain, *Nat. Photon.*, Vol. 10, pp. 167–170.

Rytov, S. M., Kravtsov, Yu. A. and Tatarskii, V. I. (1989). *Principles of Statistical Radiophysics. 4. Wave Propagtion through Random Media* (Springer, Berlin Heidelberg).

Sahin, S. and Korotkova, O. (2009). Effect of the pair-structure factor of a particulate medium on scalar wave scattering in the first Born approximation, *Opt. Lett.*, Vol. 34, pp. 1762–1764.

Sahin, S. Korotkova, O. and Gbur, G. (2011). Scattering of light from particles with semisoft boundaries, *Opt. Lett.*, Vol. 36, pp. 3957–3959.

Sahin, S. and Korotkova, O. (2012). Light sources generating far fields with tunable flat profiles, *Opt. Lett.*, Vol. 37, pp. 2970–2972.

Salem, M., Korotkova, O. and Wolf, E. (2006). Can two planar sources with the same sets of Stokes parameters generate beams with different sets of Stokes parameters? *Opt. Lett.*, Vol. 31, pp. 3025–3027.

Santarsiero, M., Piquero, G., de Sande, J. C. G. and Gori, F. (2014). Difference of cross-spectral densities, *Opt. Lett.*, Vol. 39, pp. 1713–1716.

Schell, A. C. (1961). *The multiple plate antenna*, PhD Dissertation, Massachusetts Institute of Technology, Cambridge.

Schmidt, J. M. and Kumar, G. (1996). Turbulent nature of refractive-index variations in biological tissue, *Opt. Lett.*, Vol. 21, pp. 1310–1312.

Sharma, K. A., Brown, T. G. and Alsonso, M. A. (2016). Phase-space approach to lensless measurements of optical field correlations, *Opt. Exp.*, Vol. 24, pp. 16099–16110.

Sheppard, C. J. R. (1996). Scattering by fractal surfaces with an outer scale, *Opt Commun.*, Vol. 122, pp. 178–188.

Shirai, T. and Wolf, E. (2007). Correlations between the intensity fluctuations in stochastic electromagnetic beams of any state of coherence and polarization, *Opt. Commun.* Vol. 272, pp. 289–292.

Shirai, T. and Asakura, T. (1995). Spectral changes of light induced by scattering from spatially random media under the Rytov approximation, *J. Opt. Soc. Am A*, Vol. 12, pp. 1354–1363.

Shirai, T. and Asakura, T. (1996). Multiple light scattering from spatially random media under the second-order Born approximation, *Opt. Commun.*, Vol. 123, pp. 234–249.

Shchepakina, E. A. and Korotkova, O. (2013). Spectral Gaussian Schell-model beams, *Opt. Lett.*, Vol. 38, pp. 2233–2236.

Simon, R. and Mukunda, N. (1993). Twisted Gaussian Schell-model beams, *J. Opt. Soc. Am. A*, Vol. 10, pp. 95–109.

Siviloglou, G. A. and Christodoulides, D. N. (2007). Accelerating finite energy Airy beams, *Opt. Lett.* Vol. 32, pp. 979–981.

Sparrow, C. M. (1916). On spectroscopic resolving power, *Astrophys. J.*, Vol. 44, pp. 76–86.

Stokes, G. G. (1852). On the composition and resolution of streams of polarized light from different sources, *Trans. Cambridge Phil. Soc.*, Vol. 9, pp. 399–416.

Stoletov, A. (1888). *Sur un sorte de courants electrique provoques par les rayons ultraviolets*, Comptes Rendus, Vol. 106, pp. 1149–1152.

Strutt, J. (1871a). On the light from the sky, its polarization and colour, *Phil. Mag.*, series 4, Vol. 41, pp. 107–120, 274–279.

Strutt, J. (1871b). On the scattering of light by small particles, *Phil. Mag.*, Series 4, Vol. 41, pp. 447–454.

Strutt, J. (1881) On the electromagnetic theory of light, *Phil. Mag.*, Series 5, Vol. 12, pp. 81–101.

Strutt, J. (1899). On the transmission of light through an atmosphere containing small particles in suspension, and on the origin of the blue of the sky, *Phil. Mag.* Series 5, Vol. 47, pp. 375–394.

Silverman, R. A. (1958). Scattering of plane waves by locally homogeneous dielectric noise, *Proc. Cambridge Phil. Soc.*, Vol. 54, pp. 530–537.

Tatarskii, V. I. (1961). *Wave Propagation in Turbulent Medium* (McGraw-Hill).

Tatarskii, V. I. (1971). *The Effects of the Turbulent Atmosphere on Wave Propagation* (Israel Program for Scientific Translation).

Tervo, J., Setälä, T. and Friberg, A. T. (2003). Degree of coherence for electromagnetic fields, *Opt. Exp.*, Vol. 11, pp. 1137–1143.

Thompson, B. J. (1969). Image formation with partially coherent light, in *Progress in Optics*, Vol. VII, E. Wolf (ed.) (North Holland Publishing Company, Amsterdam).

Thorpe, S. A. (2012). The turbulent ocean. Cambridge University Press, UK.

Tong, Z., Cai, Y. and Korotkova, O. (2010). Ghost imaging with electromagnetic stochastic beams, *Opt. Commun.*, Vol. 283, pp. 3838–3845.

Tong. Z. and Korotkova, O. (2010). Theory of weak scattering of stochastic electromagnetic fields from deterministic and random media, *Phys. Rev. A*, Vol. 82, 033836.

Tong. Z. and Korotkova, O. (2011a). Momentum of light scattered from collection of scatterers, *Phys. Rev. A*, Vol. 84, 043835.

Tong, Z. and Korotkova, O. (2011b). Pair-structure matrix of random collection of particles: Implications for light scattering, *Opt. Commun.*, Vol. 284, pp. 5598–5600.
Tong Z. and Korotkova, O. (2012a). Electromagnetic nonuniformly correlated beams, *J. Opt. Soc. Am. A*, Vol. 29, pp. 2154–2158.
Tong, Z. and Korotkova, O. (2012b). Beyond the classical Rayleigh limit with twisted light, *Opt. Lett.*, Vol. 37, pp. 2595–2597.
Toselli, I., Andrews, L. C., Phillips, R. L. and Ferrero, V. (2008). Free-space optical system performance for laser beam propagation through non-Kolmogorov turbulence, *Opt. Eng.*, Vol. 47, 026003.
Toselli, I., Agrawal, B. and Restaino, S. (2011). Light propagation through anisotropic turbulence, *J. Opt. Soc. Am A*, Vol. 28, pp. 483–488.
Toselli, I, Korotkova, O., Xiao, X. and Voelz, D. J. (2015). SLM-based laboratory simulations of Kolmogorov and non-Kolmogorov anisotropic turbulence, *Appl. Opt.*, Vol. 54, pp. 4740–4744.
Tyndall, J. (1869). On the blue colour of the sky, the polarization of skylight, and of polarization of light by cloudy matter generally, *Phil. Mag.*, Vol. 37, pp. 384–394.
van Cittert, P. H. (1934). *Die Wahrscheinliche Schwingungsverteilung in Einer von Einer Lichtquelle Direkt Oder Mittels Einer Linse Beleuchteten Ebene*, *Physica*, Vol. 1, pp. 201–210.
Verdet, E. (1865). Étude sur la constitution de la lumière non polariseé et de la lumière partiellement polariseé, *Annales Scientifiques de l'E.N.S.*, Vol. 2, pp. 291–316.
Voelz, D., Xiao, X. and Korotkova, O. (2015). Numerical modeling of Schell-model beams with arbitrary far-field patterns, *Opt. Lett.*, Vol. 40, pp. 352–355.
Volkov, S. N., James, D. F. V., Shirai, T. and Wolf, E. (2008). Intensity fluctuations and the degree of cross-polarization in stochastic electromagnetic beams, *J. Opt. A.: Pure and Appl. Opt.*, Vol. 10, 055001.
Wan, L. and Zhao, D. (2018). Optical coherence grids and their propagation characteristics, *Opt. Exp.*, Vol. 26, pp. 2168–2180.
Wang, F. and Korotkova, O. (2016a). Random sources for beams with azimuthal intensity variation, *Opt. Lett.*, Vol. 41, pp. 516–519.
Wang, F. and Korotkova, O. (2016b). Random optical beam propagation in anisotropic turbulence along horizontal links, *Opt. Exp.*, Vol. 24, pp. 24422–24433.
Wang, F. and Korotkova, O. (2016c). Convolution approach for beam propagation in random media, *Opt. Lett.*, Vol. 41, pp. 1546–1549.
Wang, F. and Korotkova, O. (2017). Circularly symmetric cusped random beams in free space and atmospheric turbulence, *Opt. Exp.*, Vol. 25, pp. 5057–5067.

Wang, F., Toselli, I., Li, J. and Korotkova, O. (2017a). Measuring anisotropy ellipse of atmospheric turbulence by intensity correlations of laser light, *Opt. Lett.*, Vol. 42, pp. 1129–1132.

Wang, F., Li, J., Martinez-Piedra, G. and Korotkova, O. (2017b). Propagation dynamics of partially coherent crescent-like optical beams in free space and turbulent atmosphere, *Opt. Exp.*, Vol. 25, pp. 26055–26066.

Wang, T. and Zhao, Z. (2010a). Scattering of scalar light wave from a Gaussian-Schell model medium, *Chin. Phys. B*, Vol. 19, 084201.

Wang, T. and Zhao, Z. (2010b). Scattering theory of stochastic electromagnetic light waves, *Opt. Lett.*, Vol. 35, pp. 2412–2414.

Wang, J., Yu, R., Xin, Y., Shao, Y., Chen, Y. and Zhao, Q. (2016). Ghost imaging with different speckle sizes of thermal light, *J. Opt. Soc. Korea*, Vol. 20, pp. 8–12.

Wolf, E. (1954). Optics in terms of observable quantities, *Il Nuovo Cimento*, Vol. 12, pp. 884–888.

Wolf, E. (1955). A microscopic theory of interference and diffraction of light from finite sources. II. Fields with a spectral range of arbitrary width, *Proc. Roy. Soc. London*, Vol. 230, pp. 246–255.

Wolf, E. (1982). A new theory of partial coherence in the space-frequency domain. Part 1: Spectra and cross-spectra of steady-state sources, *J. Opt. Soc. Am.*, Vol. 72, pp. 343–351.

Wolf, E. (1986). Invariance of spectrum on propagation, *Phys. Rev. Lett.*, Vol. 56, pp. 1370–1372.

Wolf, E., Foley, J. T. and Gori, F. (1989). Frequency shifts of spectral lines produced by scattering from spatially random media, *J. Opt. Soc. Am. A*, Vol. 6, pp. 1142–1149.

Wolf, E. (2003). Unified theory of coherence of polarization of random electromagnetic beams, *Phys. Lett. A*, Vol. 312, pp. 263–267.

Wolf, E. (2007a). *Introduction to the Theory of Coherence and Polarization of Light* (Cambridge University Press, Cambridge, UK).

Wolf, E. (2007b). *The influence of Young's interference experiment on the development of statistical optics*, in *Prog. Opt.*, Vol. 50, E. Wolf (Ed.), (Elsevier, Amsterdam), pp. 251–273.

Wu, D., Wang, F. and Cai, Y. (2018). High-order nonuniformly correlated beams, *Opt. Laser Tech.*, Vol. 99, pp. 230–237.

Xiao, X., Voelz, D., Toselli, I. and Korotkova, O. (2016). Gaussian beam propagation in anisotropic turbulence along horizontal links: theory, simulation and laboratory implementation, *Appl. Opt.*, Vol. 55, pp. 4079–4084.

Yaacoub, R., Pujol, O. and Dubuisson, P. (2019). Tunneling optical resonances in light-droplet interactions: Simulations of spaceborne cloud droplet observations, *J. Opt. Soc. Am. A*, Vol. 36, pp. 2076–2088.

Yakovlev, D. D. and Yakovlev, D. A. (2019). Scattering patterns of orthogonally polarized light components for statistically rotationally invariant mosaic birefringent layers, *Opt. Spectrosc.*, Vol. 126, pp. 245–256.

Yamazoe, K. (2012). Coherency matrix formulation for partially coherent imaging to evaluate the degree of coherence for image, *J. Opt. Soc. Am. A*, Vol. 29, pp.1529-1536.

Yao, J., Zhang, H., Wang, R., Cai, J., Zhang Y. and Korotkova, O. (2019). Wide-range Prandtl/Schmidt number power spectrum of optical turbulence and its application to oceanic light propagation, *Opt. Exp.*, Vol. 27, pp. 27807–27819.

Yao, J., Elamassie, M. and Korotkova, O. (2020). Spatial power spectrum of natural water turbulence with any average temperature, salinity concentration and light wavelength, *J. Opt. Sos. Am. A*, Vol. 37, pp. 1614–1621.

Young, T. (1804). The Bakerian Lecture. Experiments and calculations relative to physical optics, *Phil. Trans. Roy. Soc. Lond.*, Vol. 94, pp. 1–16.

Young, T. (1807). *A Course of Lectures on Natural Philosophy and the Mechanical Arts*, 2 Vols. (Johnson, London, 782 pages).

Yura, H. T. (1972). Mutual coherence function of a finite cross section optical beam propagating in a turbulent medium, *Appl. Opt.*, Vol. 11, pp. 1399–1406.

Zhang, Z. and Fan, D. (1992). Temporal diffraction integration of optical system and its applications, *Acta Opt. Sin.*, Vol. 12, pp. 179–182.

Zhang, Y., Ding, C., Hyde, M. W. IV and Korotkova, O. (2020a). Non-stationary pulses with complex-valued temporal degree of coherence, *J. Opt. A*, Vol. 22, pp. 105–607.

Zhang, Y., Korotkova, O., Cai, Y. and Gbur, G. (2020b). Correlation-induced orbital angular momentum changes, *Phys. Rev A*, Vol. 102, 063513.

Zernike, F. (1938). The concept of degree of coherence and its application to optical problems, *Physica*, Vol. 5, pp. 785–795.

Zhuang, F., Du, X. and Zhao, D. (2011). Polarization modulation for a stochastic electromagnetic beam passing through a chiral medium, *Opt. Lett.*, Vol. 36, pp. 2683–2685.

Zilberman, A., Golbraikh, E. and Kopeika, N. S. (2008). Propagation of electromagnetic waves in Kolmogorov and non-Kolmogorov atmospheric turbulence: Three-layer altitude model, *Appl. Opt.*, Vol. 47, pp. 6385–6391.

Printed in the USA
CPSIA information can be obtained
at www.ICGtesting.com
LVHW021153301024
794945LV00011B/133

9 789811 234972